# Three-Phase Electrical Power

# Three-Phase Electrical Power

Joseph E. Fleckenstein

## CRC Press
Taylor & Francis Group
Boca Raton   London   New York

CRC Press is an imprint of the
Taylor & Francis Group, an **informa** business

CRC Press
Taylor & Francis Group
6000 Broken Sound Parkway NW, Suite 300
Boca Raton, FL 33487-2742

© 2016 by Taylor & Francis Group, LLC
CRC Press is an imprint of Taylor & Francis Group, an Informa business

No claim to original U.S. Government works

Printed on acid-free paper
Version Date: 20150626

International Standard Book Number-13: 978-1-4987-3777-7 (Hardback)

**Library of Congress Cataloging-in-Publication Data**

Fleckenstein, Joseph E.
  Three-phase electrical power / Joseph E. Fleckenstein.
    pages cm
  Includes bibliographical references and index.
  ISBN 978-1-4987-3777-7
  1. Electric currents, Alternating--Three-phase. 2. Electric power
distribution--Multiphase. 3. Electric power transmission--Alternating current. I. Title.

TK1165.F54 2016
621.319'13--dc23                                                                    2015016379a

**Visit the Taylor & Francis Web site at**
**http://www.taylorandfrancis.com**

**and the CRC Press Web site at**
**http://www.crcpress.com**

# Contents

# Preface

The sudden availability of electrical power at the end of the nineteenth century was followed by an array of inventions that took advantage of this wondrous discovery. Today, electricity is used to illuminate buildings, to operate air conditioners, and to power trains. Electricity powers computers, copiers, medical devices, radios, televisions, and an ever-increasing variety of devices that continue to greatly improve the quality of life for millions of people the world over. At the core of it all is three-phase electrical power.

In the early years of electricity, there were two new promising and competing forms of energy. For a number of years, it was unclear as to which form, direct current (dc) or alternating current (ac), would prove to be more valuable or predominant. Eventually, ac electricity became the preferred choice for most applications in part because it is a form of power that can be run through a transformer, converted to a high voltage, and transmitted efficiently over long distances. There is still a place for dc electricity, for example, as power for commuter trains, but by far ac has become the more commonly applied form of electricity.

It is fair to say there are two general types of ac: single phase and polyphase. In the United States, single-phase electricity is the commonly used form at the residential level. As many devices, such as lights and computers, can use only single-phase electricity, there will long be a need for this type of electricity. Some single-phase electricity originates in small single-phase generators, but most single-phase electricity is derived from a three-phase supply.

The category of polyphase alternating electricity includes two-phase electricity and three-phase electricity. Two-phase electricity had its advocates early in the acceptance of ac. Today, there remains a limited use of two-phase electricity, primarily in the Great Lakes and the upper New York areas of the United States. The development of three-phase electricity in the early days was hindered in part because engineers at the time had difficulty with the related mathematics. A General Electric employee, Charles Steinmetz, who had extensive training in mathematics, is credited with the development of the mathematics that assisted in the adoption of three-phase electricity. In particular, Steinmetz developed a method of using complex numbers and phasors to calculate three-phase currents and voltages. His first paper on the subject was presented in a conference in 1893 (Reference 1.1). To this day, Steinmetz's phasors remain a convenient tool for analyzing three-phase circuits, and phasors are used extensively in this textbook.

In time, the merits of three-phase ac electricity became more apparent to those who were involved in the use and application of electricity. Today, three-phase electrical circuits remain the most effective means of transmitting electrical current from the source where it is generated to the places where it is to be consumed. And the three-phase motor is the most effective and practical means of converting electrical energy into mechanical power.

Public utilities that generate electricity have encountered a number of problems over the years in their quest to deliver economical, affordable, and reliable power to their customers. First, the expanding global demand for energy has resulted in the continually rising fuel costs of acceptable sources. Deregulation in some states has resulted in affected utilities reexamining the way they plan, build, and finance new power plants. Increasingly restrictive environmental laws have brought about a new set of problems and challenges to public utilities. In response to an ever-changing world, utilities are continually devising programs basically to provide a fair return on their investments. The objectives of these programs are mostly to more fully utilize existing plants while postponing the construction of new plants. The rising cost of electricity has also encouraged users to examine the ways they use electricity, and many are embarking on programs and methods to reduce the monthly bills. Internal energy audits have consequently become more popular. All of these ongoing changes require a cadre of personnel who will need to know and understand three-phase electricity as well as all of its peculiarities.

This textbook is designed primarily as an educational guide to those individuals who are interested in understanding and applying three-phase electrical power. In many ways, this textbook can fairly be considered as contemporary as it focuses on the present-day status of three-phase electrical power—the gear that is used today and the various programs and the issues that impact the generation and use of three-phase electrical power. The theory behind most concepts is explained, and practical equations are introduced. The following is a brief summary of the subjects encompassed by this textbook.

The first section of the textbook provides a review of ac in general and an introduction to the symbols and terms that are used in the following chapters of the book. This section presents some concepts that might be new to those readers who, say, might not have previously found the need to use phasors.

The textbook touches on the subject of generation, transmission, and distribution. If a facility is using electricity, it will be helpful if cognizant personnel understand the process and problems associated with the production and delivery of that product.

The grounding of an electrical circuit is critically important not only for the safety of personnel but also for the prevention and moderation of damage to property. With three-phase circuits, there are a number of unique considerations that do not appear with single-phase circuits. If a person is to be concerned with the use or installation of three-phase circuits, knowledge of good grounding practices applicable to three-phase circuits is a necessity.

Persons who become involved with three-phase electricity, whether for a proposed installation or an existing one, will, at times, find a need to calculate currents in three-phase circuits. A typical need could be for the purpose of sizing new electrical apparatus such as circuit breakers, conductors, conduits, and the like. Or the need might be for the purpose of demonstrating that an existing installation is in accordance with governing electrical codes. A circuit under consideration might be balanced or it could be unbalanced. Perhaps it is a three-wire or a four-wire circuit. There could be a need to add the currents of several branch circuits. In fact, there are numerous reasons why a person might encounter a need to conduct three-phase current calculations.

As with electrical currents, often there will be a requirement to calculate the power consumption of a three-phase circuit. The subject circuit might be delivering power to a balanced three-phase motor, or it could be an unbalanced circuit that serves a variety of users with different currents and power factors. Knowing the currents, voltages, and power factors is, of course, necessary. In addition, one must also know how to use these parameters to arrive at accurate and reliable values of power.

The term "demand" has long been used in association with three-phase electrical services. However, in more recent years it has taken an entirely new significance. More and more utilities have begun searching for alternative means to add peak generating capacity. Adding generating capacity is expensive and often unnecessary. Utilities are finding that rather than add generating capacity, a better policy is to move existing electrical consumption away from peak periods. In consequence, there are new business models for the billing for electricity, new meters, and new billing policies. The federal government has entered the picture with the intent of minimizing the future potentials for blackouts. Potentially large savings are available to users who come to understand the technical intricacies of new billing policies.

As long as utilities deliver power to customers, there will be need for a meter to measure the electricity that is being provided. Also, customers will often wish to measure the power usage at various locations within a facility. To correctly measure the power of a three-phase circuit, certain knowledge of the subject is required. First, the type of circuit must be determined. Is the circuit a three-phase, three-wire circuit or a three-phase, four-wire circuit? What type of meter should be used? Finally, where and how are the potential and current measurements made in order to obtain meaningful data? Today, there is an increasing need to measure and record not only power consumption but other electrical parameters as well. Reactive power, time of day usage, and demand are commonly measured values. Smart meters are being introduced and used by a number of public utilities in conjunction with a variety of pricing policies. Often, both the meters and the associated pricing policies bring welcome advantages to both utilities and customers.

When dealing with any form of electricity—dc, ac, single phase, three phase, or otherwise—circuit protection is critically important. Cables as well as almost all electrical devices must be protected from overheating that might be a result of a short circuit or merely an overcurrent condition. So, all electrical circuits and motors require protective means.

In motors larger than the fractional horsepower sizes (or roughly 1 kW), three-phase motors are almost always preferred to single-phase motors. In many regards, three-phase electricity is much better suited to powering a motor than single-phase power. So, anyone concerned with the use of electricity would be well advised to understand and appreciate the merits of three-phase motors. Globalization has added a new requirement, and complication, to the methods that have been used to describe and size motors in North America. For example, the long-used unit of "horsepower" is being used less and less, and the term "kilowatt" is used more in reference to motors. Also, in recent years there has been a developing awareness of motor efficiency and a gradual shift toward more efficient motors. In short, there have been a number of important changes affecting three-phase motors and their use.

Lagging power factors are a common problem worldwide. Mostly, the causes are due to the broad use of induction motors. Most utilities monitor the power factor of their three phase customers and generally charge a penalty above and beyond a charge for energy usage for a lagging power factor. For this reason, many three phase users find that efforts to correct a lagging power factor can pay large dividends.

Tariffs are naturally a subject of concern to any customer of electricity. Which user of electricity would not wish to reduce the monthly charge for electricity? The various types of charges for an electrical service vary greatly from one area to another, and many of the listed charges on a bill are not readily understood. It usually behooves a user to carefully examine the bills and to understand the various charges. Very often action may be taken to reduce certain parts of a bill.

Drawings are an integral and important part of three-phase electrical power. The industry uses a variety of drawings to represent a broad spectrum of conditions. There are one-line diagrams, three-line diagrams, logic diagrams, ladder diagrams, as well as a variety of other types. Most of these drawings are unique to the industry and cannot be readily understood without prior training and exposure.

A number of different types of relays are used in the control of three-phase electricity. There are control relays, time delay relays, and protection relays. In fact, there are a great variety of relays that are important to three-phase circuits and the operation of electrical gear.

This textbook addresses all of the issues pertinent to three-phase circuits: ac circuits, generation, transmission, distribution, grounding, currents, power, demand, metering, circuit protection, motors, motor protection, power factor correction, tariffs, drawings, and relays. All possible types of three-phase circuits are considered for all possible applications: balanced, unbalanced, leading, lagging, three wire, and four wire. Concepts are explained in simple terms, in detail, and in straightforward language. To better convey the concepts pertinent to three-phase electricity, numerous examples, illustrations, and photographs are used throughout the textbook.

It is pertinent to note that worldwide there are numerous electrical codes that prescribe how electrical devices are to be selected, sized, and installed. In the United States, the National Electrical Code (NEC) is the most recognized code that sets the guidelines for the installation of electrical systems. To be sure, the criteria of the NEC are not required for every building and every facility throughout the nation. The NEC is a very long and detailed textbook. In areas where the NEC is mandated, experienced electricians will know the requirements of the NEC intimately as knowledge of the code is required almost daily. Many electrical engineers, too, will find that in time a detailed understanding of the NEC will be necessary. This textbook does not in any way duplicate portions of the NEC, although some of the definitions of the NEC are repeated. The primary intent of this textbook is to present a more fundamental understanding of the state of the art as applicable to three-phase electrical power.

# *Author*

**Joseph E. Fleckenstein** graduated from Carnegie Mellon University, Pittsburgh, Pennsylvania, with a BS degree in mechanical engineering and a minor in electrical engineering. Postgraduation, he completed more than 32 courses in a variety of subjects, including graduate level courses at the University of Wisconsin and Pennsylvania State University. He has been a registered professional engineer for more than 30 years and throughout this period has been involved in a variety of roles with manufacturers and engineering firms. Over the years, he has written a number of technical documents, including technical papers, specifications, system descriptions, and prepared online courses for registered professional engineers. He is well experienced in electrical power generation, transmission, circuit protection, and controls. He is a member of IEEE and ISA.

# 1

# *Alternating Current*

The common forms of alternating electricity are single phase and polyphase. In the polyphase category, there are two types, namely, the two phase and the three phase. Two-phase electricity was popular in some areas years ago, but its popularity has dwindled, and today, the two common forms of alternating electricity are the single-phase and the three-phase types.

Alternating current (ac) originates at a generator that could be a single-phase or three-phase generator. A single-phase generator is generally of a smaller capacity and usually situated on site. Typical single-phase generators are in the range of 1–10 kW, although a few are larger.

Three-phase generators at an electrical utility may be as large as 1,300,000 kW (1,300 MW). The current generated at utility generators is commonly at a potential in the proximity of 35,000 V. The output from the three-phase generators is stepped up at transformers (SU transformers) to a high voltage that could be in the range of 140,000–500,000 V or higher. The high voltage facilitates transmission of current over long distances. Obviously, the higher the voltage, the lower the current for a given power level. And lower currents permit the use of smaller conductors. Close to the area where the electricity is to be used, the high voltage, in turn, is reduced to medium or low voltage for local distribution.

Users of large amounts of electrical power, as commercial facilities or industrial installations, are generally supplied with a three-phase electrical service. Within the installations, circuits would typically be divided into both single-phase and three-phase circuits. The three-phase circuits are used to power motors and possibly other large industrial-type loads. The single-phase branches of the three-phase service would typically be used for lighting, heating, fractional horsepower motors, and the like. A common North American electrical service to commercial and industrial users is the 480/3/60 ac service. (By convention, the "480" designates 480 V, the "3" designates three phase, and the "60" designates 60 Hz.) If the user has individual motors greater than, say, 500 HP, the voltage of the electrical service might very well be greater than 480 V and could be as high as 13,800 V.

Three-phase alternating current is essentially comprised of three single phases that are connected together but which peak at equally spaced time intervals. In order to establish the pertinent symbology for three-phase electricity, it is appropriate to first review some of the principles that define single-phase electricity.

Phasors are commonly used in the study and analysis of three-phase electricity. For this reason, the concept of phasors is introduced in this section along with an explanation of their use. As will be apparent, phasors can be very helpful in an understanding of three-phase electricity.

---

## 1.1 Single-Phase Alternating Current

In the United States, electrical utilities typically supply single-phase electricity to residential houses. A common service would be a three-wire 120/240 VAC configuration, and within the residence, the wiring would be separated into both 120 VAC and 240 VAC circuits. (According to common usage, the acronym "VAC" designates "voltage-alternating current.") The 240 VAC branch circuits are used to power larger electrical appliances as ranges, air conditioners, and heaters. The 120 VAC circuits are for wall outlets to power small loads as lamps, televisions, personal computers, small window air conditioners, and the like. In many instances, the electrical supply to apartments and condominiums will be three phase. Where a large number of residential users are grouped close together, a three-phase electrical supply is often the better choice.

### 1.1.1 Instantaneous Voltage and Instantaneous Current

Consider in the way of illustration a typical single-phase electrical service to a residence. In the United States, a 120/240 VAC service to a residence would normally be similar to the schematic representation in Figure 1.1. The service consists of three conductors. The potential of the neutral conductor would be very near or equal to ground potential and would typically be connected to ground at the utility transformer and at some point near the residence. If the guidelines of the National Electrical Code (NEC) are observed, the neutral conductor within the building must be colored either gray or white in color to distinguish those conductors from *hot* conductors. There is no requirement in the NEC for color coding of the two *hot* (120 VAC to ground) conductors, but these conductors are often colored red and black, one phase being designated as the black phase and the other as the red phase.

If an oscilloscope is used to view the instantaneous voltages of a single-phase residential service, a trace of the voltages would resemble the depiction of Figure 1.2. A pair of leads from the oscilloscope would be connected to the neutral wire and a black phase conductor with the (−) lead common to a neutral conductor and the (+) lead common to a black phase conductor. The oscilloscope would show a trace similar to the $v_{NB}^i$ trace in Figure 1.2. If another set of the oscilloscope leads is connected to the neutral and the red

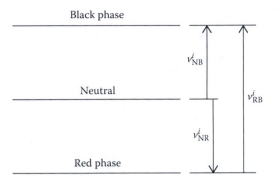

**FIGURE 1.1**
Typical single-phase service.

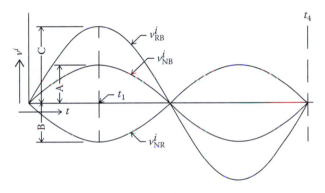

**FIGURE 1.2**
Trace of voltage with time.

phase, with the (–) lead on the neutral and the (+) lead on the red conductor, the trace would be similar to that shown for $v^i_{NR}$ in Figure 1.2. With the (–) lead on the red conductor and the (+) lead on the black conductor, the trace would be that shown as $v^i_{RB}$ in Figure 1.2. The traces in Figure 1.2 represent a single cycle. (The "$i$" notation is used to distinguish instantaneous values of voltage or current from root mean square [rms] values, as explained in the following.) The representations in Figure 1.2 show a true sinusoidal trace. To be sure, not all electrical services deliver a voltage that is a true sin wave function with time. A voltage that is not a true sin wave can cause a variety of problems. Among other problems, watt-hour meters may not correctly register the actual power consumption.

In Figure 1.2, the time period from $t = 0$ to $t = t_4$ is the time for a single cycle. In the United States, the common frequency of ac is 60 Hz (60 cycles/s). Thus, the time for a single cycle would be 1/60 s or 0.0166 s. The time from $t = 0$ to $t = t_1$, the time to the first peak voltage, would be (1/4) (1/60) s or 0.004166 s.

As mentioned earlier, the trace of $v^i_{NB}$ or $v^i_{NR}$ in Figure 1.2 is a representation of the instantaneous voltage of a typical 120 VAC service. The trace of $v^i_{NB}$ can be described by the algebraic relationship

$$v^i_{NB} = A \sin \omega t,$$

where, A is the value of voltage $v^i_{NB}$ at time $t_1$.

Since the value of $v^i_{NB}$ at time $t_1$ is equal to the peak value of $v^i_{NB}$,

$$v^i_{PK} = A$$

$$v^i_{BN} = (v^i_{PK}) \sin \omega t,$$

where
   $\omega = 2\pi f$ (rads/s)
   f is the frequency (Hz)
   $t$ is the time (s)

Similarly,

$$v^i_{NB} = B \sin \omega t$$

$$v^i_{RB} = C \sin \omega t.$$

In general,

$$v^i = (v_{PK}) \sin \omega t, \tag{1.1}$$

where
   $v^i$ is the instantaneous value of voltage (V)
   $v_{PK}$ is the peak value of voltage (V)

Much as with instantaneous voltage, instantaneous current can also be described as a function of time by the general relationship

$$i^i = i_{PK} \sin (\omega t + \theta_{SP}), \tag{1.2}$$

where
   $i^i$ is the instantaneous value of current (amps)
   $i_{PK}$ is the peak value of current "$i$" (amps)
   $\theta_{SP}$ is the angle of lead or angle of lag (degrees or radians) of current with respect to voltage in a single-phase circuit (subscript "SP" designates single phase)

NOTE: As indicated, the units of the parameter $\omega$ are rads/s. If the parameter $t$ is in seconds, as is typically the case, the units of $\omega t$ are rads. The lead or lag of current with respect to voltage is typically stated in degrees (°). So, to compute the value of $\sin(\omega t + \theta_{SP})$, either $\theta_{SP}$ must be converted to rads or $\omega t$ must be converted to degrees.

It may be noted that if the current lags voltage, $\theta_{SP} < 0$, and if the current leads voltage, $\theta_{SP} > 0$.

### 1.1.2 RMS Voltage and RMS Current

A trace of instantaneous voltage as obtained with an oscilloscope, as mentioned earlier, is of interest and educational. An oscilloscope trace provides a true visual picture of voltage and current as a function of time. Nevertheless, it is the values of rms voltage and rms current that are of the most practical use. This is due to the common and useful analogy to dc circuits. In dc circuits, power dissipation is calculated by the relationship

$$P = I^2R, \text{ or}$$

since $V = IR$ and $I = V/R$,

$$P = [V/R] (IR) = VI.$$

By common usage, these same formulas are also used for determining power as well as other parameters in single-phase ac circuits. This is possible only by the use of the rms value of voltage and, likewise, the rms value of current. The relationship between rms and peak values in ac circuits is given by the relationship

$$V = \left(1/\sqrt{2}\right)(v_{PK}) = (0.707)(V_{PK}) \quad \text{(Reference 1.1)}$$

$$I = \left(1/\sqrt{2}\right)(i_{PK}) = (0.707)(I_{PK}) \quad \text{(Reference 1.2)},$$

where
    V is the rms voltage
    I is the rms current

In general, rms voltage can be described as a function of time by the equation

$$V(t) = V \sin(\omega t), \tag{1.3}$$

where
    $V(t)$ is the voltage expressed as a function of time (rms volts)
    V is the numerical value of voltage (rms)

Equation 1.3 assumes that at $t = 0$, $V(t) = 0$.

Current, in a circuit, may lag by the amount $\theta_{SP}$. So, the rms current may be described as a function of time by the relationship

$$I(t) = I \sin (\omega t + \theta_{SP}), \tag{1.4}$$

where

$I(t)$ is the current expressed as a function of time (rms amps)

$I$ is the numerical value of current (rms)

$\theta_{SP}$ is the angle of lead or angle of lag (radians) of current with respect to voltage in a single-phase circuit

NOTE: The expressions for rms voltage and rms current stated in Equations 1.3 and 1.4, respectively, are consistent with common industry practice and are generally helpful in understanding electrical circuits. However, these expressions are not true mathematical descriptions of the rms values of voltage and current as a function of time. Expressed as a function of time, the curves of the rms values of voltage and current would have an entirely different appearance.

### Example 1.1

**Problem**: Mathematically express as a function of time the typical residential voltages of Figures 1.1 and 1.2, given that the source voltage is a nominal "120/240 VAC" service.

**Solution**

If the nominal (i.e., rms voltage) is "120 VAC," then the "peak" value of voltage would be

$$v_{PK} = (1.414) \, V = (1.414) \, (120) = 169.7 \text{ V}.$$

Therefore, the instantaneous value of voltage described as a function of time is given by the equation

$$v^i = v_{PK} \sin \omega t = (169.7) \sin \omega t \text{ (V)}.$$

The 120 VAC (rms) voltage as a function of time is

$$V(t) = (120) \sin \omega t \text{ (V)}, \text{ and}$$

the 240 VAC voltage as a function of time is

$$V(t) = (240) \sin \omega t \text{ (V)}.$$

At $t_1$, $t = 0.004166$ (as stated here), and

$$\omega t = (2\pi f t) = (2\pi) \, (60) \, (0.004166) = 1.570 \text{ rad}.$$

So, sin $\omega t = 1.0$, and

$$v(t) = v_{PK} \sin \omega t = 169.7 \text{ V (peak value)}$$

$$v(t) = (120)\ (1)\ (V) = 120 \text{ V}$$

$$v(t) = (240)\ (1)\ (V) = 240 \text{ V.}$$

Thus, the computation is confirmed.
It may also be seen that

$$v^i_{NR} = -v^i_{NB} \quad \text{and}$$

$$v^i_{RB} = 2v^i_{NB}.$$

Since $v^i_{NR}$ is the mirror image of $v^i_{NB}$, the instantaneous voltage would be described by the relationship

$$v^i_{NR} = (-169.7) \sin \omega t \text{ (V)},$$

and the voltage $v_{RB}$ would be described by the relationship

$$v^i_{RB} = (1.414)\ (240) \sin \omega t = (339.4) \sin \omega t \text{ (V).}$$

Unless specifically stated otherwise in documents, voltages and currents are always assumed to be rms values. This would also be true, for example, of a multimeter that reads voltage or current unless the meter is set to read "peak" values. (Some multimeters have the capability to read peak voltages and peak currents as well as rms values.)

### 1.1.3 Single-Phase Circuits

Application of a single-phase voltage to a load can be represented as depicted in Figure 1.3. (The ac voltage source is represented by a symbol that approximates a single cycle of a sine wave. The load is represented by the letter "L.")

#### 1.1.3.1 Resistive Loads

If an ac load is purely resistive, the current is in phase with the voltage. A typical resistive load would be one consisting of an incandescent lighting or an electrical resistance heater. A trace of an applied (rms) voltage and the typical resultant (rms) current for a purely resistive load, expressed as a function of time, is represented in Figure 1.4.

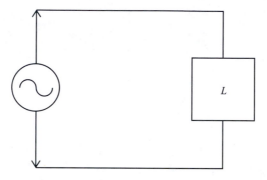

**FIGURE 1.3**
ac voltage source and load.

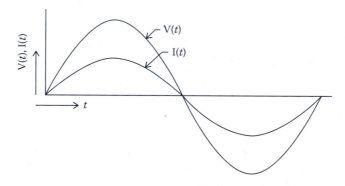

**FIGURE 1.4**
RMS voltage and current with time.

### Example 1.2

**Problem**: Assume a nominal single-phase source voltage of 240 VAC and a load that consists solely of a 5 kW heater. Describe as a function of time: instantaneous voltage, instantaneous current, rms voltage, and rms current.

**Solution**

For a single-phase application,

$$P = I^2R = VI \text{ watts,}$$

where
   R is the resistance of load (ohms)
   V is the voltage (rms)
   I is the current (rms)
   P is the 5000 (W), which is equal to (240) I
   I is the 5000/240 = 20.83 amp

So, the instantaneous voltage and the instantaneous current would be described by the relationships of Equations 1.1 and 1.2.

$$v^i = [(240)\ (1.414)]\ \sin \omega t \ (V) \ \text{or}$$

$$v^i = (339.5)\ \sin \omega t \ (V), \ \text{and}$$

$$i^i = [(20.83)\ (1.414)]\ \sin \omega t \ (\text{amps}) \ \text{or}$$

$$i^i = (29.46)\ \sin \omega t \ (\text{amps}).$$

According to common practice, the expression for the rms value of voltage as a function of time is represented by the expression

$$V(t) = (240)\ \sin \omega t \ (V).$$

Since the load is restive, $\theta_{SP} = 0$, and per Equation 1.4, the expression for rms current becomes

$$I(t) = (20.83)\ \sin \omega t \ (\text{amps}).$$

### 1.1.3.2 Leading and Lagging Power Factor

The load represented in Figure 1.3 could be resistive, inductive, capacitive, or any combination of these possibilities. In the case of a resistive load, the current is in phase with the voltage. Inductive or capacitive elements in a circuit may cause the current to either lag or lead the respective phase voltage. Common inductive loads include transformers, relays, motor starters, and solenoids. A capacitive load would commonly be capacitors or electrical components that emulate a capacitor. The insulation of an electrical system can act as a capacitor.

For a configuration of the type represented in Figure 1.3 where the load consists mostly of capacitive elements, the current would lead the voltage as shown in Figure 1.5. The current leads by the value of $\theta_{SP}$.

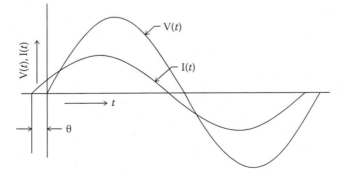

**FIGURE 1.5**
Leading power factor.

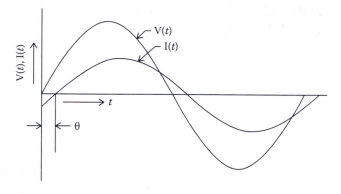

**FIGURE 1.6**
Lagging power factor.

For a configuration of the type represented in Figure 1.3 where the load consists of mostly inductive elements, the current would lag the voltage as shown in Figure 1.6. The current lags by the amount $\theta_{SP}$.

### Example 1.3

**Problem**: Assume that for the traces of voltage and current of Figure 1.6, the value of $\theta_{SP}$ is −20° (i.e., lagging current).
Find the actual value of time lag.

**Solution**

$$180° = \pi \text{ rad}$$

$$\theta_{SP} = - (20°/180°) \, \pi \text{ rad, or}$$

$$\theta_{SP} = -0.3491 \text{ rad}$$

Then, from Equation 1.4,

$$I(t) = I \sin (\omega t + \theta_{SP}).$$

At

$$I(t) = 0, \sin (\omega t + \theta_{SP}) = 0, \text{ and}$$

$$\omega t = 2\pi f t = -\theta_{SP} = -0.3491 \text{ rad,}$$

$$t = \left| -0.3491/2\pi f \right| = 0.000926 \text{ s.}$$

As explained here, for an ac voltage, 60 Hz application, a single cycle (360°) would be 0.0166 s in length. The trace of $i$ would cross the abscissa at

$$t = (20°/360°) \, (0.0166) \text{ s or}$$

$$t = 0.000926.$$

The computation of power consumed by a reactive electrical circuit requires the use of a term commonly known as the circuit's "power factor," which is often abbreviated merely as "PF."

In a single-phase, linear circuit,

$$P = VI \cos \theta_{SP}, \text{(Reference 1.3)},$$

where
  P is the power (watts)
  V is the voltage (rms)
  I is the current (rms)

$\theta_{SP}$ is the angle of lead or angle of lag of current with respect to voltage (the subscript "SP" designates "single phase")

$$\cos \theta_{SP} \text{ is the power factor} = PF.$$

In single-phase circuits, the term "$\theta_{SP}$" is a measure of the lead or lag of current with respect to the applied voltage, and the value of $\cos \theta_{SP}$ is equal to the power factor. The term "$\theta_{SP}$" is negative for a lagging power factor, positive for a leading power factor, and equal to zero for a purely resistive circuit. The value of power factor ($\cos \theta_{SP}$) is always a positive number (i.e., PF > 0). This is necessarily the case since the angle of lead or lag in a single-phase circuit can be no less than $-90°$ and no more than $+90°$, and within that range of values, the cosine of the angle of lead or lag is only positive. (So, the often used industry term of a "negative" power factor as applied to a single-phase circuit is a misnomer. Power and power factor in three-phase circuits are discussed in greater detail later.)

As indicated, the earlier stated expression for power is applicable to a single-phase linear circuit. A linear circuit is one that has a voltage trace and a current trace that may be described by a mathematical sine wave of the shape shown in Figure 1.2. Power factor as used in the expression for linear circuits is also known as the "displacement power factor." When a circuit is nonlinear, the determination of power factor becomes a more complicated task, and other terms become applicable to circuit power factor. Nonlinear circuits and the applicable power factors are treated in detail in Section 5.3.3.

## 1.1.4 Phasor Diagrams of Single-Phase Circuits

In the study of ac circuits, and particularly three-phase circuits, it is a common practice to use phasor diagrams to depict the relationship between voltages and currents (Reference 1.4). A phasor diagram uses vectors that emulate the vectors common to the mathematics of complex variables. As used in the analysis of electrical circuits, phasors represent currents and voltages. A phasor is a line segment the length of which represents the magnitude of voltage or current. The angle of the phasor's line segment with

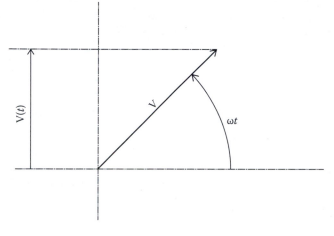

**FIGURE 1.7**
Voltage vector.

respect to a referenced axis conveys the phase angle information. The phasor diagram becomes a visual analogy of the voltages and currents in an electrical circuit. A good visualization of the currents and voltages helps to ensure that any determinations of currents will be correct. Phasors can be used in the analysis of either single-phase circuits or three-phase circuits.

In the way of illustration, consider Equation 1.1, which is for a single-phase circuit. The circuit can be represented by a vector (or "phasor") as shown in Figure 1.7. In Figure 1.7 it is apparent that if the value of $t$ or $\omega t$ is increased, the vector V would rotate counterclockwise about its base, which is at the intersection of the abscissa and the ordinate axes. The projection on the $y$-axis (ordinate) would then equal the term [V sin $\omega t$] as defined by Equation 1.3.

A vector representative of current can likewise be included in a phasor diagram as shown in Figure 1.8. In Figure 1.8, the projection on the $y$-axis equals the mathematical term [I sin $\omega t$] as defined by Equation 1.4.

If current leads the applied voltages as described by Equation 1.4, and which is represented by the plot in Figure 1.5, the associated voltage and current vectors would be similar to that shown in Figure 1.9. The representation of Figure 1.9 is for a particular point in time, namely, at $t = 0$. When the current leads the applied voltage, $\theta_{SP} > 0$.

If current lags the voltage as described by Equation 1.4, and which is represented by the plot of Figure 1.6, the associated voltage and current vectors, or phasors, would be similar to that shown in Figure 1.10. When the current lags the applied voltage, the measure of lag is $\theta_{SP} < 0$.

Here, it is shown that the projections of the rotating vector V with time are the mathematical equivalent of the (rms) current with time. Phasor diagrams

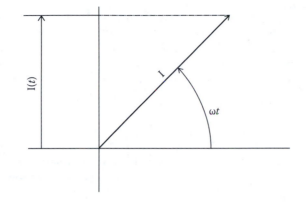

**FIGURE 1.8**
Current vector.

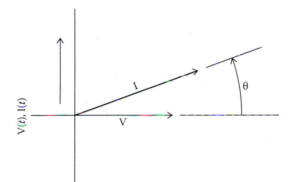

**FIGURE 1.9**
Current and voltage vectors.

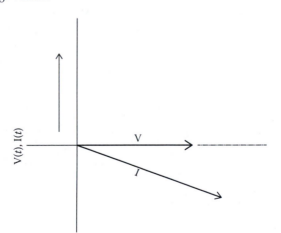

**FIGURE 1.10**
Lagging current—voltage and current vectors at $t = 0$.

are less concerned with the time variable and examine electrical properties at a selected time. So to speak, phasor diagrams look at electrical values with electrical properties frozen in time.

Phasor diagrams are especially helpful in assisting a person to visualize ac circuits, both single phase or three phase, that have currents leading or lagging the voltage. As demonstrated in the following paragraphs, phasor diagrams are particularly valuable when analyzing three-phase circuits. A phasor diagram not only provides a visual picture of the relationships between the voltage and current in a selected phase of a three-phase circuit, but it also helps a person to understand the relationship of current and voltage in one phase to the voltage and current in another phase of the circuit.

### 1.1.5 Parallel Single-Phase Loads

The merits of a phasor diagram as used in the analysis of single-phase circuits become especially apparent when considering parallel circuits with different power factors. A typical single-phase ac circuit with parallel loads is represented in Figure 1.11. One of the loads is represented by $L_1$, and the second, parallel load is represented by $L_2$. The current of load $L_1$ is assumed to be $I_1$ at power factor $\cos \theta_1$, and the current of load $L_2$ is $I_2$ at power factor $\cos \theta_2$. Since both $L_1$ and $L_2$ are subject to the same voltage ($v_{ab}$), the currents may both be referenced to that voltage as shown in Figure 1.12.

In order to construct a phasor diagram, a few rules are borrowed from the mathematics of complex variables. This is not to say that an understanding of complex variables is required. Essentially, only two relatively simple

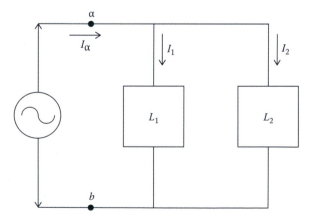

**FIGURE 1.11**
Parallel loads.

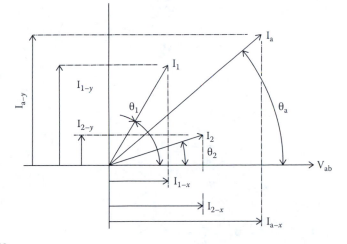

**FIGURE 1.12**
Vector addition.

rules are need to be observed. One rule pertains to the addition of vectors. The second rule is that $-\underline{I}_{XY} = \underline{I}_{YX}$, that is, the negative of vector $\underline{I}_{XY}$ is a vector that is of equal magnitude but pointing in a direction 180° from that of vector $\underline{I}_{XY}$. (Note that by convention, an underlined variable designates a vector quantity.)

When adding vectors, both magnitude and direction are material. Vectors are added by adding, first, the abscissa components to determine the abscissa component of the resultant vector. The ordinate components are added to determine the ordinate component of the resultant vector. A typical vector addition is represented in Figure 1.12. The angle, $\theta$, of any vector with respect to the positive abscissa is shown as positive in the counterclockwise direction. The vectors $\underline{I}_1$ and $\underline{I}_2$ are added to determine vector $\underline{I}_a$. The abscissa component of Ia is $I_{a-x}$, which is the sum of the abscissa quantities $I_{1-x}$ and $I_{2-x}$.

The ordinate component of $I_a$ is $I_{a-y}$, which is the sum of $I_{1-y}$ and $I_{2-y}$. For the vectors of Figure 1.12,

$$I_{a-x} = I_1 \cos \theta_1 + I_2 \cos \theta_2, \text{ and}$$

$$I_{a-y} = I_1 \sin \theta_1 + I_2 \sin \theta_2$$

$$(I_a)^2 = (I_{a-x})^2 + (I_{a-y})^2, \text{ or}$$

$$|I_a| = \{(I_{a-x})^2 + (I_{a-y})^2\}^{\frac{1}{2}}. \tag{1.5}$$

Also, $\sin \theta_a = I_{a-y} \div I_a$, or

$$\theta_a = \sin^{-1} (I_{a-y} \div I_a).$$

Obviously, if there are more than two currents comprising a line current,

$$I_{a-x} = I_1 \cos \theta_1 + I_2 \cos \theta_2 + \dots I_n \cos \theta_n$$

$$I_{a-y} = I_1 \sin \theta_1 + I_2 \sin \theta_2 + \dots I_n \sin \theta_n.$$

**Example 1.4**

**Problem**: Assume that for the configuration of Figure 1.11

$$I_1 = 10 \text{ amp @ PF} = 0.60, \text{ leading}$$

$$I_2 = 4 \text{ amp @ PF} = 0.90, \text{ leading.}$$

Determine $I_a$ and its power factor (PF).

**Solution**

From Equation 1.5,

$$I_a = \{(I_{a-x})^2 + (I_{a-y})^2\}^{1/2}$$

For $I_1$, $\cos \theta_1 = 0.60$, $\theta_1 = 53.13°$, and $\sin \theta_1 = 0.800$.
For $I_2$, $\cos \theta_2 = 0.90$, $\theta_2 = 25.84°$, and $\sin \theta_2 = 0.4358$.
$I_a$ is determined by first adding the abscissa and ordinate components.

$$I_{a-x} = I_1 \cos \theta_1 + I_2 \cos \theta_2$$

$$I_{a-x} = (10) (0.600) + (4) (0.900) = 6.000 + 3.600 = 9.600$$

$$I_{a-y} = [I_1 \sin \theta_1 + I_2 \sin \theta_2]$$

$$I_{a-y} = [(10) (0.80) + (4) (0.4358)] = [8.00 + 1.7432] = 9.743$$

$$I_a = \{(I_{a-x})^2 + (I_{a-y})^2\}^{1/2} = \{(9.600)^2 + (-9.743)^2\}^{1/2} = 13.678 \text{ amp}$$

$$\sin \theta_a = I_{a-y} \div I_a = 9.744 \div 13.678 = .712$$

$$\theta_a = \sin^{-1} (I_{a-y} \div I_a) = \sin^{-1} (9.744 \div 13.678) = 45.43°$$

$$PF = \cos 45.43° = 0.702$$

### 1.1.6 Polar Notation

As stated in the previous section, complex variables are not used in this text-book to calculate current values. Nevertheless, polar notation that is used to describe the position of a complex vector is worth mentioning. Polar notation is commonly found in textbooks that treat three-phase currents and three-phase voltages. Accordingly, a familiarity with polar notation could be helpful at times. The polar form of complex variable notation can be used to describe the position of a voltage or a current on a phasor diagram. In polar notation, a current (amperage) or a potential (voltage) is described by the magnitude of the variable and angle in the CCW (counterclockwise) direction from the positive abscissa. In the way of illustration, consider the currents of Example 1.4. Currents $I_1$ and $I_2$ are described in the example as

$$I_1 = 10 \text{ amp @ PF} = 0.60, \text{ leading } (\theta_1 = 53.13°)$$

$$I_2 = 4 \text{ amp @ PF} = 0.90, \text{ leading } (\theta_2 = 25.84°).$$

In polar notation, these two currents are represented, respectively, as

$$I_1 = 10\underline{/53.13°}$$

$$I_2 = 4\underline{/25.84°}.$$

If the currents were lagging instead of leading, the representation would be

$$I_1 = 10\underline{/-53.13°}$$

$$I_2 = 4\underline{/-25.84°}.$$

Since, $360° - 53.13° = 306.87°$ and $360° - 25.84° = 334.16°$, the lagging currents could also be expressed as

$$I_1 = 10\underline{/306.87°}$$

$$I_2 = 4\underline{/334.16°}.$$

In Example 1.4, current $I_a$ is determined by adding the abscissa and ordinate components of $I_1$ and $I_2$. In vector notation, the addition is represented by the expression

$$\underline{I_a} = \underline{I_1} + \underline{I_2}.$$

(The underscored characters indicate that the respective variables are vectors and not algebraic values.) Expressed in polar notation,

$$\underline{I_a} = 10\underline{/53.13°} + 4\underline{/25.84°}.$$

In Example 1.4, the voltage source is single phase, and the positive abscissa would naturally be taken as the single-phase voltage source. The angle of lead or lag would be in reference to the applied voltage. In three-phase delta circuits, the phase voltage is the same as the line voltage, and the positive abscissa is taken as that voltage. So, in three-phase delta circuits, the phase currents or the line currents are stated in reference to the line or phase and voltage. In the case of wye circuits, there are line voltages as well as phase voltages, that is, the line-to-neutral voltages. Therefore, when using polar notation to describe phase or line currents, care must be taken to separately indicate that the angle stated in the polar notation is in reference to either the phase voltage or the line voltage.

## 1.2 Three-Phase Alternating Current: Basic Concepts

### 1.2.1 Three-Phase Alternating Current Source

The three phases of a three-phase electric service originate in a three-phase generator. A three-phase generator has coils positioned within its stator to deliver three separate voltages that have peaks equally spaced in time. Three-phase electrical power generated at an electric utility is generally transmitted to an electrical grid, and that grid eventually provides three-phase electricity, after it is reduced to a lower voltage, to users. The eventual connections to a user would be by means of either a three-conductor circuit or a four-conductor circuit. If there are four conductors, the fourth conductor would be connected to ground. As a group, the conductors comprise the electrical "source," and the electricity provided by a utility to a user is considered as the "service." The nongrounded conductors are commonly designated as the phase A conductor, the phase B conductor, and the phase C conductor. The voltage between the phase A conductor and the phase B conductor is a single-phase voltage. Likewise, the voltage B–C is a single-phase voltage and of the same magnitude as voltage A–B. The same is true of the voltage C–A. The more common voltage sequence is A–B–C. In other words, phase B peaks after phase A, and phase C peaks after phase B. A time trace of the three voltages, as obtained with an oscilloscope, would be similar to the representation in Figure 1.13.

If the incoming conductors of the electrical service are at, say, 4,120 VAC (a common distribution voltage in the United States) or 13,800 VAC (also a common voltage in the United States), and the user is to provide for a transformer, the service will most probably be three wire. If the user is not providing a transformer, the service will probably be a four-wire service. All users generally require a 120 VAC source for convenience (wall) outlets, and this voltage source may be either provided by the utility or derived by the user from a user-provided transformer.

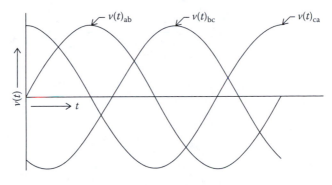

**FIGURE 1.13**
Three-phase voltages.

Within a user's premises, the establishment of the sequence of the incoming three-phase conductors is of importance. This is particularly true when the three-phase source is used to power motors since the phase sequence determines the direction of motor rotation. Motor control centers and power distribution panels will almost always have incoming terminals designated as "1," "2," and "3." These terminal designations assume the sequence 1–2–3 that corresponds to the sequence A–B–C. In consequence, motors connected to 1–2–3 will rotate in a predictable direction. If the sequences of the voltages connected to a motor are reversed, the motor will rotate in an unintended direction. Depending on the application, considerable property damage or personnel injury could result when a motor is started, and it rotates in reverse direction to the intended rotation.

### 1.2.2 Delta and Wye Circuits Defined

Three-phase electricity is provided for use as either a three-wire or four-wire configuration. The manner in which it is used is commonly in the form of a delta circuit or a wye circuit. Of course, a three-phase electrical source may also be used to serve single-phase loads. The "delta" configuration is so named because of the resemblance of the configuration to the Greek symbol "Δ." The "wye" configuration is also called the "Y" circuit and sometimes the "star" circuit. A delta circuit is a three-wire circuit, and a wye circuit can be either a three-wire circuit or a four-wire circuit.

A three-phase circuit consists of three separate single-phase voltages that are interconnected one to the other. A typical three-wire three-phase delta circuit is represented in Figure 1.14. Single-phase loads may be taken from any two of the conductors of a three-phase delta circuit. In practice, single-phase loads are more commonly derived from a four-wire wye circuit.

A typical three-phase wye circuit is shown in Figure 1.15. If a wye circuit is a four-wire configuration, point "d" of Figure 1.15 might be connected to the ground. Not all wye circuits, however, have the neutral point connected

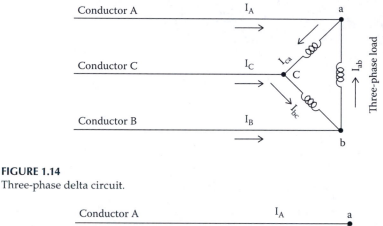

**FIGURE 1.14**
Three-phase delta circuit.

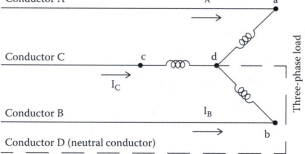

**FIGURE 1.15**
Three-phase wye circuit.

to the ground. A wye wound motor, for example, would not have the neutral point connected to the ground. In a motor, the currents in each of the three phases are equal under normal circumstances. Accordingly, there is no need to connect the neutral point to ground. Variations of the delta and wye circuits are discussed in the following paragraphs.

### 1.2.3  Common Service Voltages

The types of circuits used around the world vary greatly. Likewise, the magnitudes of voltages are often different from one country to another.

Table 1.1 presents a tabulation of common service voltages found in some areas of the world.

Following the table is a brief summary of the U.S. three-phase electrical services together with a brief description of the common usages of the services.

#### 1.2.3.1  240/120 VAC Three-Phase Four-Wire Delta (Figure 1.16)

This service is commonly known as a high-leg delta or a center-tap delta. The arrangement was more popular years ago than it is today. It was used

**TABLE 1.1**

Some Common International Three-Phase Voltages

| Description | L–NVAC | L–LVAC | Countries | Watt Node Models (Wye or Delta–Voltage) |
|---|---|---|---|---|
| Single-phase, two-wire 120 V with neutral | 120 | — | USA | 3Y-208 |
| Single-phase, two-wire 230 V with neutral | 230 | — | EU, others | 3Y-400 |
| Single-phase, two-wire 208 V (no neutral) | — | 208 | USA | 3D-240 |
| Single-phase, two-wire 240 V (no neutral) | — | 240 | USA | 3D-240 |
| Single-phase, three-wire 120/240 V | 120 | 240 | USA | 3Y-208 |
| Three-phase, three-wire 208 V delta (no neutral) | — | 208 | USA | 3D-240 |
| Three-phase, three-wire 230 V delta (no neutral) | — | 230 | Norway | 3D-240 |
| Three-phase, three-wire 400 V delta (no neutral) | — | 400 | EU, Others | 3D-400 |
| Three-phase, three-wire 480 V delta (no neutral) | — | 480 | USA | 3D-480 |
| Three-phase, three-wire 600 V delta (no neutral) | — | 600 | USA, Canada | None |
| Three-phase, four-wire 208Y/120 V | 120 | 208 | USA | 3Y-208, 3D-240 |
| Three-phase, four-wire 400Y/230 V | 230 | 400 | EU, others | 3Y-400, 3D-400 |
| Three-phase, four-wire 415Y/240 V | 230 | 415 | Australia | 3Y-400, 3D-400 |
| Three-phase, four-wire 480Y/277 V | 277 | 480 | USA | 3Y-480, 3D-480 |
| Three-phase, four-wire 600Y/347 V | 347 | 600 | USA, Canada | 3Y-600 |
| Three-phase, four-wire delta 120/208/240 Wild phase | 120, 208 | 240 | USA | 3D-240 |
| Three-phase, four-wire delta 240/415/480 Wild phase | 240, 415 | 480 | USA | 3D-480 |

*Source:* Table courtesy of Continental Control Systems, LLC, Boulder, CO. All rights reserved.

primarily at small industrial facilities that had a relatively large (240 VAC) motor load but only a small need for (120 VAC) convenience outlets and lighting.

### 1.2.3.2 480 VAC Three-Phase Two-Wire Delta or Three-Phase Three-Wire Delta (Figure 1.17)

This service would typically be used to provide electric service to a user as a remote water pump or a remote sewer pump. Wiring to the user could be made with either two conductors or three conductors. Obviously, the merits

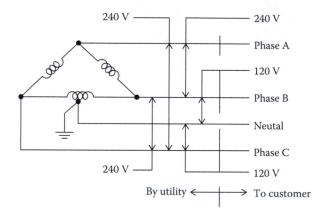

**FIGURE 1.16**
Three-phase four-wire delta service.

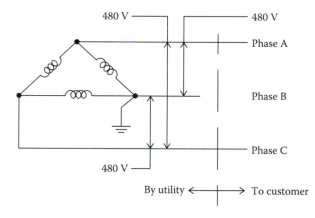

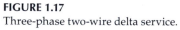

**FIGURE 1.17**
Three-phase two-wire delta service.

of the two conductor configuration are that the branch circuit requires only two conductors instead of three. The voltage in Figure 1.17 is shown as 480 VAC, but it could be less if the current is expected to be small. A large motor in the range of 500 HP or larger would probably be served with a voltage of 4120 VAC or higher.

### 1.2.3.3 208/120 VAC Wye (Three-Phase Four-Wire) (Figure 1.18)

This service in the 208/120 VAC configuration is best suited to users whose requirements are mostly single-phase 120 VAC convenience outlets and overhead lighting but who also have some need for small three-phase motors as central air-conditioning units. A typical user would be a small machine shop, a small retail store, condominiums, or an apartment

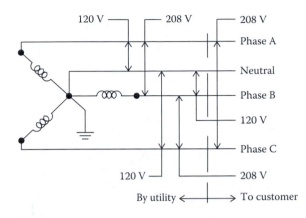

**FIGURE 1.18**
Three-phase four-wire wye service.

complex. This service is used less for new installations today than it was years ago. Nevertheless, many of the installations made years ago with this service are still in use. More recently, the configuration in 480/277 VAC (as shown in Figure 1.19) has become popular for medium-sized commercial and industrial users.

### 1.2.3.4 480 VAC Wye (Three-Phase Four-Wire) (Figure 1.19)

This is a very popular and common electrical service in North America. It is an ideal service for medium-sized commercial and industrial buildings. A typical user would be a multistory commercial building or a medium-sized manufacturing facility. The 480/3/60 VAC would serve air-conditioning units, chillers, or other motors. The 277/3/60 VAC would be

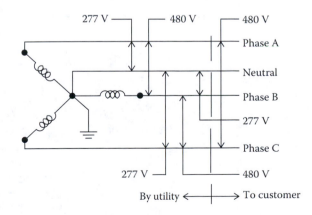

**FIGURE 1.19**
Three-phase four-wire wye service.

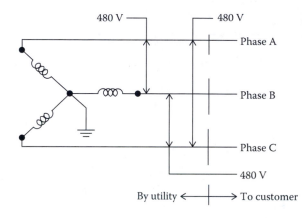

**FIGURE 1.20**
Three-phase Three-wire service.

used for overhead fluorescent lighting. A need for 120 VAC convenience outlets would be satisfied by either a separate single-phase supply or by 120 VAC derived from an in-house transformer.

### 1.2.3.5 480 VAC Wye (Three-Phase Three-Wire) (Figure 1.20)

This service would be more suited to a single user as a remote pumping station.

## 1.2.4 Phases and Lines

According to common usage, the term "phase" as used in reference to three-phase electricity can have two different connotations that can result in some misunderstandings. By common usage, the three conductors of a three-phase circuit are commonly designated as the "phase A" conductor, the "phase B" conductor, and the "phase C" conductor. Depending on the grounding arrangement, some or all of these conductors could be at a potential above the ground. In a four-wire circuit, the fourth conductor is termed the "neutral" conductor. (*Note:* The neutral conductor is at or very near the ground potential, but according to the NEC, it is not to be identified as the "ground" conductor. The subject of grounds and ground conductors is treated in greater detail in Chapter 3.) According to common usage, the term "phase" is also commonly used to designate a voltage potential as in "phase a–b," "phase b–c," or "phase c–a." In a wye circuit, the term "phase" could designate, say, phase a–d, phase b–d, or phase c–d or it could designate "phase a–b", "phase b–c," or "phase c–a."

In a delta circuit, phase currents (Figure 1.14) are currents $I_{ab}$, $I_{bc}$, and $I_{ca}$. In a wye circuit, phase currents (Figure 1.15) are $I_A$, $I_B$, and $I_C$.

The term "line" is commonly used to designate the conductors that bring voltage and current to a "circuit," a "load," or a "user." For example, in the delta circuit of Figure 1.14, the line voltages are $V_{ab}$, $V_{bc}$, and $V_{ca}$; the line currents are $I_A$, $I_B$, and $I_C$. In the wye circuit of Figure 1.15, the line voltages, as with the delta circuit, are $V_{ab}$, $V_{bc}$, and $V_{ca}$; the line currents are $I_A$, $I_B$, and $I_C$. In Figure 1.15, voltages $V_{ad}$, $V_{bd}$, and $V_{cd}$ are the phase voltages.

It is germane to note that in a delta circuit, the line voltages are the phase voltages, but the phase currents are not necessarily the same as the line currents. In a wye circuit, the phase currents are the line currents, but the phase voltages are not equal to the line voltages.

## 1.2.5 Balanced and Unbalanced Three-Phase Circuits

A three-phase circuit can be balanced or unbalanced. By definition, a balanced circuit is one in which all three currents and the leads or lags of the respective currents are identical. Conversely, an unbalanced circuit is one in which either the currents or the lead and lags of the currents are not identical. It is assumed in this textbook that in all circuits, balanced or unbalanced, all three potentials (Figures 1.14 and 1.15), $V_{ab}$, $V_{bc}$, and $V_{ca}$, are identical. Likewise, it is assumed that in a wye circuit, the potentials $V_{ad}$, $V_{bd}$, and $V_{cd}$ are identical and, further, that point "d" (Figure 1.15) is at or near the ground potential.

## 1.2.6 Lead, Lag, and Power Factor in a Balanced Three-Phase Circuit

The term "power factor" is commonly used with regard to both single-phase and (balanced) three-phase ac circuits. It is a value that is used to compute the power of a circuit, and it is commonly used in the description of motor characteristics. Power factor in linear circuits are determined by the lead or lag of current with regard to voltage. For balanced three-phase loads, delta or wye, power may be calculated by using power factor in the equation

$$P = (\sqrt{3})\ V_L\, I_L\, (PF), \qquad \text{(1.6) (Reference 1.5)}$$

where
P is the power (watts)
$V_L$ is the voltage (rms voltage)
$I_L$ is the current (rms amperage)
PF is the power factor = $\cos \theta_P$

$\theta_P$ is the angle of lead or angle of lag of phase current with respect to phase voltage (degrees or radians). (The subscript "p" designates "phase," i.e., $\theta_P$ designates lead/lag in a phase.) By normal convention, the quantity $\theta_P$ is considered positive if the current leads voltage and negative if current lags voltage. Since $\theta_P$ will always be between $-90°$ and $+90°$, the cosine will always be positive. Nevertheless, a power factor of a lagging current is sometimes called a negative power factor, and a power factor of a leading current is sometimes called a positive power factor.

Three-phase power and the applicable equations are treated in detail in Chapter 5.

Power factor is often stated as the ratio of "real power" to "imaginary power." (*Note:* Imaginary power is also known as "total power.") In those instances where power, line voltage, and line current are known, power factor may be computed for balanced three-phase loads by the expression

$$PF = P \div (V_L \cdot I_L \cdot \sqrt{3}), \text{ also}$$

$$PF = P \div (\text{volts–amps}).$$

NOTE: In single-phase circuits, the term "volt-amps" is defined as the circuit potential (voltage) times the circuit current (amperage). In other words, in single-phase circuits, volt-amps $= (V_L \cdot I_L)$. However, in three-phase balanced circuits, volt-amps $= (V_L \cdot I_L \cdot \sqrt{3})$. In balanced three-phase circuits, the term "volts–amps" is equal to the volts times the amps times the square root of 3. Thus, the term "volt-amps" as commonly used with reference to three-phase circuits is a misnomer and can lead to some confusion. If care is not taken, the use of the value "volts–amps" in three-phase computations can also lead to incorrect computations. In unbalanced three-phase circuits, the term "volts–amps" has no significance.

As with single-phase circuits, the phase current in resistive loads is in-phase with phase voltage. Capacitive elements cause phase current to lead phase voltage, and inductive elements cause phase current to lag phase voltage. Typical resistive loads would be heaters or incandescent lighting. A capacitor bank installed to counteract a lagging power factor would present a capacitive load. Common inductive loads include repulsion induction motors, which represent the most common form of three-phase motors and which always have a lagging power factor.

The equation that determines power in a balanced, linear three-phase circuit involves use of the power factor. The power factor by definition is the cosine of the angle between phase voltage and phase current (and not line current). If the angle of lead or lag is known for the line current with respect to line voltage and the phase lead/lag is desired, the angle of line lead/lag in a phase is determined by use of the relationship

$$\theta_P = \theta_L - 30°, \tag{1.7}$$

where
  $\theta_L$ is the line lead/lag angle between line current and line voltage (degrees or radians) (for lagging current, $\theta_L < 0$; for leading current, $\theta_L > 0$)
  $\theta_P$ is the phase lead/lag angle between phase current and phase voltage (degrees or radians) (for lagging current, $\theta_P < 0$; for leading current, $\theta_P > 0$)

Specifically for balanced delta circuits,

$$\theta_{P-CA} = \theta_{L-A/CA} - 30°,$$

where
$\theta_{P-CA}$ is the lead/lag of current in phase C–A with respect to voltage C–A
$\theta_{L-A/CA}$ is the lead/lag of current in conductor A with respect to voltage C–A

$$\theta_{P-AB} = \theta_{L-B/AB} - 30°,$$

where
$\theta_{P-AB}$ is the lead/lag of current in phase A–B with respect to voltage A–B
$\theta_{L-B/AB}$ is the lead/lag of current in conductor B with respect to voltage A–B

$$\theta_{P-BC} = \theta_{L-C/BC} - 30°,$$

where
$\theta_{P-BC}$ is the lead/lag of current in phase B–C with respect to voltage B–C
$\theta_{L-C/BC}$ is the lead/lag of current in conductor C with respect to voltage B–C

For balanced or unbalanced wye circuits,

$$\theta_{P-A/AD} = \theta_{L-A/CA} - 30°,$$

where
$\theta_{P-A/AD}$ is the lead/lag of current in phase A–D with respect to voltage A–D
$\theta_{L-A/CA}$ is the lead/lag of current in conductor A with respect to voltage C–A

$$\theta_{P-B/BD} = \theta_{L-B/AB} - 30°,$$

where
$\theta_{P-B/BD}$ is the lead/lag of current in phase B–D with respect to voltage B–D
$\theta_{L-B/AB}$ is the lead/lag of current in conductor B with respect to voltage A–B

$$\theta_{P-C/CD} = \theta_{L-C/BC} - 30°,$$

where
$\theta_{P-C/CD}$ is the lead/lag of current in phase C–D with respect to voltage C–D
$\theta_{L-C/BC}$ is the lead/lag of current in conductor C with respect to voltage B–C

### 1.2.7 Phasor Diagrams of Three-Phase Circuits

Phasors are handy tools that can be used to calculate currents and power in three-phase circuits. The features offered by phasor diagrams are especially useful when dealing with unbalanced three-phase circuits.

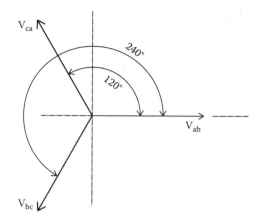

**FIGURE 1.21**
Three-phase voltage phasors.

Construction of a phasor diagram for a three-phase circuit essentially follows the rules applicable to single-phase circuits. Phasors can represent both voltages and currents. Typical phasors representative of the voltages of the three-phase circuits of Figures 1.13 through 1.15 are shown in Figure 1.21. As is the case with a single-phase circuit, projections of the three vectors on the ordinate (*y*-axis) describe the absolute values of voltages with time for the three phases. The phasors of Figure 1.21, in accordance with the rules for the construction of phasor diagrams, are shown at $t = 0$. Note that the assumed sequence of voltage as depicted in Figure 1.13 is A–B–C. According to the rules for the construction of phasor diagrams, the phasor for $V_{ca}$ is positioned 120° counterclockwise from $V_{ab}$, and $V_{bc}$ is positioned 240° counterclockwise from $V_{ab}$. The voltage phasors of Figure 1.21 represent line voltages and therefore would be applicable to both a delta circuit and a wye circuit.

A phasor diagram with phasors for the currents in a typical, balanced delta circuit with resistive loads is shown in Figure 1.22. Each current vector is shown as in phase with the phase voltage, which is the line voltage in a delta circuit.

A phasor diagram for a wye circuit showing both the line voltage phasors and the phase voltage phasors is represented in Figure 1.23. The phase voltages (viz., $V_{db}$, $V_{da}$, and $V_{dc}$) in a wye circuit lead the respective line voltages ($V_{ab}$, $V_{ca}$, and $V_{bc}$) by 30°. Also, the phase voltages in a wye circuit are necessarily of a magnitude

$$\left| V_{db} \right| = \left| V_{da} \right| = \left| V_{dc} \right| = \left| V_{ab} \div \sqrt{3} \right|$$

$$= \left| V_{ca} \div \sqrt{3} \right| = \left| V_{bc} \div \sqrt{3} \right| \text{ (Reference 1.6).}$$

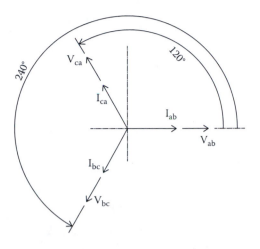

**FIGURE 1.22**
Phasor diagram of balanced delta circuit with resistive load.

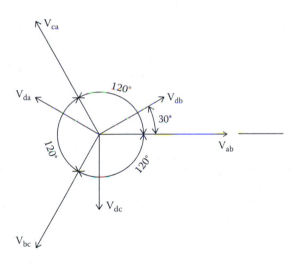

**FIGURE 1.23**
Wye circuit phasor diagram showing phase and line voltages.

A phasor diagram showing the voltage and current phasors of a typical, balanced wye circuit with resistive loads is represented in Figure 1.24. As in the case of a delta circuit, the phasors representative of the currents ($I_{db}$, $I_{da}$, and $I_{dc}$) in the phases ($V_{db}$, $V_{da}$, and $V_{dc}$) are in line with the phasors representative of the phases. It may be noted that the currents in either a balanced delta or a balanced wye circuit with resistive loads lead line voltages by 30°. Phasor diagrams for a large variety of configurations of delta and wye circuits are treated in Chapter 4.

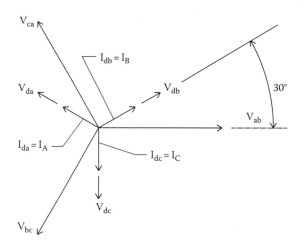

**FIGURE 1.24**
Wye circuit phasor diagram of balanced resistive load.

## 1.2.8 Advantage of Three-Phase Electricity over Single-Phase Electricity

Why use three-phase electricity instead of single-phase electricity? In comparison to single-phase electricity, three-phase electricity offers several advantages including

1. Smaller diameter transmission cables and therefore lower capital costs for transmission gear
2. Lower transmission power losses
3. Fewer number of transmission cables (three cables for three-phase circuits instead of six cables for the equivalent single-phase circuits)
4. Better suited to powering motors (the features of three-phase motors are explained in greater detail in Chapter 9)

The features of three-phase electricity explain the greater suitability of three-phase circuits to transmission and the operation of motors. These merits of three-phase power for transmission can be illustrated by means of the following simple example.

**Example 1.5**

Assume for the purposes of illustration that a given amount of power is to be transmitted, first, by three single-phase circuits and, second, by means of a single three-phase circuit. The merits of each arrangement are considered.

Single-phase power transmission:
 Assume three single-phase circuits are each to transmit 10 kW of power at 480 V and unity power factor over copper conductors for a distance of 100 m.

Each of the single-phase loads will require two conductors to transmit the electrical power over the distance of 100 m. In total, six conductors will be required to transmit the single-phase power.

$$P = VI$$

$$I = P \div V = 10,000 \div 480 = 20.833 \text{ amps}$$

So, six conductors are required for the three single-phase circuits, and each conductor will carry 20.833 amps.

Assume a voltage drop of 3% (maximum allowable voltage drop per the U.S. NEC requirement for branch circuits):

$$03 \times 480 = 14.4 \text{ V or } 7.2 \text{ V per conductor}$$

$$V = IR$$

$$R = V \div I = 7.2 \div 20.833 = 0.3456 \ \Omega$$

$$R = \rho \ (l \div A),$$

where
R is the resistance (ohms)
$\rho$ is the resistivity of copper conductor $= 1.68 \times 10^{-8}$ (ohmmeter)
*l* is the distance (meters)
A is the cross-sectional area of conductor (sq. meter)

$$A = \rho \ (l \div R) = 1.68 \times 10^{-8} \text{ (ohmmeter) } (100 \div 0.3456) \ (\text{m}^2)$$

$$A = 1.68 \times 10^{-8} \ (289.35) = 486.11 \times 10^{-8} = 4.611 \times 10^{-6} \ (\text{m}^2)$$

$$A = (\pi \div 4) \ d^2 = 4.611 \times 10^{-6}$$

$$d^2 = (4.611 \times 10^{-6}) \ (4 \div \Pi) = 5.870 \times 10^{-6}$$

$$d = 2.422 \times 10^{-3} \text{ m} \cdot 1000 \text{ mm}/1 \text{ m}$$

$$d = 2.422 \text{ mm}$$

The net volume of copper becomes

$$\text{Vol} = (6) \ (4.611 \times 10^{-6}) \ (100) = 2766.6 \times 10^{-6} \text{ m}^3 = 2.7666 \times 10^{-3} \text{ m}^3.$$

Calculate weight of copper:

$$\text{Wt.} = 8960 \ (\text{kg/m}^3) \cdot 2.7666 \times 10^{-3} \text{ m}^3 = 24.783 \text{ kg}$$

The power loss becomes

$$P = (6)\ I^2R = (6)\ (20.833^2)\ (0.3456) = 899.9\ \text{W}.$$

Next, consider a circuit that transmits the same amount of power (30 kW) by a three-phase circuit.

Three-phase power transmission:
  According to Equation 1.6,

$$P = (\sqrt{3})\ V_L\ I_L\ (\text{PF}).$$

As stated, PF = 1.0.
Therefore,

$$I_L = P \div (\sqrt{3}V_L) = (30{,}000) \div (\sqrt{3} \cdot 480) = 30{,}000 \div 831.384 = 36.084\ \text{amp}.$$

As with the single-phase circuits, assume a voltage drop of 3%, namely, 7.2 volts per conductor.

$$V = IR$$

$$R = V \div I = 7.2 \div 36.084 = 0.1995\ \Omega$$

$$A = \rho\ (l \div R) = 1.68 \times 10^{-8}\ (\text{ohmmeter})\ (100 \div 0.1995)\ (\text{m}^2)$$

$$= (1.68 \times 10^{-8})\ (501.25)$$

$$= 842 \times 10^{-8} = 8.42 \times 10^{-6}$$

$$d^2 = (8.42 \times 10^{-6})\ (4 \div \Pi) = 10.72 \times 10^{-6}$$

$$d = 3.27 \times 10^{-3} = 3.27\ \text{mm}$$

The net volume of copper becomes

$$\text{Vol} = (3)\ (8.42 \times 10^{-6})\ (100) = 2526 \times 10^{-6}\ \text{m}^3 = 2.526 \times 10^{-3}\ \text{m}^3.$$

Calculate weight of copper:

$$\text{Wt} = 8960\ (\text{kg/m}^3) \times 2.526 \times 10^{-3}\ \text{m}^3 = 22.632\ \text{kg}$$

The power loss becomes

$$P = (3)\ I^2R = (3)\ (36.084^2)\ (0.1995) = 779.2\ \text{W}.$$

The following is a summary of the calculated values:

| Circuit | Net Wt Copper (kg) | Net Power Loss (W) | Number of Conductors |
|---|---|---|---|
| Three single phase | 24.783 | 899.9 | 6 |
| One three phase | 22.632 | 779.2 | 3 |

The use of a three-phase circuit instead of three single-phase circuits would result in

1. An 8.7% reduction in the weight of copper
2. A 13.4% reduction in power losses
3. A 50% reduction in the number of cables

The example demonstrates that a three-phase circuit would be the better choice for the transmission of electrical power.

## 1.3 ac Impedance and Admittance

### 1.3.1 Impedance

The topic of "impedance" follows naturally after the subject of "phasors." Use of an impedance value permits the convenient solution of some ac circuit problems that would otherwise become relatively challenging. The concept of ac impedance is typically combined with the aforementioned phasor theory and a few simple complex number basics to arrive at values of current, voltage, and current lead/lag.

Basically, there are three types of elements that may be found in ac circuits. These may be classified as resistive, inductive, and capacitive. The value of resistance is independent of frequency, but the value of both an inductive circuit and a capacitive circuit is dependent on voltage frequency.

If a circuit contains only resistive elements, the value of current in an ac circuit can be calculated by the relationship

$$I = V/R,$$

where R is the resistance (ohms).

In a purely resistive circuit, the current is in phase with the applied voltage.

If a circuit contains only inductive elements, the value of current can be determined by the relationship

$$I = V/\omega L,$$

where
   V is the volts (ac)
   $\omega = 2\pi f$ (rads/s)
   L is the inductance (H)

In a purely inductive circuit, the current lags the applied voltage by 90° in a V–I phasor diagram.

If a circuit contains only capacitive elements, the value of current can be determined by the relationship

$$I = V\omega C.$$

where C is the capacitance (F)

In a V–I phasor diagram for a purely capacitive circuit, the current leads the applied voltage by 90°.

The aforementioned relationships of voltage and current are interesting and of value. It will be found that when computing voltages and currents that have a mix of these elements, the concept of impedance will have a very practical use.

In order to use impedance values, a person must first understand a few, simple rules applicable to complex numbers. Complex numbers are usually expressed in one of the following three forms:

Polar form: $z = |Z|/\beta$

Exponential form: $z = e^{j\beta}$

Rectangular form: $z = x + iy$

NOTE: The exponential form is not used in this textbook.

The rules for addition, subtraction, multiplication, and division of complex numbers can be performed with relatively simple algebraic and trigonometric equations. A brief review of these procedures is presented here.

Traditionally, the real numbers "$x$" and "$y$" define a complex number "$z$." The number $x$ is called the real component, and $y$ is called the imaginary component. A typical complex diagram is shown in Figure 1.25. According to the convention of representing complex numbers, the complex vector, $z$, has abscissa component $x$ and ordinate component $y$. The vector $z$ can be expressed in several ways. The polar form may be converted to the rectangular form and vice versa.

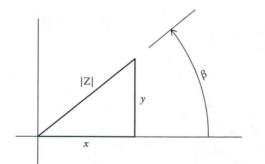

**FIGURE 1.25**
Complex number.

If $z = x + iy$, then $z = |z|/\beta$, where $|z| = [x^2 + y^2]^{\frac{1}{2}}$, $\beta = \tan^{-1}(y/x)$, and $i^2 = -1$.

If $z = |z|/\beta$, then $z = x + iy$, where $x = |z| \cos \beta$ and $y = |z| \sin \beta$.

Addition of complex numbers:

If $z_3 = z_1 + z_2$, where $z_1 = x_1 + iy_1$ and $z_2 = x_2 + iy_2$, then
$z_3 = (x_1 + x_2) + i(y_1 + y_2)$.

Subtraction of complex numbers:

If $z_3 = z_1 - z_2$, where $z_1 = x_1 + iy_1$ and $z_2 = x_2 + iy_2$, then
$z_3 = (x_1 - x_2) + i(y_1 - y_2)$.

Multiplication of complex numbers:

If $z_3 = z_1 \cdot z_2$, where $z_1 = x_1 + iy_1$ and $z_2 = x_2 + iy_2$, then
$z_3 = (x_1 x_2 - y_1 y_2) + i(x_1 y_2 + x_2 y_1)$.

Division of complex numbers:

If $z_3 = z_1 \div z_2$, where $z_1 = x_1 + iy_1$ and $z_2 = x_2 + iy_2$, then
$z_3 = [(x_1 x_2 + y_1 y_2) \div (x_2^2 + y_2^2)] + i [(x_2 y_1 - x_1 y_2) \div (x_2^2 + y_2^2)]$.

By tradition, impedance is represented by a system of symbols and letters that emulate the symbols of complex numbers. When used to describe impedance, the letter "Z" is used, and it simulates the $z$ used in complex notation. The letter "R" is used in place of "$x$," and "X" is used in place of "$y$." (The traditional use of "X" for impedance is in some regard unfortunate since it represents a value parallel to the ordinate. In complex numbers, an "$x$" is always parallel to the abscissa. Nevertheless, this is the common practice.) The letter "$j$" is used instead of "$i$" to designate the imaginary value. If these conversions are made, then the rules for impedance become

If $Z = R + jX$, then $Z = |Z|/\beta$, where $|Z| = [R^2 + X^2]^{\frac{1}{2}}$ and $\beta = \tan^{-1}(X/R)$.

If $Z = |Z|/\beta$, then $Z = R + jX$, where $R = |Z| \cos \beta$ and $X = \sin \beta$.

Addition of impedances:

If $Z_3 = Z_1 + Z_2$, where $Z_1 = R_1 + jX_1$ and $Z_2 = R_2 + jX_2$,
then $Z_3 = (R_1 + R_2) + j(X_1 + X_2)$.

Subtraction of impedances:

If $Z_3 = Z_1 - Z_2$, where $Z_1 = R_1 + jX_1$ and $Z_2 = R_2 + jX_2$,
then $Z_3 = (R_1 - R_2) + j(X_1 - X_2)$.

Multiplication of impedances:

If $Z_3 = Z_1 \cdot Z_2$, where $Z_1 = R_1 + jX_1$ and $Z_2 = R_2 + jX_2$,
then $Z_3 = (R_1R_2 - X_1X_2) + j(R_1X_2 + R_2X_1)$.

Division of impedances:

If $Z_3 = Z_1 \div Z_2$, where $Z_1 = R_1 + jX_1$ and $Z_2 = R_2 + jX_2$,
then $Z_3 = [(R_1R_2 + X_1X_2) \div (R_2^2 + X_2^2)] + j[(R_2X_1 - R_1X_2) \div (R_2^2 + X_2^2)]$.

An understanding of these relatively simple rules is all that is needed to perform the computations related to impedance as well as any other subjects treated in this textbook. A typical impedance diagram is shown in Figure 1.26. As represented in the diagram, the abscissa component of the vector Z is "R," and the ordinate component is "X."

Basically, ac impedance is the complex (not the scalar) ratio of volts to amps in an ac circuit. Impedance represents the ability of an ac circuit to resist the flow of current. Impedance is also the ratio of two phasors, but it is not a phasor. It is a complex number that connects one phasor to another phasor. Expressed as "Z," impedance may be stated mathematically as

$$\underline{Z} = \underline{V}/\underline{I}. \tag{1.8}$$

NOTE: An underlined variable, as "$\underline{Z}$," designates that the variable is a vector quantity and not a scalar quantity. However, the practice of underlining variables to designate a vector quantity is not rigorously followed. In most instances, it is understood that a stated variable is a vector quantity and not necessarily underlined. Mostly, the underline serves as a reminder that the variable is a vector quantity. A variable that is designated as an absolute value, as $|Z|$, specifically designates that the variable is a scalar quantity and not to be confused with a vector quantity.

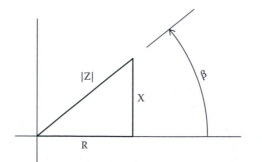

**FIGURE 1.26**
Impedance diagram.

For a purely resistive circuit, $Z_R = R$, where R is the value in ohms of the resistance.

It may also be shown that,

- For a purely inductive circuit, the impedance is $Z_L = j\omega L$.
- For a purely capacitive circuit, the impedance is $Z_C = 1/j\omega C$.

The merits of impedance in computing circuit values can best be demonstrated with some specific examples. Consider, first, a resistive circuit.

Resistive:
In the phasor domain, or what is known as the frequency domain,

$$\underline{V} = \underline{I} \cdot \underline{Z}R.$$

### Example 1.6

Consider a case in which the voltage across a resistor is given by the time domain equation

$$v = 120 \sin (\omega t + \theta_{SP}).$$

Then by definition,

$$i = 120 \, [\sin (\omega t + \theta_{SP})/R.$$

Let $\theta_{SP} = 135°$. The current in a resistor is necessarily in phase with the voltage. In the frequency domain, and written in polar notation, the voltage is

$$\underline{V} = 120\underline{/135°}.$$

Since

$$\underline{I} = \underline{V}/R,$$

$$\underline{I} = (120\underline{/135°})/R.$$

If, say, $R = 10 \, \Omega$, then $\underline{I} = (120\underline{/135°})/10 = 12\underline{/135°}$.

The phasor representation of the voltage and current of Example 1.6 is shown in Figure 1.27.

Inductive:
In the frequency domain, $\underline{V} = j\omega L\underline{I}$, where L is the inductance (in henries).

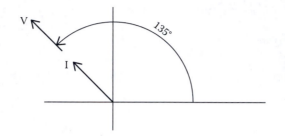

**FIGURE 1.27**
Voltage and current phasors, Example 1.6.

### Example 1.7

Consider a purely inductive circuit in which

$$v = 20 \sin (\omega t + 40°)$$

$$f = 50 \text{ Hz}$$

$$H = 1.5 \text{ H.}$$

Calculate I:
Expressed in polar notation,

$$\underline{V} = 20\underline{/40°}$$

$$\omega = 2\pi f \text{ (rads/s)}$$

$$\omega = 2\pi(50) = 314.2$$

$$\underline{I} = \underline{V}/(j\omega L) = \underline{V}/(314.2 \cdot 1.5 \cdot j) = \underline{V}/471.3j.$$

In polar notation,

$$471.3j = 471.3\underline{/90°}.$$

Thus,

$$\underline{I} = 20\underline{/40°} \div 471.3\underline{/90°}.$$

Here,  $\underline{I} = \underline{Z_1} \div \underline{Z_2}.$

Let

$$Z_1 = R_1 + jX_1 \text{ \& } Z_2 = R_2 + jX_2$$

$$R_1 = 20 \cos 40° = 15.320$$

$$X_1 = 20 \sin 40° = 12.855$$

$$R_2 = 0$$

$$X_2 = 471.3$$

$$\underline{I} = \underline{Z}_1 \div \underline{Z}_2 = [(R_1 R_2 + X_1 X_2) \div (R_2^2 + X_2^2)] + [(R_2 X_1 - R_1 X_2) \div (R_2^2 + X_2^2)]j$$

$$\underline{I} = [(15.32 \cdot 0) + 12.855 \cdot 471.3) \div (0^2 + 471.3^2)] + [0 - 15.320 \cdot 471.3) \div (0^2 + 471.3^2)]$$

$$\underline{I} = [(12.855 \div 471.3) - (15.320 \div 471.3)\,j$$

$$\underline{I} = .02727 - 0.03250\,j$$

$$|\underline{I}| = .0424.$$

Let $\theta$ be the angular position of I with respect to the positive abscissa, the angle being positive in the CCW direction from the positive abscissa.

$$\theta = -\tan^{-1}(.03250/.02727) = -50°$$

Expressed in polar notation,

$$\underline{I} = .0424\underline{/-50°}.$$

The position of I with respect to the voltage is $40° - (-50°) = 90°$ CW. The computation indicates that the current lags the voltage by 90°, which is what is expected of a purely inductive element. The vectors of the voltage and current phasors are shown in Figure 1.28.

Capacitive:
In the frequency domain, $I = j\omega CV$, where C is the inductance (in henries).

### Example 1.8

Consider a purely capacitive circuit in which the voltage across a capacitor is given by the time domain equation

$$v = 35 \sin (\omega t - 60°)$$

$$f = 60 \text{ Hz}$$

$$C = 20,000\ \mu F = 0.002\ F.$$

Expressed in polar notation,

$$\underline{V} = 35\underline{/-60°} = 17.5 - j30.31$$

$$\omega = 2\pi f \text{ (rads/s)}$$

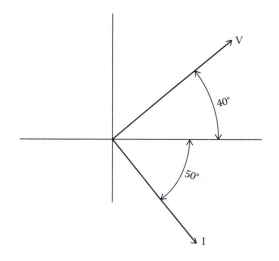

**FIGURE 1.28**
Current lagging voltage, Example 1.7.

$$\omega = 2\pi(60) = 377$$

$$Z_C = V/I = 1/j\omega C$$

$$\underline{I} = V \, j\omega C$$

$$= [17.5 - 30.31 \, j] \, [(377)(0.002)j]$$

$$= [17.5 - 30.31 \, j] \, [j \, (0.754)]$$

$$\underline{I} = 13.19j - j^2(22.85)$$

$$\underline{I} = 22.85 + 13.19j, \text{ or}$$

$$\underline{I} = 26.38\underline{/30}°.$$

The computation indicates that the current leads the voltage by 30° − (−60°) = 90° CCW, which is what is expected for a pure capacitive element. The vectors of the voltage and current phasors are shown in Figure 1.29.

## Combined Resistance and Reactance

In practice, most circuits contain a mix of resistive components along with reactive components that could be inductive or capacitive. In order to analyze these types of circuits, the use of impedance proves to be an especially helpful tool. As stated before, the impedance of a circuit, regardless of the nature of the components, can be described by the relationship

$$\underline{Z} = \underline{V}/\underline{I}.$$

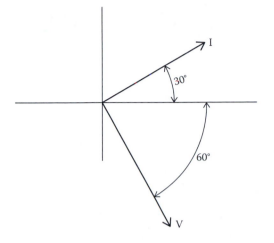

**FIGURE 1.29**
Leading current, Example 1.8.

In a circuit that contains both resistive and reactive elements, the impedance in the complex plane can be described by the general relationship depicted in Figure 1.25, where, in the rectangular form,

$$\underline{Z} = R + jX$$

R is the resistive component of impedance (ohms)
X is the reactive component of impedance (ohms)

$$|Z| = [R^2 + X^2]^{\frac{1}{2}}$$

The value of R is always R > 0°.

In this textbook, impedance diagrams follow the rule that for an inductive circuit, X > 0° and X = $X_L$. Also in this textbook an impedance diagram follows the rule X < 0° and X = $X_C$. Not all textbooks, however, follow this practice.

Stated in polar form, $\underline{Z} = |Z|\underline{/\beta}$, where $\beta = \tan^{-1}(X/R)$.

If the circuit is inductive, in the impedance diagram, β > 0° but β < 90°.

If the circuit is capacitive, in the impedance diagram, β < 0° but β > –90°.

(In this textbook, the Greek symbol θ is used to define the lead or lag of current with respect to voltage. As will be demonstrated, the symbols θ and β are related but not identical. To avoid confusion, the variable β is used to define the ratio [X/R] in an impedance diagram and θ to define the lead or lag of current with respect to voltage.)

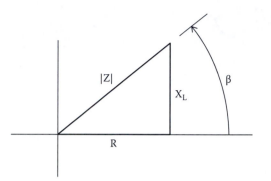

**FIGURE 1.30**
Impedance of an RC circuit.

Briefly stated,
the positive direction of both β and θ is CCW,

$$\theta \neq \beta, \quad \text{but}$$

$$\theta = -\beta$$

$$\cos \theta = \cos \beta = PF.$$

If the circuit components are only resistive and capacitive elements, the impedance diagram would resemble the representation of Figure 1.30. Since impedance is not a vector, the lines representative of impedance are not shown as vectors, that is, the lines do not have arrowheads.

### Example 1.9

A voltage of 240 VAC is imposed on a RL (resistive–inductive) circuit that contains resistance of 40 Ω and an inductive reactance of 50 Ω. Compute the current, power factor, and lead/lag of current with respect to voltage.

$$\underline{Z} = R + jX$$

$$R = 40; X = X_L = +50$$

$$\underline{Z} = 40 + j50$$

$$|Z| = [R^2 + X^2]^{1/2}$$

$$|Z| = [40^2 + 50^2]^{1/2} = 64.03$$

$$Z = V/I$$

$$I = V/Z = 240/64.03 = 3.75 \text{ amps}$$

$$\beta = \tan^{-1}(X/R) = \tan^{-1}(50/40) = 51.34°$$

If the lead/lag of current with respect to voltage is θ, then

$$\theta = -\beta = -51.34°$$

$$PF = \cos -51.34° = .624, \text{ lagging.}$$

The current lags voltage by 51.34°.

The associated impedance diagram is represented in Figure 1.31, drawn to scale, and the phasor diagram of the voltage–current relationship is shown in Figure 1.32. Note that in the impedance diagram, the reactance, which is inductive, is shown in Quadrant I, whereas in the associated phasor diagram, the current vector is shown in Quadrant IV.

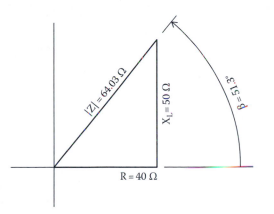

**FIGURE 1.31**
Impedance diagram of a RL circuit.

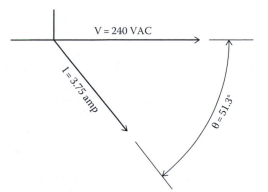

**FIGURE 1.32**
Phasor diagram of lagging current.

**Example 1.10**

A voltage of 120 VAC is imposed on a RC (resistive–capacitive) circuit that contains resistance of 35 Ω and a capacitive reactance of 45 Ω. Compute the current, power factor, and lead/lag of current with respect to voltage.

$$\underline{Z} = R + jX$$

$$R = 35; X = X_C = -45$$

$$\underline{Z} = 35 + j(-45) = 35 - j45$$

$$|Z| = [R^2 + X^2]^{\frac{1}{2}}$$

$$|Z| = [35^2 + 45^2]^{\frac{1}{2}} = 57.00$$

$$Z = V/I$$

$$|I| = V/|Z| = 120/57.00 = 2.10 \text{ amps}$$

$$\beta = \tan^{-1}(X/R) = \tan^{-1}(-45/35) = -52.12°$$

$$\theta = -\beta = 52.12°$$

$$PF = \cos 52.12° = .614$$

The current leads voltage by 52.12°.
It may also be seen that

$$\underline{I} = \underline{V}/\underline{Z}, \text{ or}$$

$$\underline{I} = (120 + j0) \div (35 - j45)$$

$$\underline{I} = 1.292 + j1.661 = 2.10\underline{/52.12°}.$$

The associated impedance diagram is represented in Figure 1.33, drawn to scale, and the phasor diagram of the voltage–current relationship is shown in Figure 1.34. (Note that in the impedance diagram in Figure 1.33, the reactance, which is capacitive, is shown in Quadrant IV, whereas in the phasor diagram, the current vector is shown in Quadrant I.)

## Components in Series
Impedance values are especially useful when the need arises to analyze electrical components connected in series. A typical ac circuit containing several electrical components in series is represented in Figure 1.35. If the impedance of each component is known, the respective current, voltages, and phase lag

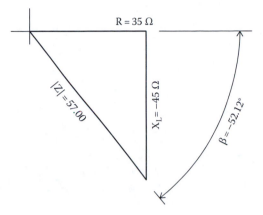

**FIGURE 1.33**
Impedance diagram of a RC circuit.

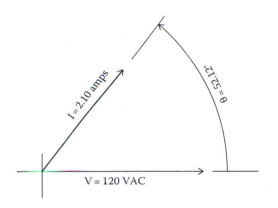

**FIGURE 1.34**
Phasor diagram of leading current.

or lead of each component may readily be determined. The current in each component is equal at every instant. Also, the sum of the voltages must equal the applied voltage. In Figure 1.35, for example,

$$V_T = V_1 + V_2 + V_3.$$

Dividing by the common current, I,

$$V_T/I = V_1/I + V_2/I + V_3/I.$$

By definition, Z = V/I
Therefore,

$$Z_T = Z_1 + Z_2 + Z_3.$$

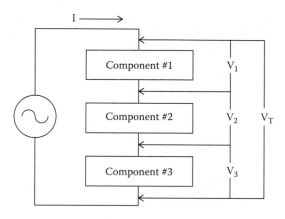

**FIGURE 1.35**
Components in series.

More generally,

$$Z_T = Z_1 + Z_2 + \cdots Z_N.$$

Once the overall impedance is known, the common current and the voltages across the respective components may readily be calculated.

### Example 1.11

Reference is made to Figure 1.36. A voltage of 480 VAC is applied to three electrical components connected in series. The components have the impedances $Z_1$, $Z_2$, and $Z_3$, which are

$$\underline{Z_1} = 10 + j8$$

$$\underline{Z_2} = 6 + j12$$

$$\underline{Z_3} = 14 - j6.$$

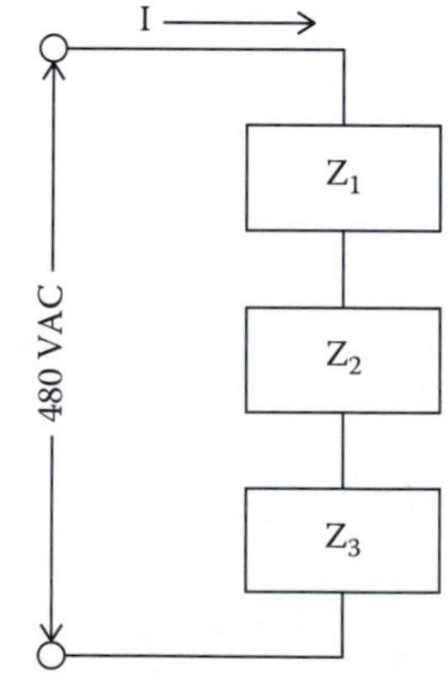

**FIGURE 1.36**
Components in series.

Determine the common current, current lead or lag, and the voltage across each component.

The overall impedance is $\underline{Z_T} = \underline{Z_1} + \underline{Z_2} + \underline{Z_3}$.

$$\underline{Z_T} = (10 + 6 + 14) + j(8 + 12 - 6)$$

$$= 30 + j14$$

$$|Z_T| = [R^2 + X^2]^{1/2}$$

$$= [30^2 + 14^2]^{1/2} = 33.105 \ \Omega$$

$$\underline{I} = V/\underline{Z} = V/(30 + j14)$$

$$= V[(30 - j14)/(30^2 + 14^2)]$$

$$= (480) \ [.02737 - j.01277]$$

$$= 13.1386 - j6.1313$$

$$\underline{I} = 14.497 \ \underline{/-25.01^\circ}$$

The circuit current, I, lags applied voltage $V_T$ by $25.016^\circ$.

$$\underline{V_1} = \underline{I}(\underline{Z_1}) = [13.1386 - j6.1313][10 + j8]$$

$$= 180.4136 + j43.795 = 185.653\underline{/13.64^\circ}$$

$$\beta_1 = \tan^{-1}(X/R) = \tan^{-1}(8/10) = 38.654^\circ$$

$$\theta_1 = -\beta = -38.654^\circ$$

The current lags voltage $V_1$ by $38.654^\circ$. (Note: According to the calculations, voltage $V_1$ is positioned $13.64^\circ$ CCW from the positive abscissa in the phasor diagram, and the current is positioned at $25.01^\circ$CW from the positive abscissa. So, the current is positioned CW from voltage $V_1$ by $25.01^\circ + 13.64^\circ$ or $38.65^\circ$ and is considered lagging. This value of lag checks with the value of $\theta$ computed from the given value of impedance.)

$$V_2 = \underline{I} \ (\underline{Z_2}) = [13.1386 - j6.1313] \ [6 + j12] = 152.4072 + j120.875$$

$$= 194.521\underline{/38.42^\circ}$$

$$\beta_2 = \tan^{-1}(X/R) = \tan^{-1}(12/6) = 63.435^\circ$$

$$\theta_2 = -\beta = -63.435^\circ$$

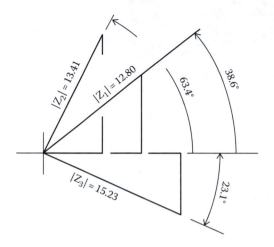

**FIGURE 1.37**
Example 2.11: Impedance.

The current lags voltage $V_2$ by 63.435°.

$$\underline{V}_3 = \underline{I}(Z_3) = [13.1386 - j6.1313][14 - j6] = 147.1526 - j164.66$$

$$= 220.832\,\underline{/-48.208°}$$

$$\beta_3 = \tan^{-1}(X/R) = \tan^{-1}(6/14) = 23.198°$$

$$\theta_3 = -\beta = -23.198°$$

The current leads voltage $V_3$ by 23.198°.

    The values of the impedances are shown in Figure 1.37. The computations may be confirmed by adding the voltages $V_1$, $V_2$, and $V_3$.
Determine: $\underline{V}_1 + \underline{V}_2 + \underline{V}_3$

$$\underline{V}_1 + \underline{V}_2 + \underline{V}_3$$

$$= 480 + j0$$

$$|V_T| = 480 \text{ VAC}$$

Thus, the computations confirm that the initial determination of the current and the voltages of the components are correct. The computed voltages, drawn to scale, as well as the amperage are shown in Figure 1.38.

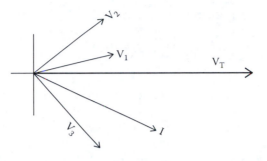

**FIGURE 1.38**
Example 2.11: voltages.

### 1.3.2 Admittance

In the previous section, the concept of impedance is defined and explained. As demonstrated, values of impedance are very helpful when dealing with components connected in series. With components connected in parallel, the value of "admittance" is more suited. Admittance is defined as

$$\underline{Y} = 1/\underline{Z} \tag{1.9}$$

where
  $\underline{Y}$ = admittance
  $\underline{Z}$ = impedance

If the impedance is known and in the form

$$Z = R + jX$$

$$\underline{Y} = 1/(R + jX) = (R - jX)/(R^2 + X^2).$$

Generally, admittance is expressed as

$$\underline{Y} = G + jB, \text{ whereby}$$

$$G = R/(R^2 + X^2) B = -X/(R^2 + X^2).$$

It follows that

$$Y_R = 1/R$$

$$Y_C = j\omega C$$

$$Y_L = -j/\omega L.$$

If the admittance is known, then the impedance may be determined from the relationship

$$\underline{Z} = 1/(G + jB) = (G - jB)/(G^2 + B^2).$$

By common usage, the term "G" is known as the conductance, and the term "B" is known as susceptance. The units of both G and B are called "siemens." As noted, the impedance is $Z = R + jX$. If the value of X, the imaginary component, is positive, the element is inductive, and if the value of X is negative, the element is capacitive. The converse is true in the case of admittance. The element is capacitive if imaginary component B of the admittance is positive and inductive if B is negative.

Consider a typical circuit, as represented in Figure 1.39, of two components connected in parallel. Each element is subjected to the same voltage but the currents may be different. As represented in Figure 1.39, the net current consists of the sum of the currents in each of the elements connected in parallel. So,

$$\underline{I_T} = \underline{I_1} + \underline{I_2},$$

where
  $I_T$ is the total current
  $\underline{I_1}$ is the current flowing through Component #1
  $\underline{I_2}$ is the current flowing through Component #2

Since all currents are subjected to the same potential,

$$\underline{I_T} / \underline{V} = \underline{I_1} / \underline{V} + \underline{I_2} / \underline{V} \quad \text{or}$$

$$1/Z_T = 1/Z1 + 1/Z2,$$

where $Z_T$ is the net impedance.

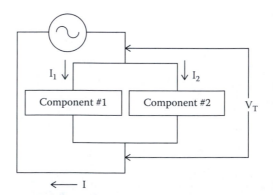

**FIGURE 1.39**
Parallel circuits.

By definition, $Y = 1/Z$.
So,

$$Y_T = Y_1 + Y_2.$$

More generally,

$$Y_T = Y_1 + Y_2 + \cdots Y_N.$$

### Example 1.12

Consider a circuit similar to that represented in Figure 1.39 wherein two components are connected in parallel. The applied voltage is 240 VAC. The admittances have been determined to be as follows:

$$Y_1 = .00591 - j(.0142)$$

$$Y_2 = .00666 - j(.0200)$$

Calculate the net current, $I_T$, and the current in each component.

Answer:

$$Y_T = Y_1 + Y_2$$
$$= [.00591 - j(.0142)] + [.00666 - j(.0200)]$$
$$= .01257 - j(0.0342)$$

$$I_T = V/Z_T = VY_T = (240)(.01257 - j0.0342)$$
$$= 3.0168 - j8.208 = 8.744 \underline{/-69.81°}$$

$$I_1 = V/Z_1 = VY_1 = (240)[.00591 - j(.0142)]$$
$$= 1.418 - j3.408 = 3.69 \underline{/-67.40°}$$

$$I_2 = V/Z_2 = VY_2 = (240)[.00666 - j(.0200)]$$
$$= 1.598 - j4.80 = 5.06 \underline{/-71.58°}$$

Check to confirm that $I_T = I_1 + I_2$

$$I_1 + I_2 = (3.69 \underline{/-67.40°}) + (5.06 \underline{/-71.58°})$$
$$= (1.418 - j3.408) + (1.598 - j4.80)$$
$$= 3.016 - j8.208 = 8.744 \underline{/69.81°}$$

Thus, the addition of the currents in Component #1 and Component #2 checks with the computed net current. This example demonstrates how the use of admittance values can simplify the computations associated with components connected in parallel.

## Components in Parallel and in Series

Here, the concepts of impedance and admittance are treated. As demonstrated, impedance is a tool that is well suited to components connected in series. Admittance is a better starting point for elements connected in parallel. Of course, a circuit may contain a number of components; some of which are connected in parallel and others connected in series. A typical series–parallel circuit is represented in Figure 1.40. How to treat a circuit of this mix? The answer may suggest itself. The impedance of the components connected in parallel is determined first, and then the mix is treated as components in series.

### Example 1.13

Consider the configuration of Figure 1.40. Assume the component properties to be as follows:

$$Z_1 = 17 + j25; \quad Z_2 = 22 + j40; \quad Z_3 = 12 + j36; \quad V = 480 \text{ VAC}$$

Determine current I, expressed in polar notation.

**Answer:**

Let Y be the admittance of Components #2 and #3, where

$$Y_1 = (R - jX)/(R^2 + X^2)$$

$$= (17 - j25)/(17^2 + 25^2)$$

$$= .01859 - j(.02735)$$

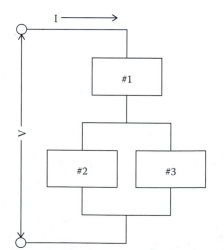

**FIGURE 1.40**
Parallel and series circuit.

$$Y_2 = (R - jX)/(R^2 + X^2)$$

$$= (22 - j40)/(22^2 + 40^2)$$

$$= .04545 - j(.08264)$$

$$Y = Y_1 + Y_2 = .06404 - j(.1099), \quad \text{where}$$

$$G = .06404 \quad \text{and} \quad B = -.1099.$$

Next, the equivalent impedance of the two components connected in parallel is determined.
For Components #2 and #3,

$$\underline{Z_{2,3}} = (G - jB)/(G^2 + B^2) = (.06404 + j.1099)/(.06404^2 + .1099^2)$$

$$= 15.615 + j26.797.$$

Let $Z_T$ be the net circuit impedance

$$\underline{Z_T} = \underline{Z_1} + \underline{Z_{2,3}} = [17 + j25] + [15.615 + j26.797] = 32.615 + j51.797$$

$$\underline{I} = \underline{V}/\underline{Z}$$

$$\underline{I} = 480/(32.615 + j51.797) = 4.17 - j6.63 = 7.83\underline{/-57.83°}.$$

---

## Problems

1. One phase of a three-phase, 50 Hz circuit has a lagging current and a power factor of 0.60. What is the lag in seconds of current with respect to voltage?

2. At a hotel in France, management advises that the voltage of the convenience outlets is "230 volts, 50 cycle." What is the value of voltage measured from positive peak to negative peak?

3. A balanced three-phase delta circuit has a lagging power factor of 0.90. State the lead or lag of line current with respect to line voltage and the angle of lead or lag with respect to line voltage.

4. A balanced wye circuit has phase currents of 20 amps and a lagging PF of 0.80. State line current and lead or lag with respect to line voltage.

5. A 208/3/60 induction motor has a nameplate current of 5.3 FLA (full load amps) and a nameplate power factor of 0.75. What is the lead or lag of line current with respect to voltage?

6. Readings on the line to a three-phase motor determined that the current was lagging line voltage by 20°. What is the power factor?

8. The readings on the line going to a three-phase motor indicated V = 480 VAC, I = 12, and P = 8000 W. Assuming the load to be balanced and that the motor draws a lagging current, what is the lead/lag of line current with respect to line voltage?

9. In a balanced wye circuit with all resistive loads, what is the lead or lag of line currents with respect to line voltages?

10. The abscissa vector components determining the phasor of the current in conductor A of a three-phase circuit are $12 + 7 - 9 + 3$. The ordinate components of the phasor determining the current in conductor A are $9 - 14 + 3 - 2$. Determine the magnitude of the current in conductor A.

11. Let the phasor for current $I_1$ have abscissa component $I_{1-x}$ and ordinate component $I_{1-y}$, and the phasor for $I_2$ have abscissa component $I_{2-x}$ and ordinate component $I_{2-y}$. If $I_3 = I_2 - I_1$, what is the abscissa component of $I_3$? What is the ordinate component of $I_3$?

12. A three-phase circuit with conductors A, B, and C delivers voltage to a number of three-phase loads. The voltage sequence is A–B, B–C, and C–A. The sum of the abscissa phasors for current in conductor A is –10, and the sum of the ordinate phasors is +10. Assuming that voltage C–A leads voltage A–B by 120°, determine the lead or lag of the current in conductor A with respect to voltage C–A.

13. The phasor vector for voltage $V_1$ is given by the polar notation $V_1 = 4160\underline{/27°}$. When the voltage is applied to a circuit that is entirely capacitive (except for a negligible amount of resistive components), the resultant current, $I_1$, is 285 amps. State the current in polar notation.

14. A current, $I_1$, is given as $I_1 = 133\underline{/130°}$. State the current as a complex number in rectangular form.

15. An electrical circuit has a resistance of 60 Ω and an inductive reactance of 80 Ω. What is the magnitude of the ac impedance?

16. The ac resistance of an ac solenoid was determined to be 1.1 Ω. A voltage of 120 VAC is applied to the solenoid, and a current of 6 amps is recorded. What is the power consumption of the solenoid?

17. Two electrical components, Components #1 and #2, are connected in-series. The impedances of the devices have been determined to be $\underline{Z_1} = 17 + j21$ and $\underline{Z_2} = 25 - j15$. A voltage of 480 VAC is applied. What is the resultant current and the lead or lag of the current with respect to the applied voltage?

18. A number of electrical components are connected in parallel. The admittance is characterized by a conductance of G = .010 and a susceptance of B = .020. What is the current flow, expressed in polar notation, in the common conductor to the components if the applied voltage is 208 VAC?

19. A circuit consists of Component #1 that is in series with a group of components designated as Group A, which consists of several components wired in parallel. The impedance of Component #1 is R = 23 $\Omega$ and X = +55 $\Omega$. The admittance of Group A is G = .003 siemens and B = .009 siemens. What is the current through Component #1 if a potential of 240 VAC is applied to the circuit? What is the lead/lag of current with respect to voltage across Component #1?

# 2

## Generation, Transmission, and Distribution

Three-phase electricity originates in a generating station that might be owned by a utility, an independent power producer (IPP), or, perhaps, on the premises of a facility. Electricity from a utility is delivered to customers within its jurisdiction, and in return, a customer is billed for delivered service. The transmission of electricity from an electric utility to local switchyards could be over high-tension lines owned by a conglomerate. Local distribution is usually the responsibility of the utility that sends the monthly bill to its customers within its jurisdiction.

In the United States, electric utilities are publically owned. Of course, both the federal government and the states have the authority to pass legislation that regulates the way electricity is generated and sold. In some states, electric rates are determined by the state, whereas in other states, competition is encouraged through "deregulated pricing."

A typical arrangement of electrical generation, transmission, subtransmission, and distribution is represented in Figure 2.1. As represented in the figure, electrical power is generated at a utility's plant, and from there, it is delivered to a grid. The grids are operated at very high voltages. As the electrical power is brought closer to the eventual users, the potential is reduced. While the arrangement of Figure 2.1 may be typical, it is pertinent to note that in recent years, more and more consumers of electricity are moving offline by generating electricity on the premises.

### 2.1 Generation

One of the earlier and popular means of generating electrical power was by hydro-driven generators. Water at an elevation above the water turbines is allowed to pass through the turbines and, in the process, turns the turbines that drive the generators. In time, the larger share of electrical power was generated by steam-driven turbine generators.

Although there are exceptions, three-phase power originates in a three-phase generator. The generator could be driven by a variety of prime movers. Common sources of electrical generation are listed in the following

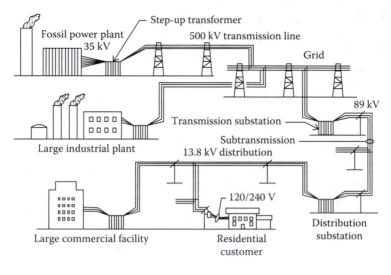

**FIGURE 2.1**
Typical transmission and distribution configuration.

paragraphs. Of course electricity generated by solar power does not involve a generator. Rather, the dc power from the solar cells is converted to ac power by means of inverters.

### 2.1.1 Fossil-Fired Steam Power Plants

In fossil-fired steam power plants, electricity is generated by the burning of a fuel in a boiler in order to generate steam at a relatively high pressure. The steam is directed to a turbine that drives a generator. In most steam power plants, steam exiting the turbine is directed to a condenser where it is cooled, condensed, and then returned to the boiler. In some industrial plants, a portion of the steam exiting a turbine might be directed to points outside the boiler house where it will be used for heating purposes. The term "fossil fuel" is something of a misnomer, since a fossil fuel is not truly a fossil. Rather, a fossil fuel is one that resulted from the anaerobic, or oxygen-free, decomposition of organic matter. The most common fossil fuels are petroleum, coal, and natural gas.

*Coal*: Throughout the twentieth century, coal was the more commonly used fuel for generating steam at electrical power plants. Coal-fired power plants still account for a large proportion of electrical power generation although at a falling rate. As of 2012, approximately 37% of all electrical power in the United States was generated at coal-fired plants (Reference 2.1). In the early part of the twentieth century, coal-fired boilers typically used rotating grates to transport coal through the boiler. As the coal burned, ash would fall through the grates to the bottom of the boiler where it would require subsequent removal. Coal was required to be of a minimum size so that it would not prematurely fall through the grates. Later, innovations as the so-called cyclone burner did away with the travelling grates. The biggest improvement in coal-fired boilers occurred around 1918, with the invention of the pulverized coal burner. Pulverized coal burners brought about a more efficient use of the boiler's furnace chamber and provided a greater thermal efficiency. The pulverized coal burner could also accommodate a wider range of coal types including fines and small-sized coal that had previously been discarded. The burning of coal results in emissions that require treatments that are both costly and limited in capabilities. The undesirable byproducts of burning coal are sulfur, mercury, particulates, and the oxides of nitrogen and carbon dioxide. Treatment of flue gases can moderate or eliminate most of these undesirable components but often at a high cost.

Starting in the 1990s, fluidized bed boilers became popular in North America for generating steam from low-grade coal, namely, coal that contains relatively high sulfur content or coal that had been mixed with earth or rocks. In fluidized bed boilers, the coal is mixed with lime, sand, or reagents in the combustion chamber where it is maintained airborne (fluidized) during the combustion process. Many of the IPPs use this type of boiler to generate electricity that is sold on an open market. One such facility is shown in Figure 2.2. The plant of Figure 2.2 is a 57.5 MW facility that was constructed adjacent to "culm" banks of coal that had been discarded over 60 years ago for the reason that it contained fines and was not marketable at the time. The plant shown in Figure 2.2 is known as the Northeast Power Plant, and it is located in Schuylkill County, Pennsylvania, near the town of McAdoo.

*Petroleum fuels*: A variety of petroleum fuels have been used over the years to fire utility boilers. Crude oil, fired directly into a boiler, has been used in past years although the use of crude oil for this purpose has decreased significantly in recent years. The most common petroleum fuel used in the boilers at electrical power plants is the bunker C grade oil, which is derived from crude oil after the lower volatiles have been extracted. Both crude oil

**FIGURE 2.2**
IPP fluidized bed steam plant. (Photograph by the author.)

and bunker C grades must be heated before injection into a boiler since both are very viscous at ambient temperatures and would not otherwise atomize adequately. Smaller boilers are usually fired with number two oil, which is much less viscous than crude oil or bunker C oil and does not require heating prior to injection into a boiler. Some petroleum products have high sulfur content and are less preferred due to the adverse SOX (sulfur–oxygen compounds) resulting emissions. In fact, many communities in the United States prohibit the firing of oil that has high sulfur content.

*Natural gas*: With the recent development of deep well drilling using fracking techniques, natural gas has become more plentiful and relatively less expensive in many parts of the world. This has become the case particularly in North America. Of all the fossil fuels, natural gas has fewer emission problems. So, with attractive pricing and lower emissions, the use of natural gas has increased. Of course, natural gas can be fired most efficiently in new boilers with furnaces specifically designed for natural gas. Existing coal-fired boilers and oil-fired gas turbine generators can also be retrofitted to accept natural gas as a fuel.

A fossil-fired steam power plant with a condenser requires a means of cooling steam delivered to the condenser. For the purpose of cooling the steam within the condenser, an abundant supply of cooling water is preferred. For this reason, most fossil power plants are located adjacent to a river, lake, or seashore. In some regions of the world, an adequate supply of water may not be available in which case an alternate means of condensing the steam is used. One such method uses an air-cooled condenser. A typical air-cooled condenser at a coal-fired power plant is shown in Figure 2.3.

**FIGURE 2.3**
Coils of the air-cooled condenser of Wygen II Unit 4 plant. (Photo courtesy of the Black Hills Corporation, Rapid City, SD.)

The photograph shows the condenser cooling coils during construction and before installation of the fans and side panels. The plant is the Wygen II Unit 4 Generating Station of the Black Hills Corporation. The plant is located in the state of Wyoming near the town of Gillette where adequate water for cooling is not available. Consequently, the need for the air-cooled condenser.

### 2.1.2 Biofuel-Fired Steam Power Plant

A biofuel is either an organic material or one that has been derived from organic material. Worldwide, the number of biofuel-fired steam plants is very small.

### 2.1.3 Nuclear Power Plant

In a nuclear power plant, heat is generated by a controlled nuclear chain reaction using fissionable fuel. The heat generated produces steam that is used to power steam turbines. Worldwide, a large number of nuclear power plants were constructed in the period from 1970 to 1990. France, which has practically no natural fossil fuel sources within its geographical boundary, was generating 75% of its electricity by nuclear power in 1990. However, nuclear accidents in Russia, Japan, and the United States caused a reevaluation of nuclear power worldwide as a reliable and safe source of electrical power. Despite these concerns, China has plans to build a number of nuclear power plants in the coming years. The large costs of nuclear plants, around $10 billion each, have also been a deterrent to new construction. Prior to the year 2000, nuclear plants were relatively large, namely, in the range of 1100–1300 MW of generated electrical power. Subsequently, smaller-sized plants have been designed.

### 2.1.4 Gas Turbine Generators

The term "gas turbine" actually is applicable to a turbine that can fire either a gaseous fuel or a liquid fuel. The two common fuels are natural gas or oil. If oil is used as the fuel, it is actually gasified by the time it arrives at the blades of the turbine. So, the operating part of the turbine actually is in fact powered by a gas. The turbines used in gas turbine generators are very nearly the same units as those used to power aircraft. Over the years, the efficiencies of gas turbines have increased in large part due to the use of materials that can withstand even higher temperatures. As a result of design improvements and the associated higher efficiencies, gas turbine generators are being used more and more to generate electricity that not many years ago would have been generated by steam power plants.

### 2.1.5 Wind Turbines

Wind turbines are one type of sustainable energy sources that has become popular as a replacement for some of the nonsustainable sources. Because of concerns over carbon emissions, global warming, and the balance of payments attributable to imported fuel, governments have been promoting and subsidizing the use of sustainable energy sources. In most cases, utilities have been required to accept power generated by independent sources located within their service areas. In Germany, 10% of the electricity is generated by wind turbines. Most German turbines are relatively large and can generate as much as 7 MW each with adequate wind speed. Outside Germany, the average size of wind turbines is smaller. A typical wind turbine farm is shown in Figure 2.4. The installation shown in the photograph is at Robin Rigg, a sand bar off the coast of Scotland. The net installation consists of 60 wind turbines, each capable of generating 3 MW of electrical power.

### 2.1.6 Solar

Much as wind-generated power and hydro-generated power, solar-generated power is a sustainable source of energy. If past trends are any indication, solar power will be an increasing and important part of electrical generation in coming years. Numerous utilities have had solar installations constructed and many are in the range of 100–500 MW.

The increasing use of solar power is attributed to several causes including government subsidies and the falling costs of the solar components. Solar power is especially adaptable to sunny parts of the world and particularly to southwest portion of the United States. In fact, authorities in the state of California maintain that they intend to have one-third of the state's electrical power generated by solar installation by the year 2020. Elsewhere, similar although less ambitious goals have been proclaimed.

The most popular form of solar-generated electricity originates in photo voltaic cells. The conversion of sunlight to electricity occurs in semiconducting elements by the photovoltaic (PV) effect. The production of electricity occurs when sunlight excites electrons within the semiconducting elements of a solar cell. In this manner, dc electricity is generated. Approximately half a dozen different materials exhibit the PV effect. Typically, a single cell will

**FIGURE 2.4**
Wind turbines. (Photo courtesy of Vestas, Aarhus, Denmark.)

be approximately 161 cm$^2$ in size and will generate approximately 2 W of power at a potential of 0.5–0.6 V. Individual cells are normally combined in an array and protected from the elements by a glass cover. Arrays are mounted in a panel suitable for installation in the field.

Inverters and controls are needed to convert the solar-generated power to usable alternating current.

### 2.1.7 Hydro

Much as power derived from solar and wind turbines, hydro power is considered a sustainable power source. In recent decades, there has been next to no growth of hydro power in developed countries. In developing countries, the growth in hydro power has been relatively brisk and is expected to continue at an accelerated rate in the future (Reference 2.2).

### 2.1.8 Geothermal

Geothermal power plants utilize naturally hot water from underground that powers steam turbines. Worldwide, the number of geothermal plants is relatively small.

## 2.2 Transmission

In the electrical industry, the term "transmission" is used to describe the transmission of electrical power from a plant where the power is generated to external destinations.

When the merits of electricity were first realized in the later part of the nineteenth century, both dc and ac electricity were initially given equal consideration. However, the invention of the transformer made apparent the superior merits of ac for the purposes of transmission and distribution. By means of a transformer, the voltage of an ac source can be efficiently elevated for the purpose of transmission and subsequently reduced for local distribution and use. A high electrical voltage is important for transmission for several reasons. First, energy losses are a function of current. The lower the current, the lower the losses. If the potential is higher, the currents and the losses are less. Also, lower currents permit smaller and less expensive cables. So, transmission potentials are generally at relatively high potential levels. Subtransmission and distribution potentials are at lower levels since very high electrical potentials are not desirable in proximity to buildings and communities. Of course, the transmission of electrical power by a three-phase circuit is more practical and efficient than transmission by single-phase circuits.

Most often, transmission lines from a power plant connect to a grid from which connections are made to substations located near the end users.

In North America, there are four grids to which generating plants may connect. Western United States has the western grid that also connects to provinces of Western Canada. In Eastern United States, there is the eastern grid that also connects to the province of Ontario in Canada. Texas has a separate grid, which is known as the Electric Reliability Council of Texas. Mexico has a separate grid. The grid connections between Canada and the United States have allowed Canada to export as much as 30% of its electrical production to the United States (Reference 2.3).

Overland transmission is conducted at high voltage over a set of three, bare conductors. An electrical transmission tower will often hold more than one set of conductors, some of which might be at different voltages. Transmission lines are almost always accompanied with at least one grounding cable positioned at the uppermost levels on the towers. The grounding cables are not part of the three-phase circuits and are intended to capture lightning strikes that would otherwise strike the current-carrying conductors. High-voltage transmission cables are always located at a high elevation above ground, away from trees and human activity. In the past years, the towers that hold the transmission cables were of the lattice type that were fabricated of galvanized and painted angle steel. Millions of these structures have been installed in the United States over the years and most may be seen to this day. Largely because of the need for periodic painting, these towers are considered as high-maintenance assets. New installations of transmission towers tend to be of a tubular steel construction. The newer types of transmission towers are mostly fabricated of a specialty steel that requires no painting and which is protected by a layer of oxidized metal that naturally forms on the metal surfaces. A typical, modern design of a high-voltage, tubular steel transmission tower is shown in Figure 2.5. In Europe, transmission towers are mostly made of concrete or wood.

**FIGURE 2.5**
Tubular steel high-voltage transmission tower. (Photograph by the author.)

In North America, the term "subtransmission" is commonly used to describe systems that are intermediate between "transmission" systems and "distribution" systems. Typical subtransmission voltages are 69 and 138 kV.

## 2.3 Distribution

A utility's distribution system reduces transmitted higher-voltage power and delivers that electrical power at a lower voltage to its retail customers. Lines on the high-voltage side of distribution transformers are called primary distribution lines, or simply "primaries." Common North American distribution voltages are 4.2, 12.5, and 13.8 kV. Older systems tended to use the lower 4.2 kV, whereas more recently, distribution potentials are mostly at a higher level.

Low-voltage connections on distribution transformers are called secondary distribution lines or "secondaries." For residential and small commercial services, transformers in North America are typically located in proximity to the customers that are served. A transformer serving residences or small commercial facilities would typically be of a size to satisfy the needs of five or more customers. Some small- and medium-sized commercial establishments often have a dedicated transformer. In North America, common secondary line voltages are 120, 208, 240, 277, and 480 VAC.

Local distribution from substations could be by either a three- or a four-wire circuit. Four-conductor circuits are characteristic of a substation transformer with a secondary wired in a wye configuration. Many medium and large commercial facilities favor the three-phase four-wire service for the reason that the four-wire system fits well with overhead fluorescent lighting used within buildings. Where a three-phase four-wire service is used, an adequate, local ground must be provided at the transformer.

Distribution systems include all of the appurtenances needed by a utility to deliver electrical power to its customers. Primary gear includes transformers, cables, poles, meters, disconnect switches, and circuit protective devices. Operation of a distribution system requires the continuous services of numerous personnel with a wide range of capabilities. Maintenance of the distribution infrastructure requires the constant attention of experienced line crews. A cadre of in-house personnel is likewise required for billing, engineering, and a variety of support services.

A typical wooden, medium-voltage distribution tower is shown in Figure 2.6.

**FIGURE 2.6**
Medium-voltage distribution tower. (Photograph by author.)

## 2.4 Electrical Gear

### 2.4.1 Generators

In a utility's power-generating station, a synchronous generator is the means whereby mechanical power is converted into electrical power. The prime mover that drives the generator could be a steam turbine, a water turbine, a gas turbine, or a wind turbine. Internal combustion engines are also used in some smaller applications. A three-phase generator creates three independent single-phase currents that peak at equally spaced intervals.

Alternating current generators, single-phase or three-phase, operate on the principle of Faraday's law of electromagnetic induction. That law states that, "A changing magnetic flux inside a loop made from a conductor material will produce a voltage in the loop." Most three-phase utility generators are of the design known as rotating field synchronous machines. In these machines, a rotating magnetic field induces current in the windings of a three-phase circuit. The principle of a three-phase synchronous generator is represented in Figures 2.7 and 2.8. The arrangement of Figure 2.7 is that of a two-pole generator with a salient pole rotor. As the rotor is rotated, it sweeps

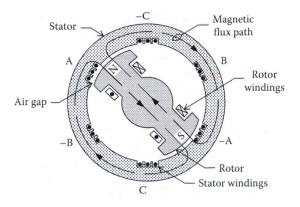

**FIGURE 2.7**
Two-pole generator.

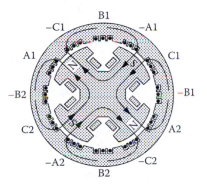

**FIGURE 2.8**
Four-pole generator.

a magnetic field over the windings located in the stator, thereby, generating electricity in those windings. A magnetic field is produced in the rotor by "field windings" through which dc current is circulated. In most designs, the dc current originates in an exciter that is adjacent to the generator. The dc current is transferred to the rotor through slip rings mounted on the shaft of the rotor. The windings in the stator are called "armature windings," and these are fixed firmly in place.

In the representation of Figure 2.7, it may be seen that if the rotor moves in a clockwise direction, it would pass, first, the positive A armature coil, then the minus C coil, and continue on the other coils represented in the depiction. The A, B, and C coils are located at positions 120° apart. Likewise, the corresponding –A, –B, and –C coils are located 120° apart and equally spaced between the A, B, and C windings. On the North American continent, the standard electrical service frequency is 60 Hz. In Europe, the frequency is

50 Hz. Elsewhere, a utility frequency could be 50 or 60 Hz. The frequency of current produced by a generator is described by the equation

$$F = NP/120,$$

where
    F is the frequency (Hz)
    N is the speed (RPM)
    P is the number of poles

This expression is applicable to either a single-phase generator or a three-phase generator. A generator driven by a steam turbine that produces 60 Hz current is commonly operated at either 1800 or 3600 RPM, whereas a gas turbine–driven generator most often operates at 3600 RPM. A generator producing 50 Hz current operates at 1500 or 3000 RPM. The 3000 and 3600 RPM machines would have two poles, and the 1500 and 1800 RPM machines would have four poles. Hydro turbines operate at approximately 200–300 RPM, and the associated generators would have many more than four poles.

Whereas a two-pole generator has one set of field windings, a four-pole generator has two sets. Likewise, a four-pole generator has twice as many armature windings as a two-pole machine. In the four-pole generator, one set of armature windings is separated by 60°, and the second set of armature coils is likewise separated by 60°. A four-pole configuration is shown in Figure 2.8 where one set of armature windings is represented by A1, B1, C1, –A1, –B1, and –C1 and the second set is represented by A2, B2, C2, –A2, –B2, and –C2. The armature windings would be interconnected externally to form either a wye connection or a delta connection. The wye connection is the preferred arrangement for generators. The armature windings of a two-pole generator would usually be connected in a wye configuration as depicted in Figure 2.9.

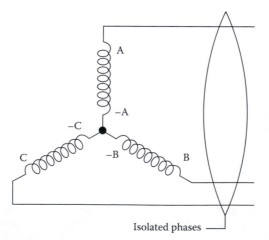

**FIGURE 2.9**
Two-pole wye connections T.

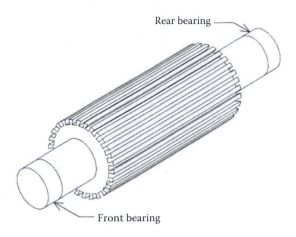

**FIGURE 2.10**
Cylindrical rotor.

In the illustrations of salient pole generators in Figures 2.7 and 2.8, it may readily be seen how a rotating magnetic field sweeps over the armature windings to generate electricity. Salient rotors are more often found in hydro generators than in steam-driven generators or gas turbine–driven generators. Large utility generators driven by steam turbines or gas turbines have what is called a cylindrical rotor. A typical cylindrical rotor is depicted in Figure 2.10. Cylindrical rotors are generally machined from a single steel billet. The field windings are subsequently installed in longitudinal groves located along the periphery of the rotor. A cylindrical rotor for a large generator might be in the range of 20–30 ft in length and 3–4 ft in diameter after machining. The armature windings are contained in longitudinal groves in the stator that run parallel to the axis of the rotor. A cross section through a cylindrical rotor and stator is represented in Figure 2.11.

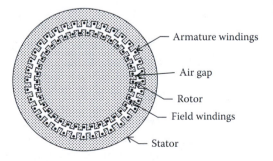

**FIGURE 2.11**
Cylindrical rotor cross section.

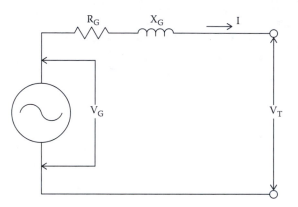

**FIGURE 2.12**
Generator equivalent circuit.

The equivalent circuit for a synchronous generator is shown in Figure 2.12. The terminal voltage of a generator may be described by the equation

$$V_T = V_G - jX_GI - R_GI,$$

where
$V_T$ is the terminal potential (voltage)
$V_G$ is the generator internal-generated potential (voltage)
$X_G$ is the generator reactance (ohms)
$R_G$ is the generator resistance (ohms)
I is the circuit current (amps)

(The generator resistance, $R_G$, is usually small compared to the generator reactance, $X_G$, and is often neglected.)

Generators produce current at a voltage that is generally no more than 35 kV. Higher voltages are impractical because of the close proximity of conductors within the generator. Most often, the output from a generator is directly connected to a step-up transformer that increases the voltage and decreases the current, for transmission overland. The three connections leading from the generator to the step-up transformer are called the isolated phases. Current in the isolated phases are conducted in isolated phase bus ducts. Some isolated bus ducts are capable of handling up to 50,000 A. Larger isolated bus ducts are cooled with forced air, whereas some smaller bus ducts are adequately cooled by the natural circulation of air over the ducts. A typical isolated phase bus duct is shown in Figure 2.13.

Although generators are very efficient machines, there are, nevertheless, internal losses that cause internal heating of the machine. If these heat losses are not removed, internal temperatures would rise to harmful levels. Large generators are most often cooled by hydrogen gas that is circulated through

**FIGURE 2.13**
Isolated bus ducts. (Photo courtesy of Crown Electric Engineering and Manufacturing LLC, Middletown, OH.)

the interior of the generator. While hydrogen gas is highly explosive if mixed with oxygen, it is the most efficient gas for extracting heat from the interior of a generator. Smaller generators can often be cooled merely by the circulation of air through the machine.

The operation of a generator is controlled by an electronic governor that determines the speed and power output of the generator. Until a generator is connected to a grid, a generator is operated independently in what is called isochronous mode, that is, speed control. During warm-up, the speed of the generator is increased slowly as it approaches synchronous speed. If a generator is driven by a steam turbine that derives steam from a fossil-fired boiler, the speed of the steam turbine and the connected generator would be slowly brought up to speed over a period of many hours. The rapid loading and unloading of a boiler must be avoided to prevent thermal stresses that could damage boiler firewall tubes. Likewise, steam turbines must be heated and cooled at a slow rate due to the different rate of growth of the rotor with respect to the turbine casing.

Once a generator is up to synchronous speed and it is ready for loading, it can be connected to a grid or an in-house load. Immediately prior to connecting to a grid or an existing in-house circuit, the phase voltages must be close in magnitude and phase to which they will be connected. When the voltages are closely matched, the generator breaker may be closed thereby connecting the generator. Once connected to a load, the mode of operation of a generator can be one of several different modes.

If a generator is connected to a grid and is driven by a steam turbine that receives steam from a boiler, the generator is generally operated in either "boiler-follow-generator" mode or "generator-follow-boiler" mode. After a generator is connected to a grid, a dispatcher will assign MW output to plant operators, and that assignment will generally call for a generator MW setting in boiler-follow-generator mode.

## 2.4.2 Transformers

Transformers are passive devices that convert ac electricity from one circuit into another through electromagnetic induction. Transformers facilitate the transmission and distribution of electrical power from the location where it is generated to the numerous points where it is used. In the most common design, a transformer contains a ferromagnetic core and two or more coils known as "windings." The windings are coupled by a mutual magnetic flux that transfers energy form one winding to another. One of the windings, the primary, is connected to an alternating current source that produces an alternating magnetic flux. The magnitude of the flux depends on the source voltage, the frequency, and the number of turns around the core. In a well-constructed transformer, all but a very small portion of the generated flux will pass through the second winding of the transformer, which is called the secondary. The magnetic flux passing through the secondary windings generates a voltage in the secondary. The voltage ratio between the primary and the secondary depends on the ratio of turns between the primary and the secondary.

The linking of the primary winding and the secondary winding can be accomplished through air but the coupling will be drastically more efficient if coupled through a core that is fabricated of ferromagnetic metal. A transformer that uses a ferromagnetic core is commonly known as an iron-core transformer. Because an iron-core transformer is the most efficient design, most power transformers are of the iron-core design. The so-called iron core of a transformer is actually fabricated of layers of laminated steel sheets that are normally in the range of 0.25–0.50 mm in thickness. The laminations reduce what are called eddy currents that would cause internal heating and a loss of efficiency. The type of laminated metal used is generally a variation of silicon steel.

There is a large variety of transformer designs that are intended for a variety of applications. Some of the better known types are the power transformer, potential transformer, current transformer, isolation transformer, and autotransformer. There are many other types as well, each intended to meet a special need. Transformers used to transmit three-phase electrical power are generally known as "power transformers." Transformers used beyond the power plant where electrical power is generated are often called "distribution transformers." Besides the power transformers, the type of transformer that is most often seen in three-power applications are the instrument transformers that are used to measure either electrical potential or electrical current.

### 2.4.2.1 Power Transformers

Three-phase electrical power can be transmitted by either a three-phase transformer or by three single-phase transformers. The use of single-phase transformers is possible since a three-phase circuit consists of three single-phase circuits that are interconnected. In fact, single-phase transformers are used in many three-phase applications. A group of three single-phase transformers interconnected to form a three-phase circuit is called a "bank of transformers." The very largest transformers used at power-generating stations to convert power from the generator to a high potential for transmission overland are the single-phase type. These large transformers are known as step-up (SU) transformers, or main power transformers (MPT), and some are capable of as much as 500 MVA. A transformer in that range would most certainly be a single-phase type. Since the voltage at the primary of a SU transformer is always lower than the secondary, the voltage is "stepped-up" by the transformer. A single-phase transformer in the 500 MW range is very large and very heavy. As it is, a single-phase transformer in the range of 500 MVA is difficult and expensive to move and ship. Larger transformers would be next to impossible to move. A failure on one single-phase transformer would be much less costly than a failure in one phase of a three-phase transformer. In short, there is considerable merit to the use of single-phase transformers in three-phase applications. In the smaller sizes, power transformers of a three-phase system could be either three-phase or single-phase types. In many sizes, the three-phase transformers are less expensive than three single-phase transformers, and that condition explains the use of most three-phase transformers.

Single-phase transformers are generally of either the core design or the shell design. The basic configuration of a core design is represented in Figure 2.14. In the core design, two separate windings are positioned on a common ferromagnetic core. The primary winding could be on one leg of the core, and the secondary winding is on the other leg of the core, as shown in Figure 2.14,

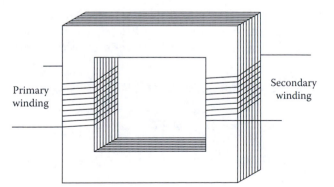

Primary winding

Secondary winding

**FIGURE 2.14**
Core-type transformer.

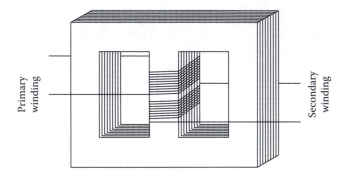

**FIGURE 2.15**
Shell-type transformer.

or both windings could be on the same leg. A shell-type design is shown in Figure 2.15. In the shell design, both the primary and the secondary windings are positioned around a center core. Three-phase transformers usually use either a three-legged core or a five-legged core. A common configuration of a five-legged core design is represented in Figure 2.16.

The single-phase configurations of Figures 2.14 and 2.15 are known as two-winding transformers. The three-phase transformer of Figure 2.16 is also called a two-winding transformer. By definition, a two-winding transformer, single-phase or three-phase, has primary windings and secondary windings. Aside from the two-winding transformers, there are also what are called three-winding transformers. Three-winding transformers have a primary winding, a secondary winding, and a third winding that is called a tertiary winding. Unless a transformer is specifically identified as a three-winding transformer, it may be assumed to be a two-winding transformer.

The mathematics that describes the parameters of transformers are sometimes classified as pertinent to an "ideal transformer." Because transformers are highly efficient, for most practical purposes, a transformer may be considered to be of the ideal type. An ideal transformer is one that is assumed to convert power with exactly 100% efficiency. This assumption greatly simplifies computations and allows results that are

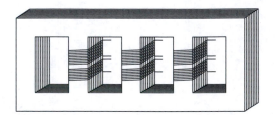

**FIGURE 2.16**
Fire-legged three-phase core design.

adequately accurate for most purposes. If a transformer is assumed to be an ideal transformer, then the following relationships are considered as applicable.

$$V_1 \div V_2 = N_1 \div N_2$$

$$V_1 \cdot I_1 = V_2 \cdot I_2$$

where,

$N_1$ is the number of turns of the primary coil
$N_2$ is the number of turns of the secondary coil
$V_1$ is the voltage of primary coil
$V_2$ is the voltage of secondary coil
$I_1$ is the primary current
$I_2$ is the secondary current

In complex form,

$$\underline{V_1} = \underline{V_2}\,(N_1 \div N_2)$$

$$\underline{V_2} = \underline{V_1}\,(N_2 \div N_1)$$

$$\underline{I_1} = \underline{I_2}\,(N_2 \div N_1)$$

$$\underline{I_2} = \underline{I_1}\,(N_1 \div N_2)$$

$$\underline{Z_1} = \underline{V_1} \div \underline{I_1}$$

$$\underline{Z_2} = \underline{V_2} \div \underline{I_2}$$

$$\underline{Z_1} = \underline{Z_2}\,(N_1 \div N_2)^2$$

$$\underline{Z_2} = \underline{Z_1}\,(N_2 \div N_1)^2$$

where

$\underline{Z_1}$ is the impedance of primary circuit
$\underline{Z_2}$ is the impedance of secondary circuit

In the study of transformers, it is often said that "the impedance is referred" from, say, the primary side to the secondary side. The converse would also be true. This is merely stating that the impedance on one side can be calculated from the impedance on the opposite side. In summary, it may be said that the ideal transformer converts voltage in proportion to the ratio of turns, currents by the inverse ratio, and impedances by the square of the turns. Power and volt–amps remain unchanged from one side of a transformer to the other side.

### Example 2.1

This example is pertinent to the practice of referring impedance form one side of a transformer to the opposite side.

Given: The circuit of Figure 2.17 has the secondary impedance

$$\underline{Z}_S = R_S + jX_S = 1.2 + j5$$

The turns ratio of the primary to the secondary is $N_1:N_2 = 6:1$.

**Problem**: (1) Show the impedance of the secondary referred to the primary circuit, (2) assume a primary voltage of 240 VAC with the secondary shorted and calculate the primary current, and (3) calculate the associated secondary current with the secondary shorted and a primary voltage of 240 VAC.

### Solution

(1) The secondary impedance referred to the primary circuit is given by the expression

$$\underline{Z}_P = (N_1 \div N_2)^2 \underline{Z}_S$$

$$= (36)\,(1.2 + j5) = 43.2 + j180$$

The impedance of the secondary circuit referred to the primary is shown in Figure 2.18.

(2) The current of the primary is given by the expression

$$\underline{I}_P = \underline{V}_P \div \underline{Z}_P = 240 \div (43.2 + j180)$$

$$= .302 - j1.260 = 1.29\underline{/\,-76.5°}\ \text{A}$$

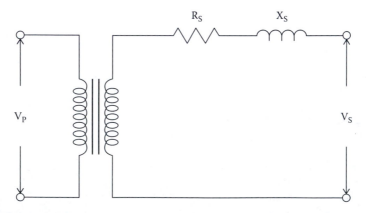

**FIGURE 2.17**
Circuit showing secondary impedance.

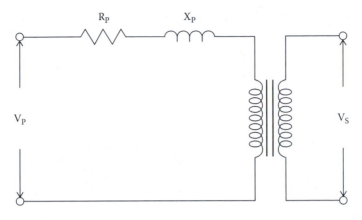

**FIGURE 2.18**
Circuit showing primary impedance.

(3) The secondary current is

$$\underline{I_S} = I_P(N_P \div N_S) = (1.29)(6) = 7.74\,A$$

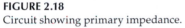

Transformers in a three-phase circuit can be interconnected in either a delta or a wye configuration. In that a transformer has both primary windings and secondary windings, there are four possible configurations. When reference is made to the arrangement of the primary windings and the secondary windings, the primary is always mentioned first. For example, a delta–wye transformer would have a delta-connected primary and a wye-connected secondary. The four possible configurations are (1) delta–wye, (2) delta–delta, (3) wye–wye, and (4) wye–delta. For a specific application, one of these configurations will generally be preferred over another possible configuration. Each configuration has specific characteristics and features.

*Delta–wye*: This is generally the preferred configuration of a step-up transformer that accepts current from a plant's generator and converts it to higher voltage. The secondary voltage is out of phase with the incoming voltage by 30°. The shift could be leading or lagging, depending on the arrangement of the interconnections. It tends to suppress the harmonics that can cause problems in other configurations. Where used in distribution applications, the secondary (wye) provides a means for local grounding. Commonly used in commercial and industrial applications.

*Delta–delta*: This is common in industrial applications. The secondary voltages are in phase with the primary voltages, which is a requirement if circuits are to be interconnected.

*Wye–wye*: This combination can cause problems due to harmonics, especially if the primary and secondary neutrals are not grounded. This configuration is often avoided due to the frequent problems with harmonics. The secondary voltages are in phase with the primary voltages.

*Wye–delta*: This is often used in step-down applications. The wye connection on the high side reduces insulation cost. It is often used in a grounding bank configuration where the wye transformer is used with a delta transformer solely to provide a grounding path. The secondary voltage is out of phase with the incoming voltage by 30°. The shift could be leading or lagging, depending on the arrangement of the interconnections.

When a transformer is loaded to some degree, there is an internal temperature rise due to the fact that no transformer is an ideal transformer. The internal losses generate heat, and that heat must be removed or damaging high temperatures will result. In transformers under approximately 1500 kVA, generated heat is removed by natural convection, and no internal oil is provided for cooling. These types are known as "dry" transformers. In power transformers over approximately 1500 kVA, heat is removed by oil that is circulated within the windings. In some sizes, oil circulates naturally, and no pumps or fans are needed to cool the oil. In larger sizes, both pumps and air-circulating fans might be required to cool the oil. The oil contained with the unit's casing also acts as a dielectric to provide insulation between current-carrying conductors. The preservation and maintenance of that oil is critical if serious damage to a transformer is to be avoided.

Large power transformers are generally fitted with an external tank that accommodates the expansion and contraction of the oil due to temperature changes. The tank is called a conservator, and it is mounted above the transformer casing. A typical transformer with a conservator tank is represented in Figure 2.19. Oil within the tank is connected to the transformer casing through piping that is called a goose neck. Expanding oil within the casing will move up the goose neck and into the bottom of the conservator. In most designs, an elastomeric bladder floats on the top

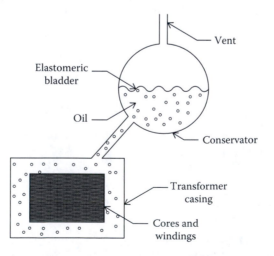

**FIGURE 2.19**
Transformer with conservator.

**FIGURE 2.20**
Step-up transformer. (Photo courtesy of WEG, Jaraguá do Sul, Brazil.)

of the oil within the conservator to separate the oil from air within the tank. The purpose of the bladder is to prevent contamination of the oil by the air and, particularly, the moisture carried by the air. Large transformers are also generally fitted with what is called a Buchholtz relay that is intended to detect gas generated in the oil that has resulted from a fault condition. In the event of the sudden appearance of gas, a trip signal is sent by the relay to the transformer's protective circuit.

A typical SU transformer is shown in Figure 2.20. The SU of Figure 2.20 contains all three transformers within a single casing. It is located at a wind farm in south Texas and is rated 160 MW at 115 kV secondary. The conservator may be seen positioned above the transformer casing.

A typical three-phase distribution transformer is shown in Figure 2.21. The manufacturer rates the transformer 30, 40, or 50 MVA depending on the type and degree of cooling. The transformer in the photograph has maximum cooling and is rated 50 MVA, 138/26.4 kV.

Prior to 1979, PCB oil was commonly used as the coolant oil in transformers as well as for other purposes. However, by 1979, the manufacture of PCB-containing oil had been banned in most countries, as well as the United States, because of the carcinogenic properties of the compound. Today, no new transformers contain PCBs.

### 2.4.2.2 Instrument Transformers

Two important types of instrument transformers used to measure circuit parameters are the potential transformer and the current transformer.

A transformer used to measure the potential of a circuit is known as a potential transformer or merely a "PT." In a well-designed potential transformer, the secondary potential is at all times in an accurate proportion to the

**FIGURE 2.21**
Three-phase distribution transformer. (Photo courtesy of SPX Transformer Solutions, Inc., Goldsboro, NC.)

primary potential, and that ratio is in an exact proportion to the turns ratio. Typically, the secondary potential might be 120 VAC and the primary as high as, say, 500,000 VAC. Accordingly, the ratio for this PT would be 500,000:120 or 4,167:1. For best accuracy, the secondary of a potential transformer should be connected to a high-impedance circuit as excessive burden (current) on the secondary will cause error in the secondary potential. The potential of the secondary must also be in phase with the primary potential.

The current transformer, commonly known as a "CT," is used to measure the current in a conductor. Very often a CT is in the shape of a toroid through which the measured current is directed. Transformers often have CTs located in the bushings that connect internal conductors to external conductors. Current transformers are identified by the ratio of the measured current to the secondary current. The most common secondary current is 5 A although CTs with a 1 A secondary are also common. For a 1000 A conductor and a 5 A secondary CT, the ratio would be 1000:5. It is worth noting that although the ratio of 1000:5 in this example is the mathematical equivalent of 200:1, this later mentioned designation is usually avoided. The presence of the "1" in the device's description might be incorrectly interpreted to suggest that the CT is intended for use with a 1 A secondary. For optimum accuracy, current

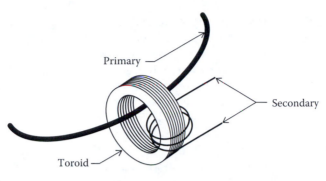

**FIGURE 2.22**
Current transformer.

transformers must be used with a low-impedance secondary. Secondary voltages in the range of 1–5 V are common. It is important that the secondary of a CT must be either across a low-impedance load or shorted anytime current is flowing through the CT. If the secondary is allowed to be open when monitored current is present, the secondary voltage will rise to a very high value. That high voltage could be dangerous to nearby personnel and might very well damage the insulation of the CT.

The design of a typical current transformer is represented in Figure 2.22. The current that is to be measured is in the conductor that is threaded through the opening of the CT toroid transformer. That conductor becomes the primary of the CT. The secondary that provides measurement of the current in the primary has loops around the toroid. For a CT with a ratio of 1000:5 and a single loop of the primary through the toroid, the secondary would have 200 turns. If the primary had 10 loops through the toroid, the secondary would have 20 turns. As with all properly constructed transformers, CTs are formed with layers of ferromagnetic sheet steel.

## 2.5 Per Unit Calculations

The per unit (pu or p.u.) method of calculations is often used as an aid in the analysis of electrical systems that operate at different voltage levels. The determination of per unit values is not the end objective of a study. Rather determined per unit values serve to assist in the calculation of other system values. More specifically, the per unit method of calculation is often used in the calculation of short circuit currents. The pu system of calculations was more prevalent in the past years. However, the availability of computer programs for system analysis has diminished to some extent the need for the pu system of calculations. Nevertheless, a person involved with three-phase electrical power would be well advised to understand the pu method, its mechanics, and how it is used.

The per unit method uses what are called base values. Selected values of parameters are compared to selected base values and assigned per unit values. The pu numbers become unitless. A pu value can readily be converted back to a voltage, power, or any other value as long as the base value is known. Although the pu system could most likely be used for systems other than electrical systems, it is probably used mostly with electrical values.

In the way of explanation of the pu system, assume that a pu system is to be used, say, to identify automobile weights. Assigning a base value would be the first step. Say, the weight of a model XYZ, which is 4200 lb, is to be taken as the base value of weight. A model KLM is noted to have a weight of 4800 lb. So the model KLM has a weight of 4800 ÷ 4200 = 1.14 pu. If a model DEF is said to have a weight of 1.4 pu, its actual weight can be determined since the base weight value is known as 4200 lb. To convert the pu value of the model DEF back to pounds, the pu value is multiplied by the weight of the base model. So the weight of model DEF would be 1.4 · 4200 or 5880 lb. This, basically, is how the pu system functions.

Applied to electrical systems, the pu system deals primarily with values of power, voltage, current, and impedance. Any two variables are selected to be base values. The selection of two base values then fixes the other values. A per unit quantity then becomes the ratio of a selected parameter to a selected base value. Basically,

Per unit = (Present value) ÷ (base value)

Per unit values are sometimes expressed as a percentage rather than a ratio. For example, if the base value of voltage had been selected as 13,800 V and the present value is 11,000 V, the value of voltage per unit is

$$V_{PU} = 11{,}000 \div 13{,}800 = 0.797 \text{ pu, or}$$

$$V_{PU} = 0.797 \, (100) = 79.7\% \text{ pu}$$

Often, nameplate values are taken as base values but this need not be the case. Some basic relationships follow and are different for single-phase systems and three-phase systems.

**Single Phase**

Typically,

$$P_{base} = 1 \text{ pu}$$

$$V_{base} = 1 \text{ pu}$$

Let

$$Q_{base} = \text{reactive power}$$

$$S_{base} = \text{apparent power}$$

The remainer of the values may be derived from the normal relationships

$$S = IV$$

$$P = S \cos \phi$$

$$Q = S \sin \phi$$

$$\underline{V} = \underline{I} \cdot \underline{Z}$$

$$\underline{Z} = R + jX$$

$$I_{base} = S_{base} \div V_{base}$$

$$Z_{base} = V_{base} \div I_{base} = V^2_{base} \div S_{base}$$

$$Y_{base} = 1 \div Z_{base}$$

**Three Phase**
In three-phase circuits, some relationships are defined differently from those used in single-phase circuits. Specifically, for three-phase systems,

$$S_{base} = (\sqrt{3}) \, V_{base} \cdot I_{base}$$

$$I_{base} = S_{base} \div [(V_{base}) \cdot (\sqrt{3})]$$

$$Z_{base} = V_{base} \div [(I_{base}) \cdot (\sqrt{3})] = V^2_{base} \div S_{base}$$

$$Y_{base} = 1 \div Z_{base}$$

In general, the per unit values are determined as

$$V_{pu} = V \div V_{base}$$

$$I_{pu} = I \div I_{base}$$

$$S_{pu} = S \div S_{base}$$

$$P_{pu} = P \div P_{base}$$

$$Z_{pu} = Z \div Z_{base}$$

$$Y_{pu} = Y \div Y_{base}$$

To change base values, the following expression is used:

$$Z_{pu-new} = Z_{pu-old} (S_{base-new} \div S_{base-old}) (V_{base-old} \div V_{base-new})^2$$

**Three-Phase Example of Determination of Per Unit Values**

Consider a three-phase transformer with a rating of 700 MVA and a secondary voltage of 145 kV. Determine $I_{base}$, $Z_{base}$, and $Y_{base}$.

There is no obligation to choose the transformer's ratings as the base values, but it is an option.

Using the transformer's rated values of apparent power and secondary voltage,

$$S_{base} = 700 \text{ MVA}$$

$$V_{base} = 145 \text{ kV}$$

The corresponding values of $I_{base}$, $Z_{base}$, and $Y_{base}$ are readily determined.

$$I_{base} = S_{base} \div [(V_{base}) \cdot (\sqrt{3})] = 700 \text{ MVA} \div [(145 \text{ kV} \cdot (\sqrt{3})] = 2.78 \text{ kA}$$

$$Z_{base} = V_{base} \div [(I_{base}) \cdot (\sqrt{3})] = 145 \text{ kV} \div [(2.78 \text{ kA} \cdot (\sqrt{3})] = 30.1 \text{ }\Omega$$

$$Y_{base} = 1 \div Z_{base} = 1 \div 30.1 = 0.0332 \text{ S}$$

If, say, a secondary voltage of 130 kV is under consideration, then the per unit value of that voltage is

$$V = V \div V_{base} = 130 \text{ kV} \div 145 \text{ kV} = 0.89 \text{ pu}$$

For either a single-phase system or a three-phase system, the per units become

$$V_{pu} = V \div V_{base}$$

$$I_{pu} = I \div I_{base}$$

$$S_{pu} = S \div S_{base}$$

$$Z_{pu} = Z \div Z_{base}$$

In all of these expressions, the numerator is a complex vector, whereas the denominator is a real number. For base values, a value of $S_{base} = 50$ or 100 MVA is often selected. For in-house electrical systems, a base of $S_{base} = 10$ MVA is more suitable as the VA values are generally smaller than the VA of overland transmission lines. For transmission lines, 100 MVA is often used. Nevertheless, any value of S may be used.

**Example 2.2**

Consider a sample calculation pertaining to per unit calculations. Reference is made to Figure 2.23. The represented circuit contains a generator, a transmission line, and a load. Assume the following values.

The generated voltage is $V_G = 4160\underline{/0°}$.
Internal generator reactance is $Z_G = j20$.
Line impedance is negligible.
The load has impedance $Z_L = 200 + j220$.

**Problem**: Assume base values 4000 VAC and 14 A. Determine per unit values of voltage, current, imaginary power, and impedance.

**Solution**

The net circuit impedance becomes

$$Z_C = (200 + j220) + (j20) = 200 + j240$$
$$= 312.4\underline{/50.194°}$$

Circuit current is

$$I_C = V_G \div Z_C = [4160 + j0] \div [200 + j240]$$
$$I_C = 8.52459 - j10.2295 = 13.315\underline{/-50.194°}$$

The apparent circuit power is

$$S_C = (\sqrt{3})\, V \cdot I$$
$$S_C = (\sqrt{3})\, V_C \cdot I_C$$
$$= (\sqrt{3})\, [4160\underline{/0°} \cdot 13.30\underline{/-50.194}]$$
$$= (\sqrt{3})\, [(4160 + j0)(8.52459 - j10.2295)]$$
$$= 61422.5 - j73705.6 = 95{,}943.93\underline{/-50.194°}$$

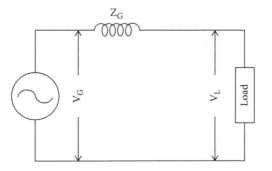

**FIGURE 2.23**
Generator and load.

The load voltage is

$V_L = I \cdot Z_L$

$\quad = (8.52495 - j10.2295)\,(200 + j\,220) = 3955.408 - j170.5 = 3959.08\underline{/-2.47°}$

Voltage attributed to the internal reactance is

$\quad V_G = I \cdot Z_G = (8.52495 - j10.2295)\,(0 + j20) = 204.59 + j170.5 = 266.2\underline{/39.8°}$

The selection of base values is entirely arbitrary. For this example, two base values are selected as

$$V_B = 4000 \text{ VAC and}$$

$$I_B = 14 \text{ A}$$

These base values are scalar values of voltage and current. Next, determine the values of apparent power, real power, system impedance, and system admittance.

To determine $S_{base}$, only scalar values are used.

$$S_{base} = (\sqrt{3})\,V_{base} \cdot I_{base} = (\sqrt{3})\,(4000)(14) = 96{,}994$$

$$Z_{base} = V^2_{base} \div S_{base} = (4{,}000)^2 \div 96{,}994 = 164.95$$

Now, the per unit values can be assigned.

$$V_{C\text{-}pu} = V_C \div V_{base} = 4160\underline{/0°} \div 4000 = 1.04\underline{/0°} \text{ pu}$$

$$I_{C\text{-}pu} = I_C \div I_{base} = 13.315\underline{/-50.194°} \div 14 = 0.951\underline{/-50.194°} \text{ pu}$$

$$S_{C\text{-}pu} = S_C \div S_{base} = 95{,}943.93\underline{/-50.194°} \div 96{,}994 = .989\underline{/-50.194°} \text{ pu}$$

$$Z_{C\text{-}pu} = Z_C \div Z_{base} = 312.4\underline{/50.194°} \div 164.95 = 1.893\underline{/50.194°} \text{ pu}$$

Example 2.2 demonstrates that once the base values of a circuit are selected, it becomes an easy matter to convert from actual values to per unit values and vice versa. The conversion of parameters to actual values is performed by merely multiplying the per unit number for a parameter by the previously determined base values for that variable.

### Example 2.3

Transformers are often involved in per unit calculations. In the way of demonstration, consider a single-phase transformer with properties

100 kVA, 4160:240 V; high-side impedance of $1.5 + j2.0$

Consider a high-side current of 5 amp.

**Problem**: Express current and impedance in pu values.

**Solution**

Use nameplate values as base values:

$$V_{base,H} = 4160 \text{ V}$$

$$V_{base,L} = 240 \text{ V}$$

$$S_{base} = 100 \text{ kVA volt–amps}$$

$$I_{base, H} = S_{base} \div V_{base,H} = 100,000 \div 4,160 = 24.03 \text{ A}$$

$$I_{base, L} = S_{base} \div V_{base, L} = 100,000 \div 240 = 416.6 \text{ A}$$

$$Z_{base, H} = 4160 \div 24.03 = 173.11 \ \Omega$$

$$Z_{base, L} = 240 \div 416.6 = 0.576 \ \Omega$$

On the high side,

$$I = 5 \text{ A} \div 24.03 = 0.208 \text{ pu}$$

On the low side,
the low-side current corresponding to 5 A on the high side is
$(4160 \div 240)(5) = 86.66$ A.
The low-side current in pu value is

$$I = 86.66 \div 416.6 = 0.208 \text{ pu}$$

The calculation shows that the pu value of current is the same on both the high side and the low side of the transformer.
The high-side impedance is

$$Z_H = 1.5 + j2.0 = 1.5/173.11 + j2.0/173.11 \text{ pu} = .00866 + j.01155 \text{ pu}$$

The low-side impedance is

$$Z_L = (Z_H)(V_L \div V_H)^2$$

$$= (1.5 + j2.0)(240 \div 4160)^2$$

$$= .00499 + j.00665$$

$$= .00499/.576 + j.00665/.576 \text{ pu}$$

$$= .00866 + j.01155 \text{ pu}$$

Expressed in per unit quantities, the low- and high-side impedances are identical.

### 2.5.1 Systems with Transformers

The features of the pu unit method of calculating system properties are often used when calculations involve a transformer. For most calculations, very little error will be encountered if the transformer is assumed to be an ideal transformer. An ideal transformer is represented in Figure 2.24. For the ideal transformer of Figure 2.24, the following relationships are assumed to be true:

1. An ideal transformer is 100% efficient.
2. Equal apparent power on both the high side and the low side.
3. $I_X \cdot V_X = I_Y \cdot V_Y$.
4. $Z_X = Z_Y (V_X \div V_Y)^2$.
5. $Z_Y = Z_X (V_Y \div V_X)^2$.

When dealing with transformers, it will be found that

1. The per unit value of impedance is identical on both sides of a transformer.
2. The per unit value of imaginary power is identical on both sides of a transformer.
3. The per unit of power is identical on both sides of a transformer.

#### Example 2.4

This example considers a circuit with two transformers as represented in Figure 2.25. Assume the following conditions applied:

$$\text{Generator G: 85 MVA, 35 kV, } X_{pu} \text{ at } 0.21$$

Transformer T1:

$$\text{60 MVA, 35 kV/145 kV, } X_{pu} = 0.07$$

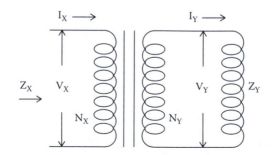

**FIGURE 2.24**
Ideal transformer.

**FIGURE 2.25**
System with two transformers.

> Line reactance impedance: 30 Ω
> Transformer T2:

$$30 \text{ MVA, } 145 \text{ kV/13.8 kV, } X_{\text{p.u}} = 0.09$$

$$\text{Motor M: 10 MVA, 13.8 kV, } X_{\text{pu}} \text{ at } 0.19$$

**Problem**: Determine the normalized impedance values for the system.

**Solution**

Resistances in a system of this type would be relatively small and for all practical purposes may be ignored. The first step, then, is to determine base values of imaginary power (S), voltages (V), and impedances (Z). As is common for systems of this type, a base value of S = 100 MVA is assumed. Base voltages are determined by the transformers. The system has three voltages (35, 145, and 13.8 kV). So, these voltages become the base values. Next, the base impedances are determined. The locations of the impedances are represented in Figure 2.26.

$$Z_{\text{BASE}} = V^2{}_{\text{BASE}} \div S_{\text{BASE}}$$

For the transmission line, $Z_{\text{BASE}} = (145)^2 \div 100 = 210.25 \ \Omega$

$$Z_{\text{L-PU}} = Z_{\text{L}} \div Z_{\text{BASE}} \text{ (where } Z_{\text{L}} = \text{actual value of resistance} = 30 \ \Omega\text{)}$$

$$Z_{\text{L-PU}} = 30 \div 210.25 = .1426 \text{ pu}$$

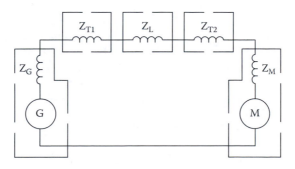

**FIGURE 2.26**
Assumed impedances (Example 2.4).

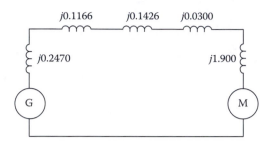

**FIGURE 2.27**
Calculated per unit impedances (Example 2.4).

For the generator, a new value of per unit reactance is determined.

$$Z_{\text{G-PU-NEW}} = Z_{\text{G-PU-OLD}} \, (S_{\text{BASE-NEW}} \div S_{\text{BASE-OLD}}) \, (V_{\text{BASE-OLD}} \div V_{\text{BASE-NEW}})^2$$

$$Z_{\text{G-PU-NEW}} = (0.21) \, (100 \div 85) \, (35 \div 35)^2 = .2470 \text{ pu}$$

For transformer T1,

$$Z_{\text{T1-PU-NEW}} = Z_{\text{T1-PU-OLD}} \, (S_{\text{BASE-NEW}} \div S_{\text{BASE-OLD}}) \, (V_{\text{BASE-OLD}} \div V_{\text{BASE-NEW}})^2$$
$$\times Z_{\text{T1-PU-NEW}}$$

$$= (.07) \, (100 \div 60) \, (35 \div 35)^2 = .1166 \text{ pu}$$

For transformer T2,

$$Z_{\text{T2-PU-NEW}} = Z_{\text{T2-PU-OLD}} \, (S_{\text{BASE-NEW}} \div S_{\text{BASE-OLD}}) \, (V_{\text{BASE-OLD}} \div V_{\text{BASE-NEW}})^2$$

$$Z_{\text{T2-PU-NEW}} = (.09) \, (100 \div 30) \, (145 \div 145)^2 = .0300 \text{ pu}$$

For the motor,

$$Z_{\text{M-PU-NEW}} = Z_{\text{M-PU-OLD}} \, (S_{\text{BASE-NEW}} \div S_{\text{BASE-OLD}}) \, (V_{\text{BASE-OLD}} \div V_{\text{BASE-NEW}})^2$$

$$Z_{\text{M-PU-NEW}} = (.19) \, (100 \div 10) \, (13.8 \div 13.8)^2 = 1.900 \text{ pu}$$

The computed values of per unit for the system are shown in Figure 2.27. These pu values would be of use if, say, a short circuit study were to be conducted to determine the levels of fault currents.

## 2.6 Cybersecurity

Originally known as "cyber security," the subject is now mostly known by the single word "cybersecurity." Cybersecurity has become an increasingly critical function worldwide. The objective of cybersecurity is to protect assets against hackers who use the Internet to gain unauthorized access to the computer files and programs of targeted organizations. The purpose of hacking might be for a variety of reasons. One, primary purpose of hacking is financial gain. Other individuals pursue the malicious destruction of the programs and files of others merely for the mental qualifications that might be found in bypassing challenging firewalls. Another, more potent, capable, and well-trained group of hackers act on behalf of their respective governments. A number of governments have found reasons for actively pursuing cyber campaigns. The theft of intellectual property has become a major reason for some organizations seeking to hack into the files of firms that might have valuable drawings and proprietary data. A manufacturer, for example, can gain considerable advantage over a competitor if it steals that competitor's information pertaining to ongoing projects, research information, and the like. Military hackers are among the most capable and advanced hackers.

National defense is one of the most important functions of cybersecurity. It is well recognized that the electrical systems within a country could be the target of an aggressor or adversary. Many governments are seriously concerned over the vulnerability of their electrical systems to cyber attacks. Potentially damaging access to critical electrical systems has come about because generation and distribution systems that in the past years were isolated from networks are now interconnected over networks. These interconnections have led to large improvements in efficiencies. At the same time, these interconnections have introduced the risk of exposure to cyber attacks. The concerns are that cyber attacks could cause damage to industrial facilities or merely the tripping of power-generating plants. The results, it is speculated, could cause troublesome brownouts or blackouts. These brownouts or blackouts alone could have associated inconveniences and costs. However, if coupled with attacks by terrorists, the resulting damage could be seriously compounded.

In consequence to the potential exposure of electrical systems, a number of governments have initiated programs to encourage and subsidize defensive measures intended to prevent or minimize the potential for damage from cyber attacks. These measures, it is commonly recognized, are designed to identify, protect, detect, respond, and recover. Particularly with regard to the protection of electrical generation and distribution facilities, the North American Reliability Corporation (NERC) has published several standards that recommend fundamental practices to be followed to ensure network security. A number of other organizations worldwide have also issued guidelines and standards with the intent of educating and preparing prospective targets of cyber intrusions.

The most common means used to guard against cyber attacks are a firewall or an intrusion prevention system that interrogates and filters incoming transmissions. A firewall, for example, would typically be used to allow or disallow transmissions interconnecting to a plant's distributed control system (DCS). A DCS is today the most common control system used in electrical power-generating stations. The DCS controls the loading and operation of boilers, generators, gas turbine generators, as well as all of the plant's auxiliary equipment. Fraudulent messages to a DCS could easily result in a plant going offline or, in the extreme, serious damage to plant equipment. A firewall at a power-generating station would consist of a separate computer and program that examines incoming transmissions and allows only legitimate transmissions to pass through.

The issues presented by cyber attacks are challenging and continually evolving. Firewalls and intrusion prevention systems installed to protect assets must be continually updated to guard against increasingly capable hackers. It is anticipated that the need for continuous and ever improving cybersecurity measures shall extend well into the future.

Some facilities, as critical military installations, have eliminated the potential of cyber attacks by installing on-site generating capabilities that are not linked in any way beyond the boundaries of the facility. Apparently, it was decided at a high level that only in this manner can operations personnel at a facility be assured with certainty that hackers will be unable to gain access to the controls of the electrical assets.

## Problems

1. Would it be correct to say that a fossil fuel is one that is the result of aerobic decomposition of organic material?
2. What is the feature of a fluidized bed boiler used for generating steam?
3. Of the possible fossil fuels, which one has the least polluting emissions?
4. Would it be correct to say that gas turbines can generate electricity with only natural gas as a fuel?
5. Name at least three types of renewable sources of electrical power.
6. If a solar panel is 152 cm by 182 cm, approximately how much power may be expected from the panel?
7. Why is it that the transmission voltages are relatively high, namely, from 140,000 V to higher than 500,000 V?
8. How many conductors are commonly used for the high-voltage transmission of three-phase electrical power?
9. Would it be accurate to say that the worldwide use of hydro power is declining because all of the possible sites have been used?
10. A hydro plant in France has turbines that rotate at 300 RPM. To generate 50 Hz current for their customer's use, how many poles must the generators have?

11. How many grids are there in North America?

12. Are the newer types of transmission towers fabricated of galvanized steel?

13. Would it be correct to say that generators typically generate electricity at a potential well over 100 kV?

14. Would a three-phase voltage of 13.8 kV normally be considered a transmission, subtransmission, or distribution voltage?

15. In a generator with a cylindrical rotor, three-phase current is produced in which windings by electromagnetic induction?

16. Do solar panels require an inverter or a converter to change dc power to ac power?

17. What is the function of an SU transformer?

18. Name the two types of ferromagnetic cores used in transformers.

19. Isolated phase ducts at a power-generating plant connect what to what?

20. Which transformer connections have no phase shift between primary and secondary voltages?

21. A 13.8 kV:480 V transformer has a primary winding consisting of 1000 turns. What is the number of turns in the secondary, rounded to the next highest whole number?

22. A CT with a rating of 100:5 is installed to measure current in a conductor that is looped twice through the toroid of the CT. If the conductor current is 40 A, what is the current in the secondary?

23. What is the function of a conservator on a power transformer?

24. A 480/120 transformer has secondary impedance $Z_S = 1.5 + j8$. Calculate the impedance referred to the primary circuit and the primary current with the secondary shorted.

25. A circuit contains a generator that has a terminal voltage of 5000 VAC, a negligible internal resistance, and an internal reactance of $j30\ \Omega$. The sole load has an impedance of $Z = 150 + j170$. Line losses connecting the generator to the load are negligible. Assume a base potential of 5000 VAC and base current of 15 A. Express load per unit voltage, per unit current, and per unit imaginary power in polar form.

# 3

## Grounding

### 3.1 Definitions

In order to understand the concepts of grounding, one would be well advised to first understand the terms commonly used with regard to grounding. Around the world, English speakers use a variety of words that have the same definition. In recent years, any English speaker involved with three-phase electricity would have encountered the words "ground," "grounding," "earth," and "earthing." A few words of explanation regarding these four words are in order.

The word "ground" is a noun, and it has many connotations in the English language. Here are a few of the definitions of "ground" according to *Webster's Dictionary* (Reference 3.1).

1. As a noun: The surface of the earth
2. As an adjective: On or near the ground
3. As a transitive verb: To set on or cause to touch the ground

The more common usage of the word "ground" in the English language is as a noun. The word "grounding," although not a noun, is also used as a noun. (A word that ends with "ing" is normally a verb but, when used as a noun, is considered a gerund.) Although the words "ground" and "grounding" are commonly used in the United States, many other English speakers use alternative terms. Outside North America, the words "earth" and "earthing" are, respectively, the equivalent of "ground" and "grounding." In general, the word "earth" has a similar but slightly different connotation in the English language. According to *Webster's Dictionary*, the definition of "earth" includes the following common usages.

1. As a noun: The planet on which we live
2. As a noun: Land as distinguished from sea or sky
3. As a noun: The soft, granular, or crumbly part of land; soil

For the purposes of this textbook, the common American words "ground" and "grounding" are used. Nevertheless, it is recognized that the equivalent words are, respectively, "earth" and "earthing."

Following are a few additional terms related to the subject of grounding. A consistent set of definitions is necessary if one is to understand the jargon of the subject.

### 3.1.1 Ground

The National Electrical Code (NEC) defines "ground" as follows.

> Ground. A conducting connection, whether intentional or accidental, between an electrical circuit or equipment and the earth or to some conducting body that serves in place of the earth.

### 3.1.2 Equipment Grounding Conductor

The NEC defines "equipment grounding conductor" as follows.

> Equipment Grounding Conductor. The conductor used to connect the non-current-carrying metal parts of equipment, raceways, and other enclosures to the system grounded conductor, the grounding electrode conductor, or both, at the service equipment or at the source of a separately derived system.

According to the NEC, ground conductors are to be colored green. An equipment grounding conductor is not necessarily insulated.

### 3.1.3 Neutral Conductor

The term "neutral conductor" is associated with grounding. The NEC indirectly defines a neutral conductor as a "grounded circuit conductor." A neutral conductor is one that may under normal circumstance be a conductor of electrical current. For example, the neutral wire of a four-wire wye circuit will be conducting current in an unbalanced circuit. A neutral conductor will be connected to a ground conductor and will be at or very near the ground potential. Yet, a neutral conductor is not considered a ground conductor. In the United States, neutral conductors are insulated and (per NEC) may be colored white or gray but never green.

### 3.1.4 Bonding

The term "bonding" is frequently used in reference to the grounding of equipment, and it often appears in wiring codes and literature that treats the subject of grounding. The NEC defines "bonding" as follows:

> Bonding: The permanent joining of metallic parts to form an electrically conductive path that will ensure electrical continuity and the capacity to conduct safely the current likely to be imposed.

A typical use of the term bonding is as "The main bonding jumper provides the connection between the equipment ground and the grounded circuit conductor."

Essentially, the purpose of bonding is to ensure that metallic components will be at the same electrical potential and that no metallic component will be allowed to be at an elevated and possibly dangerous potential. The purpose of bonding, as with equipment grounding, is to minimize or alleviate danger to personnel. Conductors are often bonded to metallic components merely with a screw. Large equipment grounding conductors are often welded or brazed to ensure a more dependable bond and one that would not be degraded by a high current or corrosion.

### 3.1.5 Ground Loops

The NEC does not define "ground loops." The NEC is written for the purpose of defining the proper practices that are to be followed to ensure a safe electrical installation, that is, one that will not introduce a risk for damage to property or hazards to personnel. The NEC does not delve into the theory behind its recommended practices, and it does not explain those consequences that may result if the stated criteria are not followed. For this reason, the subject of ground loops is not treated in the NEC. Nevertheless, ground loops should be a subject of interest to anyone dealing with large three-phase electrical installations.

Ground loops are the possibly adverse consequence of an electrical installation that is grounded at more than one point. Since two or more grounds will rarely be at precisely the same potential, an electrical current will flow between the two grounds. That current is considered a ground loop. Ground loops are generally of little consequence in residential applications. However, in industrial applications ground loops can introduce significant errors in the grounding of instrument systems. Underground ground loops in large three-phase industrial installations can also cause the gradual corroding of underground piping, metallic tanks, and other underground metallic objects.

## 3.2 Reasons for Grounding

According to common practice that was followed years ago, many electrical installations were not grounded. Perhaps the need for adequate grounding was not fully recognized, understood, or considered necessary. Today, grounding is considered important to most, but not all, electrical installations. Yet, the importance of proper grounding is often neglected. In fact, the U.S. Office of Safety and Health Administration reports that the most common safety violation is the improper grounding of equipment or circuits (Reference 3.2).

The fundamental purpose of grounding is to both protect personnel as well as property. An electrical ground prevents conductor voltages from exceeding the rating of the respective conductor insulation. Equipment grounding will prevent equipment as cabinet enclosures and motor cases from becoming charged to a potential that could be harmful to personnel or equipment.

Whereas grounding is considered necessary to most electrical circuits, there are many circuits that, for a variety of reasons, are definitely not to be grounded. Specifically, the NEC (Article 250.7) requires that (for safety reasons) the following circuits are to be ungrounded:

1. Cranes

2. Health-care facilities (some restrictions)

3. Electrolytic cells

In a system that is specifically not grounded, the first inadvertent grounding becomes the ground point for the circuit. A feature of an ungrounded system is that the first, inadvertent grounding will not trip the overcurrent protective devices. This characteristic of an ungrounded system is sometimes important to a process that would be harmed by the immediate deactivation of an electrical system. An ungrounded system also minimizes the chances of a serious arc flash that might otherwise result from a grounded system. Ungrounded systems are often fitted with a ground fault detector that will provide notification of a ground fault condition and, in some instances, may simultaneously initiate a trip of the circuit.

The two common intended functions for the grounding of an electrical system are as a non-current-carrying function or as a current-carrying function.

Symbols commonly used to indicate a ground are shown in Figure 3.1.

Besides the need for adequate grounding of an electrical system, most facilities also require grounding for the purpose of minimizing the effects of lightning.

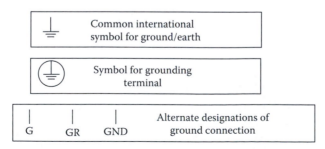

**FIGURE 3.1**
Symbols for a ground.

### 3.2.1 Non-Current-Carrying Ground

A typical non-current-carrying ground includes the "equipment ground conductor" (as defined by the NEC) that connects equipment to a ground. A typical application is a metallic cabinet that houses electrical equipment. The primary purpose for grounding of the cabinet would be to avert an electrical shock to personnel in the event that a live conductor of an electrical circuit would inadvertently contact the metallic cabinet. If a live conductor would come in contact with metallic components connected to the equipment ground conductor, the circuit protective device would be tripped thereby avoiding a condition potentially hazardous to personnel. In short, the purpose of the equipment ground is primarily personnel safety.

### 3.2.2 Current-Carrying Ground

Unlike non-current-carrying conductors that conduct current only under abnormal circumstances, current-carrying ground conductors conduct currents under normal operation of a circuit and are necessary for proper functioning of an electrical circuit. A neutral conductor is a typical current-carrying conductor.

### 3.2.3 Lightning Grounds

Grounding for lightning protection of buildings and facilities is necessary for several reasons. A facility lacking adequate lightning protection could be vulnerable to powerful lightning strikes that could cause damage in a number of ways. First, there is always the potential for a fire resulting from a strike due to the high currents that are sometimes involved. Lightning can also damage a number of support systems as CATV, fire protection, security systems, and electronic controls. Under some circumstances, there is also the risk of damage to the insulation of electrical systems.

Lightning occurs as a result of cloud masses passing over structures on earth. A moving cloud mass can be at a very high electrical potential with respect to a building or structure. When the potential rises to a value that can jump the gap between the earth and the cloud, a lightning strike can occur. A properly installed lightning grounding system will permit moderate electrical current flow between the clouds and the protected system in advance of a strike. A lightning protection system will include strike termination devices that are usually mounted high and are connected to the grounding system. The result can be a large reduction in the potential for damage that might otherwise result from a subsequent lightning strike. A properly installed lightning protection system will interconnect the grounds of different structures of a facility so as to eliminate any potential difference between the buildings.

An important characteristic of the grounding connections of a lightning system is that the path to ground must be through conductors of a relatively low resistance. The grounding conductors must connect to a ground matt or ground electrodes. If the path to ground is of a high resistance, the high current of a lightning strike will generate high temperatures because of the $I^2R$ effect and possibly cause a fire. NFPA Code 780, *Standard for the Installation of Lightning Protection Systems*, sets forth the requirements for a properly installed lightning protection system. Generally, a facility should have a single ground matt that connects to both the electrical ground system and the lightning ground system. An integral part of a properly installed lightning protection system will include adequately sized surge arrestors that divert the high potentials of a lightning strike away from the conductors of an electrical system and to ground.

High-tension electrical transmission lines usually have at least one bare ground conductor positioned above the current-carrying conductors for the purpose of absorbing at least a part of a lightning strike. A lightning strike can be at an extremely high potential and can travel along elevated transmission lines to cause damage to electrical gear not capable of withstanding the high voltages and currents. Large transformers are almost always fitted with lightning arrestors designed to divert the high potential of a lightning strike to a ground conductor. Unlike an electrical system that is usually grounded at a single point, lightning grounding conductors provided over transmission lines may connect to ground at numerous locations.

### 3.2.4 Grounding of LV, MV, and HV Cable Screens

Insulated cables and conductors intended for service above 2 kV are generally fitted with a metallic "screen" that surrounds the cable's insulation. The screen usually is fabricated of a thin copper or aluminum sheet, and it is connected to a ground conductor at some point. (*Note:* Cables intended for sound systems and instrumentation interconnections are often fitted with similar screens that are called "shields." For those applications, however, the screens are for a different purpose. More specifically, those screens are intended to prevent interference from external radio frequency interference [RFI] or electromagnetic interference [EMI].) The high-voltage cable screen depicted in Figure 3.2 will continually drain what would otherwise be an accumulation of an electrical charge. The screen has two important functions. First, it prevents a potential buildup on the outer surface of the jacket, which could be dangerous to personnel. In addition, the screen prevents a failure of the insulation that might otherwise result from stress at points where the cable comes under physical pressure. The metallic current conductor is positioned in the center of the fabrication. A layer of insulating material completely surrounds the conductor. The metallic "screen" is installed over the layer of insulation. An elastomeric "jacket," as shown in Figure 3.2, may cover the entire assembly. There are many variations in the design of cables for duty over 2 kV. The exact design of the cable will vary from manufacturer to manufacturer and will depend on a number of factors including voltage level and other characteristics of the intended application.

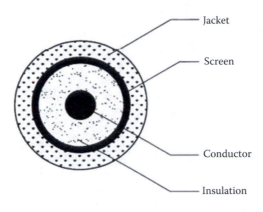

**FIGURE 3.2**
Typical cable construction for service over 2 kV.

## 3.3 Single-Phase Grounding

The grounding of a typical residential, single-phase service is represented in Figure 3.3. The local utility typically installs a transformer near a group of homes and extends 120/240 VAC electrical power to each house. The utility would be responsible for the service up to the and including the meter. Beyond the meter, the wiring installation becomes the responsibility of the installer of the electrical system and the home owner. Nevertheless, the utility might impose specific requirements for circuit protection within the building's electrical panel. The schematic representation of Figure 3.3 shows a typical electric panel that would house the circuit protective devices. The utility will commonly require a specific type of circuit protection for the two incoming live conductors. Branch circuits within the building today are

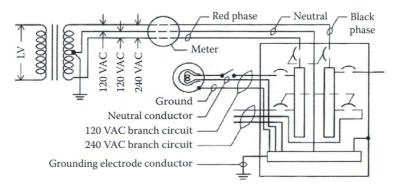

**FIGURE 3.3**
Single-phase service.

mostly protected with circuit breakers. Years ago, branch circuit protection was accomplished with fuses.

A neutral conductor at an installation will extend from the local transformer to the ground bar within the electrical panel. In the past, the neutral conductor has been bare, although today most neutral conductors are mostly insulated. According to the latest edition of the NEC, 120 VAC branch circuits must consist of a live conductor, a neutral conductor, and a ground wire. The neutral conductor must be insulated, and the ground wire may be, and usually is, uninsulated. (In the illustration of Figure 3.3, the live conductors would be the red phase and the black phase.) The representation of Figure 3.3 shows a 120 VAC circuit extended to a lighting fixture. The circuit is extended from the red phase to a light switch. The light switch controls the light. A neutral conductor of the branch circuit connects to the ground bar in the panel. A ground wire, which is part of the branch circuit, connects to the metal of the lighting fixture. Often the ground for a residence is a rod outside the house that had been driven deep into the ground. In the past years, grounding to a water pipe was allowed and is common in many older installations. However, the increasing use of plastic piping would present a problem if a ground were connected to water piping. For this reason, recent codes disallow using water piping as a ground source. Unless both the red phase and the black phase have identical electrical loads, there would normally be a current flow through the grounding electrode to ground. So, a dependable ground, other than the building's water piping, is a necessity.

Should a short fault develop in the installation represented in Figure 3.3, current would flow to the ground through the shortest path. A typical ground fault occurring in a branch circuit of the red phase is shown in Figure 3.4. As shown, current going to the ground would return to the local transformer through the two grounds that had been made available: one at the residence and one at the transformer. The largest share of the fault current would go to the building's ground since that would in most cases be the nearest ground connection. When the fault current rises to the setting of

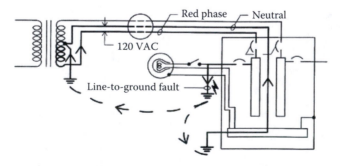

**FIGURE 3.4**
Single-phase line to ground fault.

the circuit protective device, the current flow would be interrupted and the circuit deactivated.

Here, a line-to-ground fault is treated. Of course, another possible fault would involve a line-to-line short. Line-to-line faults are less common and present less potential for injury or property loss. Nevertheless, circuit protective devices must be capable of preventing an overcurrent condition resulting from a line-to-line short. (The subject of circuit protection is treated in detail in Chapter 8.)

## 3.4 Three-Phase Grounding

There are various types of three-phase grounds. Utilities have one set of grounding needs, whereas customers have a different set of requirements. In consequence, the grounding practices of these two different groups tend to have a different appearance.

### 3.4.1 Utility Three-Phase Grounding Practices

Distribution transformers used by utilities must be grounded. Common practices followed by utilities to ground distribution transformers are illustrated in Figures 1.16 through 1.20, inclusive.

### 3.4.2 User Three-Phase Grounding Practices

The grounding of three-phase circuits at the facility of a user of electric power may have a different appearances from that of the utility's grounding practices. If the criteria of the NEC are to be followed, then grounding must follow the exact criteria of the NEC. However, many installations are not obligated to abide by the NEC. In any event, good grounding practices are always warranted. Three-phase grounding follows many of the principles applicable to single-phase circuits.

A typical three-phase four-wire service to a three-phase panel is represented in Figure 3.5. The four-wire service in the example of Figure 3.5 by definition has four conductors: phase A (represented as conductor "A"), phase B (represented as conductor "B"), phase C (represented as conductor "C"), and the neutral conductor (represented as conductor "N"). A branch circuit consisting of conductors A, B, and C are extended to a wye wound motor. An equipment grounding conductor is likewise extended to the motor and connects to the motor frame. (*Note:* The NEC allows that equipment grounding may be accomplished in lower power circuits by means of conduit rather than by a separate equipment grounding conductor.) Operation of the motor depicted in Figure 3.5 is controlled by a motor starter. (Details on motors and

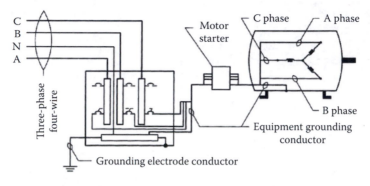

**FIGURE 3.5**
Three-phase service.

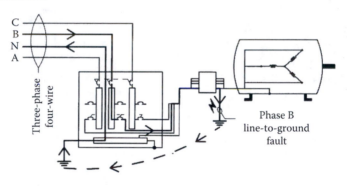

**FIGURE 3.6**
Three-phase line to ground fault.

motor starters are contained in Chapter 9.) As is the case with a single-phase electrical panel, a ground bar is provided within the panel for connections to the neutral conductors, the equipment grounding conductors, and the grounding electrode conductor. The NEC (Article 250-84) requires that the resistance to the ground must be no greater than 25 Ω. However, many engineers consider a much lower resistance necessary whenever large currents are involved. A resistance in the range of 0.1–1.0 Ω, or lower, is considered a more practical value.

The current path resulting from a typical ground fault in a three-phase circuit is represented in Figure 3.6. As represented in Figure 3.6, the ground fault is assumed to be from the B phase of Figure 3.5 to the ground. Current would flow through the short to the ground, back to the panel ground bar, to the neutral conductor, and back to the respective transformer.

### 3.4.3 Grounding Resistors

Electrical power, single-phase or three-phase, supplied to a user as a residence or a commercial building is generally grounded with a "solid ground" connection. A solid ground would typically be a conductor that connects to a ground source as a ground rod, a ground mat, or, as was common in the past, merely underground water pipes. Solid ground connections are shown in Figures 3.3 through 3.6. The purpose of the grounding electrode conductor is to minimize resistance to the flow of electrical current to the ground source. In other words, the grounding electrode conductor should be of a low resistance. On the other hand, industrial power distribution facilities involving large transformers, generators, large motors, and some other types of electrical gear are often fitted with a resistor of one type or another in the ground path. According to the *IEEE Red Book* (Reference 3.1), the reasons for limiting the current by resistance grounding may be one or more of the following:

1. To reduce burning and melting effects in faulted electrical equipment such as switchgear, transformers, cables, and rotating machines
2. To reduce mechanical stresses in circuits and apparatus carrying fault currents
3. To reduce electrical-shock hazards to personnel caused by stray ground fault currents in the ground return path
4. To reduce the arc blast or flash hazard to personnel who may have accidentally caused or who happen to be in close proximity to the ground fault
5. To reduce the momentary line-voltage dip occasioned by the occurrence and clearing of a ground fault
6. To secure control of the transient overvoltages while at the same time avoiding the shutdown of a facility circuit on the occurrence of the first ground fault (high-resistance grounding)

According to industry jargon, there are essentially two classifications of grounding resistors, namely, the low-resistance resistors and the high-resistance resistors. In the electrical industry, both are commonly called the neutral ground resistor (NGR).

### 3.4.3.1 Low-Resistance NGR

A low-resistance NGR is used to reduce fault current whenever one phase or more than one phase of a circuit shorts to the ground. If a low-resistance NGR is used, it is expected that the circuit overprotection device will trip and deactivate the circuit. A low-resistance NGR is sized to withstand a specific current for a specific period of time, typically no more than 10 s. The sizing of the resistor, of course, assumes that the overcurrent protective device will trip as a result of the fault and, so, the NGR will not be required to withstand the fault current for a period of time longer than its rating. A typical low-resistance NGR is represented in Figure 3.7. Should a short occur, say, between

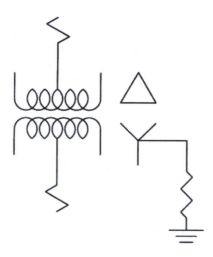

**FIGURE 3.7**
Neutral ground resistor.

phase A and the ground, current will flow to the ground and return to the circuit through the NGR. The circuit overprotection device (circuit breaker or fuse) will detect the overcurrent condition and immediately deactivate the circuit. An NGR will be of no benefit should a line-to-line short occurs.

### 3.4.3.2 High-Resistance NGR

A high-resistance ground is commonly called an HGR. An HGR serves, much as a low-resistance NGR, to limit fault current in the event of a line-to-ground short. An HGR is of a relatively high resistance and is intended to reduce fault currents to a lower value than what would be encountered with a low-resistance NGR. As a result, an HGR will have a greater effect to reduce an arc-flash hazard and will limit the potential for mechanical or thermal damage. IEEE Standard 141-1993 (Reference 3.3) recommends that HGRs not be used on systems over 5 kV except when certain grounding relaying is employed. Unlike low-resistance NGRs, HGRs are often sized to avoid a trip of the overprotection device. For this reason, a ground fault detector is sometimes needed in conjunction with an HGR. If an HGR is used with an ungrounded system, certain design measures are needed as exceedingly high, and potentially destructive, conductor voltages can otherwise result under an arcing condition. A typical installation of high-resistance grounds is shown in the photograph of Figure 9.3.

### 3.4.4 Ground Fault Detector

Ground fault detectors are commonly used with ungrounded and HGR systems. A ground fault detector typically has a CT through which all of the

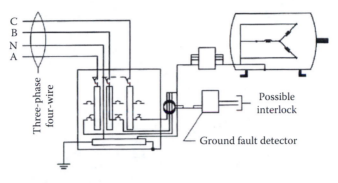

**FIGURE 3.8**
Ground fault detector and HGR.

circuit's power conductors pass. The CT may be remote from the detector or the CT could be integral to the device. Under normal conditions, the three-phase currents balance one another, and the ground fault detector does not detect a discrepancy. In the event of a short to ground, the sensed currents would become unbalanced, and the ground fault detector would detect the difference. In some circuits, a ground fault detector is used merely to initiate an alarm in the event of a ground fault condition without any action to bring about deactivation of the circuit. In other instances, a ground fault detector can be used to directly initiate action to deactivate the respective circuit. Applicable codes in some instances require ground fault detectors to deactivate circuits. More specifically, ground fault detectors are required by some codes for mine electrical installations and for some portable equipment. A typical circuit employing a ground fault detector and an HGR is represented in Figure 3.8.

### 3.4.5 Neutral Grounding Reactor

On larger generators and transformers, it is common to use a neutral grounding reactor in lieu of either a neutral ground resistor or a high ground resistor to limit ground fault currents. Unlike a neutral ground resistor or a high ground resistor that offer resistance to current flow by means of resistive elements, a neutral ground reactor resists current flow with inductive elements. A neutral ground reactor consists of coils, wound around either an air core or an iron core. A neutral ground reactor is wired between the circuit neutral and ground as represented in Figure 3.9. During normal operation current, flow through the reactor to ground is nil. With a balanced condition, current flow would be zero; under an unbalanced network condition, current flow would be slight due to the elevation of the network neutral. Much as a neutral ground resistor or a high ground resistor, a neutral ground reactor greatly reduces stress on equipment that would otherwise result from the high currents of a ground fault.

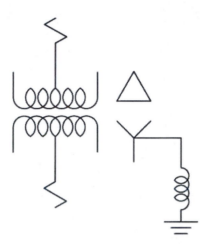

**FIGURE 3.9**
Neutral grounding reactor.

## Problems

1. What is the difference between "grounding" and "earthing"?
2. Is a neutral conductor the same as the equipment grounding conductor? Explain.
3. What is a possible adverse consequence of ground loops at large industrial installations?
4. What is the purpose of grounding?
5. What is the purpose of a ground fault detector in an ungrounded system?
6. What type of device would be used to divert the potential of a lightning strike away from the conductors of an electrical system to the ground?
7. Name two functions of screens in high-voltage cables.
8. Can a conduit be considered as a viable equipment ground path?
9. In the event of line-to-line short, would the properties of a low-resistance neutral ground resistor influence the speed of response of the circuit protective devices?
10. Would it be correct to say that where ground fault detectors are used, the detector would in all cases initiate a trip for the respective circuit?

# 4

## Calculating Currents in Three-Phase Circuits

Engineers and electricians concerned with three-phase electricity may on occasion find the need to calculate the currents in a three-phase circuit. Possibly, the values of current are needed to size electrical apparatus for a proposed new installation. Or, the values may be required to evaluate the parameters of an existing system. Some three-phase calculations can be completed with relative ease. For example, the computations pertinent to balanced three-phase circuits are comparatively straightforward. On the other hand, the currents in an unbalanced three-phase circuit require a greater degree of expertise and more attention to details.

The method of calculating currents in this chapter is by the use of vector algebra. It is the least complicated, most direct, and an inherently intuitive approach to a subject that at times can be challenging.

## 4.1 Calculating Currents in Balanced Three-Phase Circuits

Calculations applicable to balanced three-phase circuits are more easily performed than those applicable to unbalanced three-phase circuits. For this reason, the balanced circuits are treated first. The two principle types of balanced three-phase circuits are the delta and the wye circuits. In the most general sense, the currents of both types can be classified as in-line with the phase voltages, leading phase voltages or lagging phase voltages.

### 4.1.1 Calculating Currents in a Balanced Three-Phase Delta Circuit: General

A typical three-phase circuit with a delta load is represented in Figure 1.14. In three-phase delta circuits, the voltage across the load is the line voltage, but the phase current is different from the line current. As represented in Figure 1.13, the instantaneous voltages are $V(t)_{ab}$, $V(t)_{bc}$, and $V(t)_{ca}$. Each of these phase voltages is 120° apart from the adjacent phase as shown in the phasor diagram in Figure 1.16. Expressed in rms terms, the rotation would be $V(t)_{ab} - V(t)_{bc} - V(t)_{ca}$. In Figure 1.14, the line current in conductor "A" is $I_A$, the line current in conductor "B" is $I_B$, and the line current in conductor "C" is $I_C$. The current in phase "a–b" is $I_{ab}$, the current in phase "b–c" is $I_{bc}$, and the current in phase "c–a" is $I_{ca}$.

For balanced three-phase delta circuits, the line currents are determined by the relationship

$$I_L = \left(\sqrt{3}\right) I_P \dots \qquad (4.1) \text{ (Reference 4.1)}$$

where
  $I_L$ is the line current (subscript "L" designates "line")
  $I_P$ is the phase current (subscript "P" designates "phase")

### 4.1.1.1 Resistive Loads

Considered separately, each of the phases of a three-phase delta circuit is a single-phase circuit. Accordingly, if the three loads in the delta circuit of Figure 1.15 are all resistive, the phase currents would be in phase with the phase voltages as represented in Figure 4.1. Obviously, for each of the three phases, PF = 1.0. In Figure 4.1, the respective voltage vectors are represented by the symbols $V_{ab}$, $V_{bc}$, and $V_{ca}$. Since the phase voltages for delta circuits are the same as the line voltages, vectors $V_{ab}$, $V_{bc}$, and $V_{ca}$ are for both the line and the phase voltages. The current vectors are represented by the symbols $I_{ab}$, $I_{bc}$, and $I_{ca}$. As represented in Figure 1.14, the current in conductor B is the current entering point "b" from phase "a–b" less the current that flows from phase "b–c." Stated in mathematical terms,

$$\underline{I_B} = \underline{I_{ab}} - \underline{I_{bc}},$$

where
  $\underline{I_B}$ is the current in conductor B
  $\underline{I_{ab}}$ is the current in phase a–b
  $\underline{I_{bc}}$ is the current in phase b–c

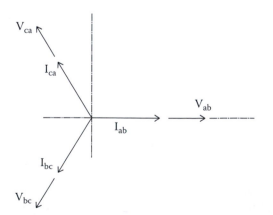

**FIGURE 4.1**
Voltage and current phasors for balanced resistive delta load.

Similarly,

$$I_C = I_{bc} - I_{ca},$$

where
  $I_C$ is the current in conductor C
  $I_{bc}$ is the current in phase b–c
  $I_{ca}$ is the current in phase c–a

and

$$I_A = I_{ca} - I_{ab},$$

where
  $I_A$ is the current in conductor A
  $I_{ca}$ is the current in phase c–a
  $I_{ab}$ is the current in phase a–b

The phasors for $I_{ab}$ and $I_{bc}$ are shown in Figure 4.2. To determine the phasor for the current in Line B, the negative vector of $I_{bc}$ is added to vector $I_{ab}$. The negative vector of $I_{bc}$ (i.e., $-I_{bc}$) is ($+I_{cb}$) as shown in Figure 4.2. The addition of vector $I_{ab}$ and vector $I_{cb}$ generates vector $I_B$.

By vector algebra, in Figure 4.2,

$$\cos 30° = [I_{B-y} \div I_{cb}]$$

where $I_{B-y}$ is the ordinate component of current $I_B$.

$$I_{B-y} = [I_{cb}] \cos 30°$$

$$\cos 30° = [\sqrt{3} \div 2]$$

$$I_{B-y} = [I_{cb}][\sqrt{3} \div 2]$$

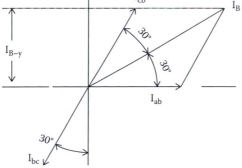

**FIGURE 4.2**
Vectors determining line currents in balanced resistive delta circuit.

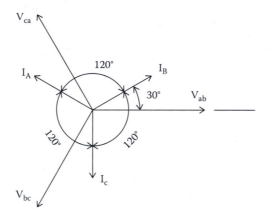

**FIGURE 4.3**
Line voltage and current phasors for balanced resistive delta load.

and

$$\cos 60° = I_{B-y} \div I_B$$

$$I_B = I_{B-y} \div \cos 60°$$

$$\cos 60° = 1/2$$

$$I_B = \left[I_{cb}\right]\sqrt{3}, \quad \text{or}$$

$$I_B = \left[I_{bc}\right]\sqrt{3}$$

This relatively simple computation for the assumed specific case is in agreement with Equation 4.1. Since the circuit was said to be balanced, $|I_B| = |I_A| = |I_C|$, and each current phasor is separated by 120° from the other current phasors. With these criteria, a phasor diagram can be generated to show the relationship between the line currents and the line voltages. This representation is shown in Figure 4.3.

### Example 4.1

**Problem**: Assume that for the three-phase delta load of Figure 1.14, each of the three loads is a heater rated 5 kW and the line voltage is 480 VAC. Find the line and phase currents.

### Solution

Since all three loads are equal, by definition, the circuit is considered balanced. The line currents, the phase currents, and the respective leads/lags of the currents with respect to voltage are all equal.

Let,

Phase current a–b $= I_{ab}$ (rms amp)

Phase current b–c $= I_{bc}$ (rms amp)

Phase current c–a $= I_{ca}$ (rms amp)

Phase a–b power $= P_{ab}$

Phase b–c power $= P_{bc}$

Phase c–a power $= P_{ca}$

Line current in conductor A $= I_A$ (rms)

Line current in conductor B $= I_B$ (rms)

Line current in conductor C $= I_C$ (rms)

For each phase,

$$P = VI$$

$$P_{ab} = 5000(W) = (480)I$$

$$I_P = I_{ab} = I_{bc} = I_{ca} = 5000/480 = 10.416 \text{ A}$$

According to Equation 5.1,

$$I_L = I_A = I_B = I_C = \left(\sqrt{3}\right)(10.416) = 18.04 \text{ A}$$

In this relatively simple example, the line currents, phase currents, and the respective powers could be determined with the well-known and straightforward formulas for a balanced three-phase circuit.

### Example 4.2

**Problem**: Confirm that for a balanced resistive load in a delta circuit, the phasor diagram for line current replicates the mathematical relationship of all three voltages and currents.

### Solution

The value of instantaneous current as represented by vector $I_{ab}$ can be represented by the general Equation 1.2.

$$i(t)_{ab} = I_{ab}\sin(\omega t + \theta),$$

where $\theta$ is the lead or lag of current with respect to voltage $V_{ab}$.

Reference is made to Figure 1.14. Since the load under consideration is a balanced, resistive, three-phase load, the currents in the phases are in phase with the respective voltages. Let the subscript "P" designate "phase" and $\theta_P$ designate the lead/lag of the phase currents with respect to voltages. Then, $\theta_P = 0$ for all three phases. Let,

$$I_{ab} = I_{bc} = I_{ca} = I_P, \quad \text{and}$$

$$i(t)_{ab} = I_{ab}\sin(\omega t + 0)$$

$$i(t)_{bc} = I_{bc}\sin(\omega t + 240°)$$

$$i(t)_{ca} = I_{ca}\sin(\omega t + 120°)$$

The expression $i(t)_{bc}$ can be stated as

$$i(t)_{bc} = I_{bc}\sin(\omega t + 240°), \text{ or}$$

$$i(t)_{bc} = I_{bc}[\sin(\omega t + 240°)] = I_P\{\sin\omega t\cos 240° + \cos\omega t\sin 240°\}$$

$$\cos 240° = -\cos 60°$$

$$\sin 240° = -\sin 60°$$

$$i(t)_{bc} = I_P\{-\sin\omega t\cos 60° - \cos\omega t\sin 60°\}$$

$$i(t)_{bc} = -I_P\{\sin\omega t\cos 60° + \cos\omega t\sin 60°\}$$

$$i(t)_{bc} = -I_P\sin(\omega t + 60°) = -I_P\left\{(1/2)\sin\omega t + \left(\sqrt{3}/2\right)\cos\omega t\right\}$$

$$i(t)_{ab} = I_P\sin\omega t$$

$$i(t)_{ab} - i(t)_{bc} = I_P\sin\omega t + I_P\left\{1/2\sin\omega t + \left(\sqrt{3}/2\right)\cos\omega t\right\}$$

$$i(t)_B = i(t)_{ab} - i(t)_{bc} = I_P\left\{(3/2)\sin\omega t + \left(\sqrt{3}/2\right)\cos\omega t\right\}$$

$$i(t)_\text{B} = I_\text{P}\left(\sqrt{3}\right)\left\{\left(\sqrt{3}/2\right)\sin\omega t + (1/2)\cos\omega t\right\}$$

$$i(t)_\text{B} = I_\text{P}\left(\sqrt{3}\right)\left\{\cos 30° \sin\omega t + \sin 30°\cos\omega t\right\}$$

$$i(t)_\text{B} = I_\text{P}\left(\sqrt{3}\right)\sin(\omega t + 30°)$$

This computation confirms the value of one of the line currents. Since the circuit is balanced, the other two line currents would be equal in magnitude and rotated by 120° from one another. The computation also confirms that the line current in a balanced resistive delta load is the product of $\left(\sqrt{3}\right)$ times the phase current and, further, that the line current leads the phase currents by 30°. Therefore, the model as defined by the vector diagram of Figure 4.2 is an accurate representation of the phase and line currents for the described example.

The computations of Example 4.2 also illustrate the difficulties in mathematically determining currents. These computations also demonstrate the merits and simplicity in using phasor diagrams to determine current values.

### 4.1.1.2 Capacitive Loads

In the previous section, it is shown how currents in a balanced delta circuit with resistive loads are determined. Unlike currents in a resistive load where the phase currents are in phase with the phase voltages, phase currents in a predominately capacitive circuit lead the phase voltages by some amount between 0° and 90°.

Commercial and industrial loads are mostly inductive because the largest part of their power usage is generally attributed to the use of three-phase repulsion induction motors. And, all induction motors operate with a lagging power factor. Nevertheless, capacitive circuits are also found at commercial and industrial users. The most common form of a capacitive circuit in power applications is a capacitor bank that is used to counteract lagging power factors resulting from the use of induction motors. Because some utilities will charge users an extra fee for a lagging power factor, capacitor banks are sometimes installed by commercial and industrial customers in parallel with the normal loads to correct power factor. A capacitor bank alone provides a leading power factor without consuming significant power and, when combined with a lagging power factor, it will bring the service lagging power factor more near unity. A typical example of the use of a capacitor bank to correct power factor is treated in Example 4.5.

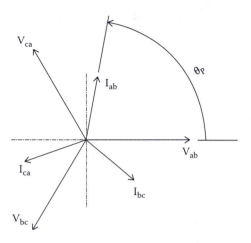

**FIGURE 4.4**
Phase voltage and current phasors for balanced capacitive delta load.

A circuit with a typical capacitive load is represented in Figure 4.4. (Actually, the circuit represented in Figure 4.4 would be only partly capacitive. The circuit would necessarily contain resistive and/or inductive element since the angle $\theta_P$ is shown as less than 90°. If the circuit were purely capacitive, the angle of lead, $\theta_P$, would be exactly +90°.) Note that the arc indicating the angle $\theta_P$ shows the positive direction of $\theta_P$ to be CCW from the positive abscissa as indicated by the arrowhead on the arc. This practice is consistent with the polar notation of complex numbers, whereby the positive angle of a vector is likewise considered the CCW direction from the positive abscissa. The vectors determining line current $I_B$, assuming the phase current vectors in Figure 4.4, are shown in Figure 4.5. The phasors representative of currents $I_A$ and $I_C$ would be determined in a similar manner. The resultant line currents, $I_A$, $I_B$, and $I_C$, are shown in Figure 4.6.

### 4.1.1.3 Inductive Loads

In a balanced three-phase inductive circuit, phase currents lag the phase voltages by some amount between 0° and 90°. For a delta circuit, Equation 4.1 remains applicable so that for each of the three currents, $I_L = \left(\sqrt{3}\right)I_P$. Using the specific notation of Figure 1.4 for current in conductor B, $I_B = \left(\sqrt{3}\right)I_{ab}$. The geometry of the phasors determining $I_B$ indicate that $\theta_P = \theta_L - 30°$. The phase currents for an inductive load lag line voltage and in the phasor diagram, the phasor for the phase current is positioned clockwise from the phasor for line voltage. For an inductive load, $\theta_P < 0°$ by definition. It may also be noted that for $\theta_P$ greater than −30° (e.g., −20° or +10°), the line current leads the line voltage and for $\theta_P$ less than −30° (e.g., −40° or −50°), the line current lags line voltage.

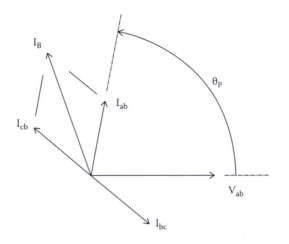

**FIGURE 4.5**
Vectors determining line current $I_B$ in balanced capacitive delta circuit.

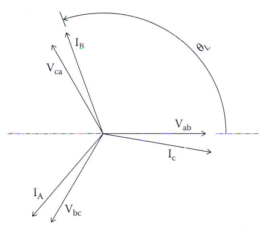

**FIGURE 4.6**
Line voltage and line current phasors for balanced capacitive delta circuit.

Much as with the phasor diagram for a capacitive load, the phasor for a particular line current in an inductive circuit is determined by adding component vectors as was shown in Figure 4.4 for current $I_B$ in a capacitive circuit. Example 4.3 demonstrates the procedure for determining line currents in a specific inductive circuit.

### Example 4.3

**Problem**: The nameplate on a delta wound induction motor states "480/3/60," "FLA" as 10A, and "PF" of 0.707. Determine the line currents, draw the phasor diagram for line and phase currents, and calculate the power.

**Solution**

The line voltage is 480 VAC, three phase, 60 Hz, and the line current is 10 A. Since induction motors have a lagging power factor, the current lags voltage by the angle $\cos^{-1} 0.707$ or 45°.

According to Equation 1.7,

$$\theta_P = \theta_L - 30°, \quad \text{or}$$

$$\theta_L = \theta_P + 30° = -15°$$

According to Equation 1.6,

$$P = \left[\sqrt{3}\right] V_L I_L \cos\theta_P$$

$$P = \sqrt{3}\,(480)(10)(0.707)$$

$$= 5877 \text{ W}$$

$$I_P = I_L/\sqrt{3} = 10/\sqrt{3} = 5.77 \text{ A}$$

Expressed in polar notation,

$$I_L = 10\,\underline{/-15°}, \text{ and } I_P = 5.77\,\underline{/-45°}$$

The phasor diagram for the phase currents and voltages is shown in Figure 4.7. The notation of Figure 1.4, which is for a delta circuit, is followed. The

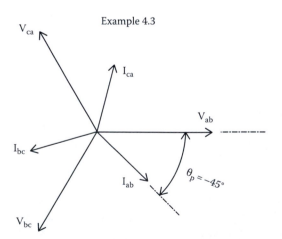

Example 4.3

**FIGURE 4.7**
Phase voltage and current phasors for balanced inductive delta load.

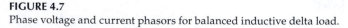

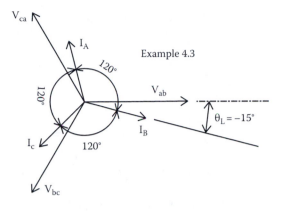

**FIGURE 4.8**
Line voltage and current phasors for balanced inductive delta load.

current phasor for phase a–b (i.e., $I_{ab}$) is rotated clockwise 45° from phase voltage $V_{ab}$, thereby indicating that the phase current lags the phase voltage. Each of the line voltages is rotated 120° from one another. Likewise, all three phase currents ($I_{ab}$, $I_{bc}$, and $I_{ca}$) are 120° apart.

The phasor diagram for the line currents and line voltages for Example 4.3 are shown in Figure 4.8 where the line current in conductor "B" is shown as lagging the respective line voltage vector ($V_{ab}$) by 15°. The three line currents are shown separated from one another by 120°. (Note that in Figures 4.7 and 4.8, the arcs indicating the positions and values of $\theta_P$ and $\theta_L$ have arrowheads at both ends. This practice is followed to avoid confusion with the positive direction of the variables. The positive direction of the variables is indicated by an arc having a single arrowhead that designates the positive direction of the respective variable.)

### 4.1.1.4 Two or More Loads

It is common to have two or more balanced delta loads on a common three-phase feeder. A typical three-phase feeder with two balanced delta loads is represented in Figure 4.9 where the three-phase feeder consists of conductors "A," "B," and "C." The feeder serves two loads, namely, load 1 and load 2.

Conductor A has branches P and S, conductor B has two branches Q and T, and conductor C has branches R and U. A common problem would be to determine the currents in conductors A, B, and C.

Obviously,

$$\underline{I_A} = \underline{I_P} + \underline{I_S},$$

$$\underline{I_B} = \underline{I_Q} + \underline{I_T}, \quad \text{and}$$

$$\underline{I_C} = \underline{I_R} + \underline{I_U}$$

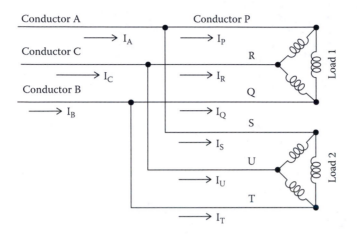

**FIGURE 4.9**
Two balanced delta loads.

(The vector $I_A$ is determined by adding the vectors $I_P$ and $I_S$, the vector $I_B$ is determined by adding the vectors $I_Q$ and $I_T$, and vector $I_C$ is determined by adding the vectors $I_R$ and $I_U$. An algebraic addition of these values would yield incorrect values.)

The phasor diagram for the two loads of Figure 4.9 is similar to that represented in Figure 4.10. The primary objective of the phasor diagram of Figure 4.10 is to determine the vales of $I_A$, $I_B$, and $I_C$. Since it was assumed that the loads are balanced, $I_A = I_B = I_C$. So, it becomes necessary to merely determine one of the currents as, say, $I_B$. (It is more convenient to first determine current $I_B$, rather than currents $I_A$ or $I_C$. This is so because the phasor for the associated reference voltage, $V_{ab}$, is positioned along the positive abscissa. The associated calculations for $I_B$ are more easily performed than those required to determine

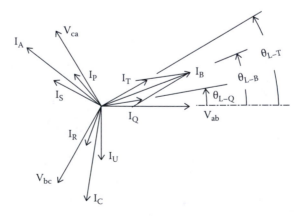

**FIGURE 4.10**
Two delta loads.

$I_A$ or $I_C$. The determinations of $I_A$ and $I_C$ follow that of current $I_B$.) The values of Figure 4.10 may be determined by general Equation 1.5, which is for a single-phase circuit. The values are substituted into Equation 1.5 as applicable to a three-phase circuit and for the purpose of determining current $I_B$.

Accordingly,

$$|I_B| = \left\{ (I_{B-y})^2 + (I_{B-x})^2 \right\}^{\frac{1}{2}} \cdots \tag{4.2}$$

where

$I_{B-x} = I_1 \cos \theta_1 + I_2 \cos \theta_2 + \cdots I_n \cos \theta_n$
$I_{B-y} = [I_1 \sin \theta_1 + I_2 \sin \theta_2 + \cdots I_n \sin \theta_n]$
$\theta_B = \sin^{-1}(I_{B-y} \div I_B)$

This section deals with balanced delta loads. Nevertheless, Equation 4.2 is in fact applicable to two or more balanced loads of any type. This subject is discussed in more detail in Section 4.1.2.4, which section is specifically applicable to balanced wye loads.

For the specific representation of Figure 4.10, which is for two three-phase loads,

$$|I_B| = \left\{ (I_{B-x})^2 + (I_{B-y})^2 \right\}^{\frac{1}{2}}, \quad \text{where}$$

$$I_{B-x} = I_Q \cos \theta_Q + I_T \cos \theta_T, \quad \text{and}$$

$$I_{B-y} = I_Q \sin \theta_Q + I_T \sin \theta_T$$

$$\theta_B = \sin^{-1}(I_{B-y} \div I_B)$$

**Example 4.4**

**Problem**: Assume there are two delta wound induction motors on a common 480 VAC circuit as represented in Figure 4.9. For load 1, FLA = 6 A, PF = 0.90, and for load 2, FLA = 7 A, PF = 0.70. Find the currents and power factor in the common three-phase feeder (currents A, B, and C in Figure 4.9).

**Solution**

For load 1, let $\theta_{P-1}$ designate the lead/lag of the (phase) current of load 1. Then, $\cos \theta_{P-1} = 0.90$.

Since the motors are induction motors, the current lags phase voltage.

$$\theta_{P-1} = -\cos^{-1} 0.90 = -25.842° \text{ and } I_1 = 6$$

$$\theta_{L-1} = \theta_{P-1} + 30° = -25.842° + 30° = +4.15° \left( \text{line current leads line voltage} \right)$$

Similarly for load 2,

$$\cos\theta_{P-2} = 0.70$$

$$\theta_{P-2} = -\cos^{-1}0.70 = -45.572° \text{ and } I_2 = 7$$

$$\theta_{L-2} = -45.572° + 30° = -15.572° \text{ (line current lags line voltage)}$$

The orientations of the vectors representative of the line currents are shown in Figure 4.11.
Reference is made to Figure 4.9.
Let

$$\theta_{L-A} = \text{lead/lag of line current "A"}$$

$$\theta_{L-B} = \text{lead/lag of line current "B"}$$

$$\theta_{L-C} = \text{lead/lag of line current "C"}$$

$$I_1 = 6 = I_P = I_Q = I_R$$

$$I_2 = 7 = I_S = I_T = I_U$$

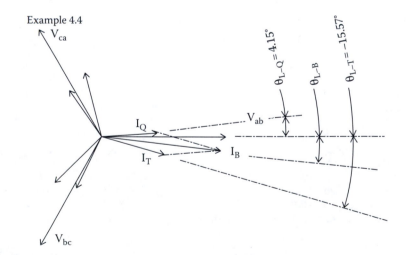

**FIGURE 4.11**
Two inductive delta loads.

Reference is made to Equation 4.2, which states in general terms,

$$I_B = \left\{ (I_{B-x})^2 + (I_{B-y})^2 \right\}^{1/2}$$

where
$$I_{B-x} = I_1 \cos \theta_1 + I_2 \cos \theta_2 + \cdots I_n \cos \theta_n$$
$$I_{B-y} = [I_1 \sin \theta_1 + I_2 \sin \theta_2 + \cdots I_n \sin \theta_n]$$
$$\theta_B = \sin^{-1}(I_{B-y} \div I_B)$$

The notation of Equation 4.2 is amended specifically for two loads, whereby the subscript "1" designates load 1 and the subscript "2" designates load 2.

$$I_B = \left\{ (I_{B-x})^2 + (I_{B-y})^2 \right\}^{1/2}, \quad \text{where}$$

$$I_{B-x} = I_1 \cos \theta_1 + I_2 \cos \theta_2, \quad \text{and}$$

$$I_{B-y} = [I_1 \sin \theta_1 + I_2 \sin \theta_2]$$

$$\theta_{L-B} = \sin^{-1}(I_{B-y} \div I_B)$$

For the line currents in the common feeder (i.e., currents A, B, and C),

$$I_{B-x} = I_1 \cos \theta_1 + I_2 \cos \theta_2 = (6)(\cos 4.15°) + (7)\,(\cos - 15.57°)$$

$$= (6)(.997) + (7)(.963) = 5.982 + 6.741 = 12.723$$

$$I_{B-y} = [I_1 \sin \theta_1 + I_2 \sin \theta_2] = [(6)(\sin - 4.15°) + (7)(\sin + 15.57°)]$$

$$= [(6)(.072) - (7)\,(.268)] = [(.432) - (1.876)] = -1.444$$

$$|I_B| = \left\{ (I_{B-x})^2 + (I_{B-y})^2 \right\}^{1/2}$$

$$I_B = \left\{ (12.723)^2 + (-1.444)^2 \right\}^{1/2}$$

$$I_B = 12.804 \text{ A}$$

Obviously, $I_B = I_A = I_C = 12.804$ amps

$$\theta_{L-B} = \sin^{-1}(I_{B-y} \div I_B)$$

$$\theta_{L-B} = \sin^{-1}(-1.444 \div 12.804) = \sin^{-1}(-0.1127)$$

$$\theta_{L-B} = \theta_{L-A} = \theta_{L-C} = -6.47° \ \text{(line current lagging line voltage)}$$

Power factor pertains to phase lead/lag and not line lead/lag. So, per Equation 1.7,

$$\theta_P = \theta_L - 30°.$$

$$\theta_P = -6.47° - 30° = -36.47° \quad \text{and} \quad PF = \cos(-36.47°) = .804$$

### Example 4.5

**Problem**: Reference is made to Figure 4.9. Assume a utility customer has a net load that would be the equivalent of load 1, Figure 4.9, with 480 VAC, 100 A, and PF = 0.50, lagging. Find the capacitor bank current size required to bring the net power factor to unity.

### Solution

The capacitor bank would be installed in parallel with the inductive load. Using the notation of Figure 4.9,

$$\text{Let } \theta_{L-A} = \text{lead/lag of (line) current "A" of electrical source}$$

$$\text{Let } \theta_{L-B} = \text{lead/lag of (line) current "B" of electrical source}$$

$$\text{Let } \theta_{L-C} = \text{lead/lag of (line) current "C" of electrical source}$$

$$\text{Let } \theta_{L-P} = \text{lead/lag of (line) current "P" of lagging load}$$

$$\text{Let } \theta_{L-Q} = \text{lead/lag of (line) current "Q" of lagging load}$$

$$\text{Let } \theta_{L-R} = \text{lead/lag of (line) current "R" of lagging load}$$

$$\text{Let } \theta_{L-S} = \text{lead/lag of (line) current "S" of capacitor bank}$$

$$\text{Let } \theta_{L-T} = \text{lead/lag of (line) current "T" of capacitor bank}$$

$$\text{Let } \theta_{L-U} = \text{lead/lag of (line) current "U" of capacitor bank}$$

$$I_1 = 100 \text{ A} = I_P = I_R = I_Q$$

For load 1 (the lagging load), $\cos \theta_{P-1} = 0.50$, $\theta_{P-1} = -60°$.
In general, from Equation 1.7,

$$\theta_P = \theta_L - 30°$$

$$\theta_L = \theta_P + 30°$$

So,

$$\theta_{L-P} = \theta_{L-R} = \theta_{L-Q} = -60° + 30° = -30°$$

Current in the capacitor bank will lead phase voltage by approximately 90°. So, for load 2,

$$\theta_{P-2} = +90°$$

$$\theta_{L-S} = \theta_{L-T} = \theta_{L-U}$$

$$= \theta_{P-2} + 30° = 120°$$

The phasor diagram for this example is shown in Figure 4.12. The objective is to have $I_A$, $I_B$, and $I_C$ at $\theta_L = +30°$, that is, to have the line currents leading line voltage by 30°. This would be the position of the line currents corresponding to a pure resistive load that would correspond to a unity power factor.

Reference is made again to Equation 4.2, which states in general terms,

$$|I_B| = \left\{ (I_{B-x})^2 + (I_{B-y})^2 \right\}^{1/2}$$

where
$$I_{B-x} = I_1 \cos\theta_1 + I_2 \cos\theta_2 + \cdots I_n \cos\theta_n$$
$$I_{B-y} = [I_1 \sin\theta_1 + I_2 \sin\theta_2 + \cdots I_n \sin\theta_n]$$
$$\theta_B = \sin^{-1}(I_{B-y} \div I_B)$$

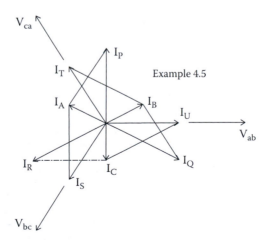

**FIGURE 4.12**
Capacitive and inductive delta loads.

With notation altered to use the notation of Figure 4.9 for two loads,

$$\left|I_B\right| = \left\{(I_{B-x})^2 + (I_{B-y})^2\right\}^{1/2}, \quad \text{where}$$

$$I_{B-x} = I_1 \cos\theta_1 + I_2 \cos\theta_2, \quad \text{and}$$

$$I_{B-y} = [I_1 \sin\theta_1 + I_2 \sin\theta_2]$$

$$\theta_{L-B} = \sin^{-1}(I_{B-y} \div I_B)$$

By trial and error computations, it was determined that a capacitor bank current of 86.61 A will satisfy the required criteria.

The computed values are readily confirmed:

$$I_2 = 86.61 \text{ A}; \quad I_1 = 100 \text{ A}; \quad \theta_1 = -30°; \quad \theta_2 = +120°$$

$$I_{B-x} = I_1\cos\theta_1 + I_2\cos\theta_2 = (100)(\cos-30°) + (86.61)(\cos 120°)$$
$$= (100)(.8660) + (86.61)(-.5) = 86.60 - 43.309 = 43.291$$

$$I_{B-y} = [I_1 \sin\theta_1 + I_2 \sin\theta_2]$$

$$I_{B-y} = [(100)(\sin-30°) + (86.61)(\sin 120°)] = [-50 + (75.00)] = 25.00$$

$$\left|I_B\right| = \left\{(I_{B-x})^2 + (I_{B-y})^2\right\}^{1/2}$$

$$I_B = \left\{(43.291)^2 + (25.00)^2\right\}^{1/2} = 50.00 \text{ As}$$

$$\theta_{L-B} = \sin^{-1}(I_{B-y} \div I_B) = \sin^{-1}(25.00 \div 50.00)$$
$$= \sin^{-1}(.500) = 30° \left(\text{i.e., line current leads line voltage by } 30°\right)$$

It may be noted that if, say, a wattmeter were installed in the line to the motor, the power would be measured at 100 A and a power factor of 0.50, resulting in a power calculation of 41.569 kW. If a capacitor bank were installed and power measured in the common feeder upstream of the capacitor bank and the motor, the meter would indicate the same power, namely, 41.569 kW, although at a current of 50.00 A and unity power factor. In other words, the addition of the capacitor bank did not alter the power consumption, but it greatly improved the power factor.

In summary, Installation of a capacitor bank with current of 86.61 A will return service line current to unity power factor (from a lagging power factor of 0.50) and reduce service current ($I_A$, $I_B$, and $I_C$ of Figure 4.9) from 100 to 50.00 A. Power consumption would remain unaltered, but the improvement in power factor might very well result in a reduced electric billing. (The subject of power factor correction is treated in detail in Chapter 10.)

## 4.1.2 Calculating Currents in a Balanced Three-Phase Wye Circuit: General

It is noted in the previous section that in a delta circuit, the phase voltage is identical in magnitude to the line voltage, but the phase current has a magnitude and lead or lag that is different from the line current. In some ways, the wye circuit is the opposite of a delta circuit. In a wye circuit, the phase voltages have a magnitude and lead or lag that is different from the line voltage, but the phase currents and the associated lead/lags are identical to the line current. A typical three-phase wye circuit is represented in Figure 1.15 where the voltage sequence is assumed to be $V_{ab} - V_{bc} - V_{ca}$ as was the case for the delta circuit of Figure 1.14. Each of the line voltages of Figure 1.14 is 120° apart from the adjacent phase as shown in Figure 1.16 for the assumed delta circuit. In Figure 1.14, the line current in conductor "A" is $I_A$, the line current in conductor "B" is $I_B$, and the line current in conductor "C" is $I_C$. In a wye circuit, there is a fourth point, namely, point "d" in Figure 1.15, which could be connected to the ground. A fourth or neutral wire is needed, connected to the ground, if it is anticipated that the three phases may not be balanced. However, all wye circuits do not have or need a neutral wire. A wye induction motor, for example, would have no neutral wire since currents in all three phases would be very nearly equal. In Figure 1.15, the current in phase "a–d" is $I_{ad}$, the current in phase "b–d" is $I_{bd}$, and the current in phase "c–d" is $I_{cd}$. If point "d" is connected to ground by neutral conductor D, current in conductor D would be nil in a balanced wye circuit but would be conducting current in an unbalanced circuit. For a balanced three-phase wye circuit, the phase voltages are related to the line voltages by the expression

$$V_L = \left(\sqrt{3}\right) V_P \dots \tag{4.3}$$

(Reference 4.2),

where
   $V_L$ is the line voltage
   $V_P$ is the phase voltage

A typical phasor diagram for a balanced wye circuit is shown in Figure 4.13. It may be seen that voltage $V_{db}$ leads voltage $V_{ab}$ by 30°, voltage $V_{da}$ leads voltage $V_{ca}$ by 30°, and voltage $V_{dc}$ leads voltage $V_{bc}$ by 30°. It is also apparent

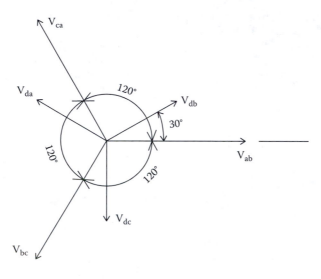

**FIGURE 4.13**
Voltages in balanced wye circuit.

that for any wye circuit, balanced or unbalanced, $\theta_P = \theta_L - 30°$, as was noted earlier to be applicable to a balanced delta circuit.

### 4.1.2.1 Resistive Loads

Resistive loads in single-phase circuits are necessarily in phase with the applied voltage. If all three phases of a three-phase circuit have resistive loads of equal magnitude, the phase current would likewise be in phase with the respective phase voltage as represented in Figure 4.14. The phase currents are equal in magnitude to the line currents but at an angle of 30° to the line voltages.

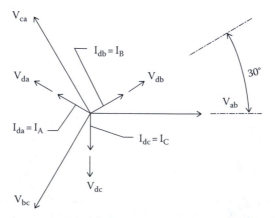

**FIGURE 4.14**
Voltage and current phasors for balanced resistive wye load.

**Example 4.6**

**Problem**: Assume that for the wye load in Figure 1.15, each of the three loads is a heater rated 5 kW (as in Example 4.1), and the line voltage is 480 VAC (also as in Example 4.1). Find the line and phase currents and the respective lead/lags.

**Solution**

The applicable current and voltage phasors are represented in Figure 4.14.

Let,

Phase current a–d = $\overline{I_{ad}}$ (rms amp)
Phase current b–d = $\overline{I_{bd}}$ (rms amp)
Phase current c–d = $\overline{I_{cd}}$ (rms amp)
Phase a–d power = $P_{ad}$
Phase b–d power = $P_{bd}$
Phase c–d power = $P_{cd}$
Line current in conductor A = $\overline{I_A}$ (rms)
Line current in conductor B = $\overline{I_B}$ (rms)
Line current in conductor C = $\overline{I_C}$ (rms)
From Equation 4.3, $V_{ad} = V_{bd} = \overline{V_{cd}} = 1/(\sqrt{3})\,(480) = 277.13$ V
For each phase, $P = V_p I_p$,

where
$V_P$ is the phase voltage
$I_P$ is the phase current

$$I_P = \frac{P}{V_P} = \frac{5000}{277.13} = 18.04 \text{ A}$$

The phase currents are in phase with the phase voltages. So,

$$\theta_P = 0$$

$$\cos\theta_P = 1.0$$

From Equation 1.7,

$$\theta_{P-A/AD} = \theta_{L-A/CA} - 30°$$

$$\theta_{P-A/AD} = 0°$$

$$\theta_{L-A/CA} = +30°$$

Since the load is balanced, the lead/lag in conductors B and C are the same as that in conductor A. Thus,

$$I_P = 18.04 \text{ A at } 0° \text{ to phase voltage, and}$$

$$I_L = 18.04 \text{ A at } +30° \text{ to line voltage (leading line voltage)}$$

### 4.1.2.2 Capacitive Loads

For capacitive loads, the phase current leads the phase voltage by angle $\theta_P$, which would be between 0° and +90°. Currents in capacitive wye circuits always lead line voltage.

### 4.1.2.3 Inductive Loads

With an inductive load, the phase current lags the phase voltage by angle $\theta_P$, which would be between $\theta_P = 0°$ and $\theta_P = -90°$. If the phase current lags by less than 30°, the line current leads line voltage. On the other hand, if phase current lags phase voltage by more than 30°, the line current would also lag line voltage.

### 4.1.2.4 Two or More Loads

In Section 4.1.1.4, it was explained that the applicable equation for two or more delta loads is Equation 4.2. As stated, Equation 4.2 is applicable when the line currents of two or more delta loads are known. It will become apparent that if the line currents are known for two or more balanced wye loads, Equation 4.2 is still applicable. The loads could be delta or wye. The manner in which the line currents were generated is immaterial. For this reason, Equation 4.2 is repeated in the summary of equations, Section B.3, where it is stated that the equation is applicable to "balanced three-phase circuits." In other words, Equation 4.2 is not restricted to delta circuits; it is applicable to any type or combination of balanced three-phase circuit.

---

## 4.2 Calculating Currents in Unbalanced Three-Phase Circuits: General

Once a person understands balanced three-phase circuits and the use of phasor diagrams to visualize the voltages and currents in those circuits, it is an easy transition into the realm of unbalanced three-phase circuits. As with balanced circuits, phasor diagrams can help a person to understand what is happening within a circuit. In the succeeding text, delta circuits are considered first, and the wye circuits are considered after the delta circuits. Because of the very large number of possible combinations of loads and phase angles in unbalanced circuits, all of these combinations cannot possibly be treated. Rather, the method for calculating line currents is described and explained. With the presented methodology, a person may then readily calculate currents in any possible, given combination of unbalanced delta and wye circuits.

By definition, an unbalanced circuit has at least one phase current that is not equal to the other phase currents either in magnitude or phase angle. Of course, all three phase currents could be of unequal magnitude. In all cases, line voltages are assumed to be of equal magnitude, separated by 120° of rotation and in the sequence A–B, B–C, and C–A.

### 4.2.1 Unbalanced Three-Phase Delta Circuits

When all three loads of a delta circuit are resistive, the phase currents are all in phase with the line voltages. Also, the power factors of all three phases are unified (PF = 1.0), and the lead/lag angle in the phases are all zero ($\theta_P = 0$). As explained earlier, in a balanced resistive delta circuit, the line currents all lead the line voltages by 30°. If the loads are resistive and unbalanced, the line currents could be at various angles to the line voltages.

A determination of line currents might be necessary if, say, the currents in the conductors of a common feeder circuit are to be calculated. Of course, other needs could be present.

#### 4.2.1.1 Unbalanced Three-Phase Delta Circuits with Resistive, Inductive, or Capacitive Loads

A generalized view of a delta circuit assumes that the current in each phase is at some angle, $\theta_P$, to the phase voltage. In other words, if the current in a phase is in phase with the phase voltage, as would be the case with resistive loads, then $\theta_P = 0$. If the load is capacitive, $\theta_P > 0$, and the current is said to be "leading." If the load is inductive, $\theta_P < 0$, and the current is said to be "lagging." (A circuit with a leading current is said to have a "negative" power factor, and circuit with lagging current is said to have a "positive" power factor.) A typical delta circuit is represented in Figure 1.14, and a generalized summary of the phase voltages and phase currents is represented in Figure 4.15.

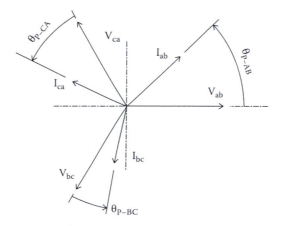

**FIGURE 4.15**
General phasors of unbalanced delta circuit.

It is assumed that none of the phase currents are equal. Earlier, it was shown how the line currents are determined. For example, the phasor of line current $I_A$ is determined by adding the phasor for $I_{ca}$ and the negative phasor of $I_{ab}$, namely, $I_{ba}$. The procedure for adding the vectors that determine $I_A$ is illustrated in the following text. For the purposes of analysis, it is assumed that no two currents are necessarily of equal magnitude or necessarily at the same lead/lag angle.

In order to develop the relationship for line current $I_A$, let

$I_{ab-x}$ be the abscissa component of vector $I_{ab}$
$I_{ca-x}$ be the abscissa component of vector $I_{ca}$
$I_{A-x}$ be the abscissa component of vector $I_A$
$I_{ab-y}$ be the ordinate component of vector $I_{ab}$
$I_{ca-y}$ be the ordinate component of vector $I_{ca}$
$I_{A-y}$ be the ordinate component of vector $I_A$

Current $I_A$ is determined as shown in Figure 4.16.
In Figure 4.16,

$$I_{ba-x} = I_{ba}\cos\gamma$$

$$\gamma = 180° + \theta_{P-AB}$$

$$\cos\gamma = -\cos\theta_{P-AB}$$

$$I_{ba-x} = I_{ba}(-\cos\theta_{P-AB})$$

$$I_{ba-x} = -I_{ba}\cos\theta_{P-AB}$$

$$I_{ba-y} = I_{ba}\sin\gamma$$

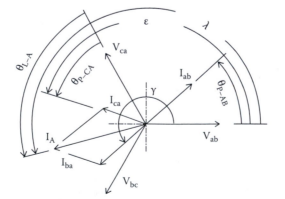

**FIGURE 4.16**
Phasors determining $I_A$ in unbalanced delta circuit.

$$\sin \gamma = -\sin \theta_{P-AB}$$

$$I_{ba-y} = -I_{ba}\sin \theta_{P-AB}$$

$$I_{ca-x} = I_{ca}\cos \varepsilon$$

$$\varepsilon = 120° + \theta_{P-CA}$$

$$\cos \varepsilon = \cos (120° + \theta_{P-CA})$$

$$= \cos 120°\cos \theta_{P-CA} - \sin 120° \sin \theta_{P-CA}$$

$$\cos 120° = -1/2$$

$$\sin 120° = \left(\sqrt{3}/2\right)$$

$$I_{ca-x} = I_{ca}\left[(-1/2) \left(\cos \theta_{P-CA}\right) - \left(\sqrt{3}/2\right)(\sin \theta_{P-CA})\right]$$

$$I_{ca-x} = -I_{ca}\left(1/2\right)\left[\left(\sqrt{3}\right)\sin \theta_{P-CA} + \cos \theta_{P-CA}\right]$$

$$I_{ca-y} = I_{ca}\sin \varepsilon$$

$$\sin \varepsilon = \sin (120° + \theta_{P-CA}) = \left[\left(\sqrt{3}/2\right)\cos \theta_{P-CA} + (-1/2) \sin \theta_{P-CA}\right]$$

$$I_{ca-y} = I_{ca}\left(1/2\right)\left[\left(\sqrt{3}\right) \cos \theta_{P-CA} - \sin \theta_{P-CA}\right]$$

$$I_{A-x} = I_{ba-x} + I_{ca-x}$$

$$I_{A-y} = I_{ba-y} + I_{ca-y}$$

$$|I_A| = \left\{\left(I_{A-x}\right)^2 + \left(I_{A-y}\right)^2\right\}^{\frac{1}{2}}$$

$$\lambda = \sin^{-1}(I_{A-y} \div I_A)$$

$$\theta_{L-A} = (\lambda - 120°)$$

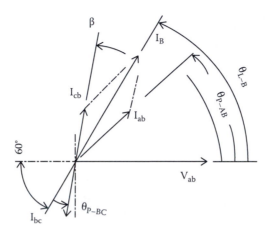

**FIGURE 4.17**
Phasors determining $I_B$ in unbalanced delta circuit.

To develop the relationship for line current $I_B$, let

$I_{ab-x}$ be the abscissa component of vector $\overline{I_{ab}}$
$I_{cb-x}$ be the abscissa component of vector $\overline{I_{cb}}$
$I_{B-x}$ be the abscissa component of vector $\overline{I_B}$
$I_{ab-y}$ be the ordinate component of vector $\overline{I_{ab}}$
$I_{cb-y}$ be the ordinate component of vector $\overline{I_{cb}}$
$I_{B-y}$ be the ordinate component of vector $\overline{I_B}$

Current $I_B$ is determined as shown in Figure 4.17.
    With reference to Figures 4.15 and 4.17, it may be noted that

$$I_{ab-x} = I_{ab}\cos\theta_{P-AB}$$

$$I_{ab-y} = I_{ab}\sin\omega$$

$$I_{ab-y} = I_{ab}\sin\theta_{P-AB}$$

$$I_{cb-y} = I_{cb}\cos\beta$$

$$\beta = 60° + \theta_{P-BC}$$

$$I_{cb-x} = I_{cb}\cos(60° + \theta_{P-BC})$$

$$I_{cb-x} = I_{cb}\left[\cos 60° \cos\theta_{P-BC} - \sin 60° \sin\theta_{P-BC}\right]$$

$$\cos 60° = 1/2$$

$$\sin 60° = \sqrt{3}/2$$

$$I_{cb-x} = I_{cb}\left[(1/2)\cos\theta_{P-BC} - \left(\sqrt{3}/2\right)\sin\theta_{P-BC}\right]$$

$$I_{cb-x} = -I_{cb}(1/2)\left[\left(\sqrt{3}\right)\sin\theta_{P-BC} - \cos\theta_{P-BC}\right]$$

$$I_{cb-y} = I_{cb}\sin\beta$$

$$I_{cb-y} = I_{cb}\sin(60° + \theta_{P-BC})$$

$$= I_{cb}\left[\left(\sqrt{3}/2\right)\cos\theta_{P-BC} + (1/2)\sin\theta_{P-BC}\right]$$

$$= I_{cb}\left[\left(\sqrt{3}/2\right)\cos\theta_{P-BC} + (1/2)\sin\theta_{P-BC}\right]$$

$$I_{cb-y} = I_{cb}(1/2)\left[(\cos\theta_{P-BC} + \sin\theta_{P-BC}\right]$$

Thus,

$$I_{B-x} = I_{ab-x} + I_{cb-x}$$

$$I_{B-y} = I_{ab-y} + I_{cb-y}$$

$$|I_B| = \left\{(I_{B-x})^2 + (I_{B-y})^2\right\}^{\frac{1}{2}}$$

$$\theta_{L-B} = \sin^{-1}(I_{B-y} \div I_B)$$

Current $I_C$ is determined as shown in Figure 4.18.
  Let

$I_{bc-x}$ be the abscissa component of vector $I_{bc}$
$I_{ac-x}$ be the abscissa component of vector $I_{ac}$
$I_{C-x}$ be the abscissa component of vector $I_C$
$I_{bc-y}$ be the ordinate component of vector $I_{bc}$
$I_{ac-y}$ be the ordinate component of vector $I_{ac}$
$I_{C-y}$ be the ordinate component of vector $I_C$

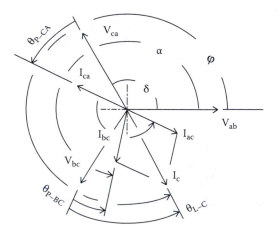

**FIGURE 4.18**

Phasors determining $I_C$ in unbalanced delta circuit.

In Figure 4.18,

$$I_{bc-x} = I_{bc}\cos \alpha$$

$$\alpha = 240° + \theta_{P-BC}$$

$$\cos \alpha = \cos (240° + \theta_{P-BC})$$

$$\cos (240° + \theta_{P-BC}) = \cos 240° \cos \theta_{P-BC} - \sin 240° \sin \theta_{P-BC}$$

$$\cos 240° = -1/2$$

$$\sin 240° = -\left(\sqrt{3}/2\right)$$

$$\cos (240° + \theta_{P-BC}) = (-1/2) \cos \theta_{P-BC} - \left[-\left(\sqrt{3}/2\right)\right]\sin \theta_{P-BC}$$

$$= (-1/2)\left[\cos \theta_{P-BC} + \left(\sqrt{3}\right)\sin \theta_{P-BC}\right]$$

$$= (1/2)\left[\left(\sqrt{3}\right)\sin \theta_{P-BC} - \cos \theta_{P-BC}\right]$$

$$I_{bc-x} = I_{bc}(1/2)\left[\left(\sqrt{3}\right)\sin \theta_{P-BC} - \cos \theta_{P-BC}\right]$$

$$I_{bc-y} = I_{bc}\sin \alpha$$

$$\sin \alpha = \sin (240° - \theta_{P-BC})$$

$$\sin (240° + \theta_{P-BC}) = \sin 240° \cos \theta_{P-BC} + \cos 240° \sin \theta_{P-BC}$$

$$= -\left(\sqrt{3}/2\right) \cos \theta_{P-BC} + (-1/2) \sin \theta_{P-BC}$$

$$= (-1/2)\left[\left(\sqrt{3}\right) \cos \theta_{P-BC} + \sin \theta_{P-BC}\right]$$

$$I_{bc-x} = -I_{bc} (1/2)\left[\left(\sqrt{3}\right) \cos \theta_{P-BC} + \sin \theta_{P-BC}\right]$$

$$I_{ac-x} = I_{ac} \cos \delta$$

$$\delta = (300° + \theta_{P-AC})$$

$$I_{ac-x} = I_{ac} \cos (300° + \theta_{P-AC})$$

$$\cos (300° + \theta_{P-AC}) = \cos 300° \cos \theta_{P-AC} - \sin 300° \sin 0$$

$$\cos 300° = 1/2$$

$$\sin 300° = -\left(\sqrt{3}/2\right)$$

$$\cos (300° + \theta_{P-AC}) = (1/2)\cos \theta_{P-AC} - \left[-\left(\sqrt{3}/2\right)\right]\sin \theta_{P-AC}$$

$$= (1/2)\left[\left(\sqrt{3}\right)\sin \theta_{P-AC} + \cos \theta_{P-AC}\right]$$

$$I_{ac-x} = I_{ac} (1/2)\left[\left(\sqrt{3}\right)\sin \theta_{P-AC} + \cos \theta_{P-AC}\right]$$

$$I_{ac-y} = I_{ac} \sin \delta$$

$$I_{ac-y} = I_{ac}\sin (300° + \theta_{P-AC})\sin(300° + \theta_{P-AC})$$

$$= \sin 300°\cos \theta_{P-AC} + \cos 300°\sin \theta_{P-AC}$$

$$= -\left(\sqrt{3}/2\right)\cos \theta_{P-AC} + (1/2)\sin \theta_{P-AC}$$

$$= -(1/2)\left[\left(\sqrt{3}\right)\cos \theta_{P-AC} - \sin \theta_{P-AC}\right]$$

$$I_{ac-y} = -I_{ac}\left(1/2\right)\left[\left(\sqrt{3}\right)\cos\theta_{P-AC} - \sin\theta_{P-AC}\right]$$

$$I_{C-x} = I_{bc-x} + I_{ac-x}$$

$$I_{C-y} = I_{bc-y} + I_{ac-y}$$

$$\left|I_C\right| = \left\{(I_{C-x})^2 + (I_{C-y})^2\right\}^{\frac{1}{2}}$$

$$\varphi = \sin^{-1}(I_{C-y} \div I_C)$$

$$\theta_{L-C} = (\varphi - 240°)$$

Recognizing

$$\left|I_{ba}\right| = \left|I_{ab}\right|, \left|I_{cb}\right| = \left|I_{bc}\right| \quad \text{and} \quad \left|I_{ac}\right| = \left|I_{ca}\right|, \quad \text{and}$$

$$\theta_{P-BA} = \theta_{P-AB}, \theta_{P-CB} = \theta_{P-BC} \quad \text{and} \quad \theta_{P-AC} = \theta_{P-CA},$$

the algebraic equations may be rewritten to exclude the terms $I_{ba}$, $I_{cb}$, $I_{ac}$, $\theta_{P-BA}$, $\theta_{P-CB}$, and $\theta_{P-AC}$.

In summary, for an unbalanced delta circuit,

$$\left|I_A\right| = \left\{(I_{A-x})^2 + (I_{A-y})^2\right\}^{\frac{1}{2}} \tag{4.4}$$

$$\theta_{L-A} = (\lambda - 120°)\ldots \tag{4.5}$$

where

$I_A$ is the current in line A

$\theta_{L-A}$ is the lead (lag) of current $I_A$ with respect to line voltage $V_{ca}$

$I_{ba-x}$ is the $-I_{ab}\cos\theta_{P-AB}$

$I_{ca-x} = -I_{ca}\left(1/2\right)\left[\left(\sqrt{3}\right)\sin\theta_{P-CA} + \cos\theta_{P-CA}\right]\sin$

$I_{A-x} = I_{ba-x} + I_{ca-x}$

$I_{ba-y} = -I_{ab}\sin\theta_{P-AB}$

$I_{ca-y} = I_{ca}\left(1/2\right)\left[\left(\sqrt{3}\right)\cos\theta_{P-CA} - \sin\theta_{P-CA}\right]$

$I_{A-y} = I_{ba-y} + I_{ca-y}$

$\lambda = \sin^{-1}(I_{A-y} \div I_A)$

Valid range of $\theta_{P-AB}$ and $\theta_{P-CA}$: $\pm 90°$; valid range of $\theta_{L-A}$: $+120°$ to $-60°$

$$|I_B| = \left\{ \left( I_{B-x} \right)^2 + \left( I_{B-y} \right)^2 \right\}^{\frac{1}{2}} \tag{4.6}$$

$$\theta_{L-B} = \sin^{-1}(I_{B-y} \div I_B) \dots \tag{4.7}$$

where

$I_B$ is the current in Line B
$\theta_{L-B}$ is the lead (lag) of current $I_B$ with respect to line voltage $V_{ab}$
$I_{ab-x} = I_{ab} \cos \theta_{P-AB}$
$I_{cb-x} = -I_{bc}(1/2) [(\sqrt{3}) \sin \theta_{P-BC} - \cos \theta_{P-BC}]$
$I_{B-x} = I_{ab-x} + I_{cb-x}$
$I_{ab-y} = I_{ab} \sin \theta_{P-AB}$
$I_{cb-y} = I_{bc}(1/2) [(\sqrt{3}) \cos \theta_{P-BC} + \sin \theta_{P-BC}]$
$I_{B-y} = I_{ab-y} + I_{cb-y}$

Valid range of $\theta_{P-AB}$ and $\theta_{P-CB}$: $\pm 90°$; valid range of $\theta_{L-B}$: $+120°$ to $-60°$

$$|I_C| = \left\{ \left( I_{C-x} \right)^2 + \left( I_{C-y} \right)^2 \right\}^{\frac{1}{2}} \tag{4.8}$$

$$\theta_{L-C} = (\varphi - 240°) \tag{4.9}$$

where

$I_C$ is the current in Line C
$\theta_{L-C}$ is the lead (lag) of current $I_C$ with respect to line voltage $V_{bc}$
$I_{bc-x} = I_{bc}(1/2) [(\sqrt{3}) \sin \theta_{P-BC} - \cos \theta_{P-BC}]$
$I_{ac-x} = I_{ca}(1/2) [(\sqrt{3}) \sin \theta_{P-CA} + \cos \theta_{P-CA}]$
$I_{C-x} = I_{bc-x} + I_{ac-x}$
$I_{bc-y} = -I_{bc}(1/2) [(\sqrt{3}) \cos \theta_{P-BC} + \sin \theta_{P-BC}]$
$I_{ac-y} = -I_{ca}(1/2) [(\sqrt{3}) \cos \theta_{P-CA} - \sin \theta_{P-CA}]$
$I_{C-y} = I_{bc-y} + I_{ac-y}$
$\varphi = \sin^{-1}(I_{C-y} \div I_C)$

Valid range of $\theta_{P-BC}$ and $\theta_{P-AC}$: $\pm 90°$; valid range of $\theta_{L-C}$: $+120°$ to $-60°$

### Example 4.7

**Problem**: With reference to Figures 1.14 and 4.15, assume the following conditions for an unbalanced delta circuit:

$I_{ab} = 5$ A at PF = 1.0
$I_{bc} = 10$ A at PF = 0.9 lagging
$I_{ca} = 15$ A at PF = 0.8 leading

Find line currents in conductors A, B, and C.

**Solution**

$$\theta_{P-AB} = \cos^{-1}1.0 = 0$$

$$\theta_{P-BC} = \cos^{-1}0.9 = -25.84°$$

$$\theta_{P-CA} = -\cos^{-1}0.8 = +36.86°$$

The respective currents may also be stated in polar notation. The lead/lag of current $I_{ab}$ is with regard to voltage $V_{ab}$. Voltage $V_{ab}$ is assumed to be in alignment with the positive abscissa. Therefore, the polar notation of current $I_{ab}$ is

$$I_{ab} = 5\angle 0°$$

The lead/lag of current $I_{bc}$ with regard to voltage $V_{bc}$ is 240° CCW from the positive abscissa. Current $I_{bc}$ lags $V_{bc}$. Therefore, the measure of angle to current $I_{bc}$ in the CCW direction from the positive abscissa is $240° - 25.84° = 214.16°$. The polar notation of $I_{bc}$ is

$$I_{bc} = 10\angle -214.16°$$

The lead/lag of current $I_{ca}$ is with regard to voltage $V_{ca}$, which is 120° CCW from the positive abscissa. Current $I_{ca}$ leads $V_{ca}$. Therefore, the measure of angle to current $I_{ca}$ in the CCW direction from the positive abscissa is $120° + 36.86° = 156.86°$. The polar notation, then, is

$$I_{bc} = 10\angle 156.86°$$

From Equations 4.4 and 4.5,

$$I_A = \{(I_{A-x})^2 + (I_{A-y})^2\}^{½}$$

$$\theta_{L-A} = (\lambda - 120°)$$

where
$I_{ba-x} = -I_{ab} \cos \theta_{P-AB} = -(5) \cos 0 = -5$
$I_{ba-y} = -I_{ab} \sin \theta_{P-AB} = I_{ba} (0) = 0$
$I_{ca-x} = -I_{ca} (1/2) [(\sqrt{3}) (\sin \theta_{P-CA} + \cos \theta_{P-CA}]$
$\quad = -(7.5) [(\sqrt{3}) \sin (36.86°) + \cos (36.86°)] = -13.794$
$I_{ca-y} = I_{ca} (1/2) [(\sqrt{3}) \cos \theta_{P-CA} - \sin \theta_{P-CA}]$
$I_{ca-y} = (15) (1/2) [(\sqrt{3}) \cos (36.86°) - \sin (36.86°)] = (7.5) [(1.385) - (0.6)] = 5.89$
$I_{A-x} = I_{ba-x} + I_{ca-x} = -5 + (-13.79) = -18.79$
$I_{A-y} = I_{ba-y} + I_{ca-y} = 0 + 5.89 = 5.89$
$|I_A| = \{(I_{A-x})^2 + (I_{A-y})^2\}^{½} = \{(-18.79)^2 + (5.89)^2\}^{½} = 19.69$ A
$\lambda = \sin^{-1} (I_{A-y} \div I_A) = \sin^{-1} [5.89 \div 19.69] = \sin^{-1}.299 = 162.59°$
$\theta_{L-A} = (\lambda - 120°) = (162.59° - 120°) = 42.59°$

In polar notation, $I_A = 19.69 / (120° \pm 42.59°) = 19.69 / 162.59°$
From Equations 4.6 and 4.7,

$$|I_B| = \{(I_{B-x})^2 + (I_{B-y})^2\}^{1/2}$$

$$\theta_{L-B} = \tan^{-1} (I_{B-y} \div I_{B-x})$$

where

$I_{B-x} = I_{ab-x} + I_{cb-x}$
$I_{B-y} = I_{ab-y} + I_{cb-y}$
$I_{ab-x} = I_{ab} \cos \theta_{P-AB} = (5) \cos 0 = 5$
$I_{ab-y} = -I_{ab} \sin \theta_{P-AB} = -(5) \sin 0 = 0$
$I_{cb-x} = -I_{bc} (1/2) [(\sqrt{3}) \sin \theta_{P-BC} - \cos \theta_{P-BC}]$
$\quad = -(10) (1/2) [(\sqrt{3}) \sin -25.84° - \cos -25.84°] = 8.27$
$I_{cb-y} = I_{cb} (1/2) [(\sqrt{3}) \cos \theta_{P-BC} + \sin \theta_{P-BC}]$
$\quad = (10) (1/2) [(\sqrt{3}) \cos -25.84° + \sin -25.84°] = 5.61$
$I_{B-x} = I_{ab-x} + I_{cb-x} = 5 + 8.27 = 13.27$
$I_{B-y} = I_{ab-y} + I_{cb-y} = 0 + 5.61 = 5.61$
$|I_B| = \{(I_{B-x})^2 + (I_{B-y})^2\}^{1/2} = \{(13.27)^2 + (5.61)^2\}^{1/2} = 14.407$ A
$\theta_{L-B} = \sin^{-1} (I_{B-y} \div I_B) = \sin^{-1} (5.61 \div 14.407) = 22.91°$

In polar notation, $I_B = 14.40 / 22.91°$
From Equations 4.8 and 4.9,

$$|I_C| = \{(I_{C-x})^2 + (I_{C-y})^2\}^{1/2}$$

where

$I_{C-x} = I_{bc-x} + I_{ac-x}$
$I_{C-y} = I_{bc-y} + I_{ac-y}$
$I_{bc-x} = I_{bc} (1/2) [(\sqrt{3}) \sin \theta_{P-BC} - \cos \theta_{P-BC}]$
$\quad = (10) (1/2) [(\sqrt{3}) \sin -25.84° - \cos -25.84°] = -8.27$
$I_{bc-y} = -I_{bc} (1/2) [(\sqrt{3}) \cos \theta_{P-BC} + \sin \theta_{P-BC}]$
$\quad = -(10) (1/2) [(\sqrt{3}) \cos -25.84° + \sin -25.84°] = -5.61$
$I_{ac-x} = I_{ca} (1/2) [(\sqrt{3}) \sin \theta_{P-CA} + \cos \theta_{P-CA}]$
$\quad = (15) (1/2) [(\sqrt{3}) \sin 36.86° + \cos 36.86°] = 13.79$
$I_{ac-y} = -I_{ca} (1/2) [(\sqrt{3}) \cos \theta_{P-CA} - \sin \theta_{P-CA}]$
$\quad = -(15) (1/2)[(\sqrt{3}) \cos 36.86° - \sin 36.86°] = -5.89$
$I_{C-x} = I_{bc-x} + I_{ac-x} = -8.27 + 13.79 = 5.52$
$I_{C-y} = I_{bc-y} + I_{ac-y} = -5.61 + (-5.89) = -11.50$
$|I_C| = \{(I_{C-x})^2 + (I_{C-y})^2\}^{1/2}$
$\quad = \{(5.52)^2 + (-11.50)^2\}^{1/2} = 12.76$ A
$\theta_{L-C} = (\varphi - 240°)$

$I_C$ is in Quadrant IV.

$$\varphi = \sin^{-1} (I_{C-y} \div I_C)$$

$$= \sin^{-1} (-11.500 \div 12.756)$$

$$= 295.64°$$

$$\theta_{L-C} = (\varphi - 240°) = 295.64° - 240° = 55.64°$$

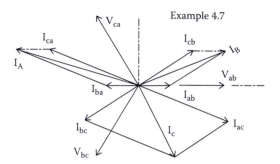

FIGURE 4.19
Unbalanced three-phase delta circuit.

In polar notation,

$$I_C = 12.756 / (240 \pm 55.64°)$$

$$= 12.756 / 295.64°$$

The values calculated in Example 4.7 are shown in Figure 4.19 where the phasors and the associated angles are drawn to scale.

### 4.2.1.2 Unbalanced Three-Phase Delta Circuit with Only Resistive Loads

A typical three-phase delta circuit is represented in Figure 1.14. With a balanced resistive load, the delta phasor diagram showing phase currents would be as represented in Figure 4.1. The line currents for the balanced circuit would be determined as shown in Figure 4.2. For an unbalanced delta circuit with resistive loads, a similar phasor diagram can be constructed. A typical phasor diagram for a resistive circuit is represented in Figure 4.20.

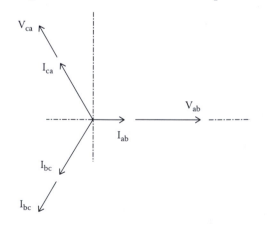

FIGURE 4.20
Phasors for unbalanced delta circuit with resistive load.

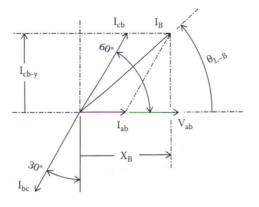

**FIGURE 4.21**
Phasors determining $I_B$ in an unbalanced resistive delta circuit.

In the figure, two of the phasors representative of phase currents $I_{bc}$ and $I_{ca}$ are shown as equal, whereas the third current, $I_{ab}$, is assumed to be smaller.

The phasor diagram for determining line current $I_B$ is shown in Figure 4.21. As was the procedure with a balanced delta circuit, vector $I_B$ is determined by adding the vector of $\underline{I_{ab}}$ and the negative vector of $\underline{I_{bc}}$, which is $I_{cb}$. According to the rules for adding vectors, the "x" (or abscissa) component of vector $I_B$ is determined by adding the "x" components of $I_{ab}$ and $I_{cb}$. Likewise, the "y" (or ordinate) component of vector $I_B$ is determined by adding the y components of vectors $\underline{I_{ab}}$ and $\underline{I_{cb}}$. Since $\underline{I_{ab}}$ is in phase with reference voltage $V_{ab}$, the vector for current $\underline{I_{ab}}$ has no "y" component.

Essentially, the equations for a delta circuit with all resistive loads becomes a special case of Equations 4.4 through 4.9, wherein $\theta_{L-A} = 0$, $\theta_{L-B} = 0$, and $\theta_{L-C} = 0$. Under these circumstances, Equations 2.4.1 through 2.4.6 simplify to the Equations 4.10 through 4.15.

$$|I_A| = \left\{ (I_{A-x})^2 + (I_{A-y})^2 \right\}^{\frac{1}{2}} \tag{4.10}$$

$$\theta_{L-A} = (\lambda - 120°) \tag{4.11}$$

where
  $I_A$ is the current in Line A
  $\theta_{L-A}$ is the lead/lag of current $I_A$ with respect to line voltage $V_{ca}$

$$I_{A-x} = -I_{ab} - (1/2)\, I_{ca}$$

$$I_{A-y} = I_{ca}\, (\sqrt{3}/2)$$

$$\lambda = \sin^{-1} (I_{A-y} \div I_A)$$

$$|I_B| = \left\{(I_{B-x})^2 + (I_{B-y})^2\right\}^{\frac{1}{2}}$$ (4.12)

$$\theta_{L-B} = \sin^{-1}(I_{B-y} \div I_B)$$ (4.13)

where
$I_B$ is the current in Line B
$\theta_{L-B}$ is the lead (lag) of current $I_B$ with respect to line voltage $V_{ab}$
$I_{B-x} = I_{ab} + (1/2) I_{bc}$
$I_{B-y} = I_{bc} (\sqrt{3}/2)$

$$|I_C| = \left\{(I_{C-x})^2 + (I_{C-y})^2\right\}^{\frac{1}{2}}$$ (4.14)

$$\theta_{L-C} = (\varphi - 240°)$$ (4.15)

where
$I_C$ is the current in Line C
$\theta_{L-C}$ is the lead (lag) of current $I_C$ with respect to line voltage $V_{bc}$
$I_{C-x} = (1/2) I_{ca} - (1/2) I_{bc}$
$I_{C-y} = -I_{bc} (\sqrt{3}/2) - I_{ca} (\sqrt{3}/2)$
$\varphi = \sin^{-1} (I_{C-y} \div I_C)$

### Example 4.8

**Problem**: With reference to Figure 1.14 (a delta circuit), given that $I_{ab}$ is 5 A, resistive, and both $I_{bc}$ and $I_{ca}$ are each 10 A, resistive. Find line currents $I_A$, $I_B$, and $I_C$ and the respective lead/lag with respect to line voltages.

### Solution

Reference is made to Equations 4.4 through 4.9.

$$I_{A-x} = -I_{ab} - (1/2) I_{ca} = -(5) - (1/2) (10) = -10$$

$$I_{A-y} = I_{ca} (\sqrt{3}/2) = (10) (\sqrt{3}/2) = 5\sqrt{3} = 8.66$$

$$|I_A| = \{(I_{A-x})^2 + (I_{A-y})^2\}^{\frac{1}{2}} = \{(-10)^2 + (8.66)^2\}^{\frac{1}{2}} = 13.22 \text{ A}$$

$$\lambda = \sin^{-1} (I_{A-y} \div I_A) = \sin^{-1} (8.66 \div 13.22) = \sin^{-1} (.655)$$

$I_A$ is in Quadrant II.

$$\lambda = 139.08°$$

$$\theta_{L-A} = (\lambda - 120°) = (139.08° - 120°) = 19.10°$$

$$I_{B-x} = I_{ab} + (1/2)\,I_{bc} = 5 + (1/2)(10) = 10$$

$$I_{B-y} = I_{bc}\,(\sqrt{3}/2) = (10)(\sqrt{3}/2) = 5\sqrt{3}$$

$$|I_B| = \{(I_{B-x})^2 + (I_{B-y})^2\}^{\frac12} = \{(10)^2 + (5\sqrt{3})^2\}^{\frac12} = 13.228 \text{ A}$$

$$\theta_{L-B} = \sin^{-1}(I_{B-y} \div I_B) = \sin^{-1}(5\sqrt{3} \div 13.228) = \sin^{-1}(.655)$$

$I_B$ is in Quadrant I.

$$\theta_{L-B} = 40.89°$$

$$I_{C-x} = (1/2)I_{ca} - (1/2)I_{bc} = (1/2)(10) - (1/2)(10) = 0$$

$$I_{C-y} = -I_{bc}\,(\sqrt{3}/2) - I_{ca}(\sqrt{3}/2) = -(10)(\sqrt{3}/2) - (10)(\sqrt{3}/2)$$

$$= -(10)\sqrt{3} = -17.32$$

$$|I_C| = \{(I_{C-x})^2 + (I_{C-y})^2\}^{\frac12}$$

$$= \{(0)^2 + (-10\sqrt{3})^2\}^{\frac12}$$

$$= (10)\sqrt{3} = 17.32 \text{ A}$$

$$\varphi = \sin^{-1}(I_{C-y} \div I_C)$$

$$= \sin^{-1}(-17.32 \div 17.32)$$

$$= \sin^{-1}(-1)$$

$I_C$ is on the negative ordinate.

$$\varphi = 270°$$

$$\theta_{L-C} = \theta_{L-C} = (\varphi - 240°)$$

$$= (270° - 240°) = 30°$$

## 4.2.2 Unbalanced Three-Phase Wye Circuit

A typical three-phase unbalanced wye circuit is represented in Figure 1.15 and the phasors for a typical unbalanced wye load are represented in Figure 4.22. For the purposes of analysis, it is assumed that no two phase currents are necessarily of equal value or necessarily at the same lead/lag

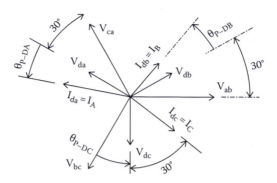

**FIGURE 4.22**
Phasors for unbalanced resistive wye circuit.

angle to the respective phase voltage. In a resistive load $\theta_P = 0$, that is, the phase current is in phase with the phase voltage. If the load is capacitive, $\theta_P > 0$, and the current is said to be "leading." (If the load is inductive, $\theta_P < 0$, and the current is said to be "lagging" phase voltage.) Let

$I_A$ be the current in conductor A
$I_B$ be the current in conductor B
$I_C$ be the current in conductor C
$I_D$ be the current in conductor D
$\theta_{L-A}$ be the lead/lag angle of vector $I_A$ with respect to voltage $V_{ca}$
$\theta_{L-B}$ be the lead/lag angle of vector $I_B$ with respect to voltage $V_{ab}$
$\theta_{L-C}$ be the lead/lag angle of vector $I_B$ with respect to voltage $V_{bc}$

(Note that in the case of an unbalanced wye circuit, the current in the neutral conductor [conductor D in Figure 1.15] is of importance and the size of the neutral conductor is likewise important. Conductor D must be sized to carry the largest combination of currents that may result from the combination of currents in conductors A, B, and C. It may be shown that the largest possible current in conductor D will be twice the maximum current in conductor A, conductor B, or conductor C.)

In comparison to the delta circuit, determining line currents in a wye circuit is a simple process since the phase currents are equal in magnitude to the line currents. Furthermore, the phase voltages lead line voltages by 30°. Consequently, the phase currents lead line voltages by 30° plus or minus the respective lead/lag angle of current with respect to voltage. More specifically,

$$\theta_{P-DA} = \theta_{L-A} - 30°$$

$$\theta_{P-DB} = \theta_{L-B} - 30°$$

$$\theta_{P-DC} = \theta_{L-C} - 30°$$

**Example 4.9**

**Problem**: With reference to Figures 1.15 and 4.22, assume the following conditions for an unbalanced wye circuit:

$$I_{da} = 15 \text{ A at PF} = 0.8 \text{ leading}$$

$$I_{db} = 5 \text{ A at PF} = 1.0$$

$$I_{dc} = 10 \text{ A at PF} = 0.9 \text{ lagging}$$

Find line currents in conductors A, B, and C.

**Solution**

The phase currents are the line currents and are repeated since the magnitude of the phase currents is the same as that of the line currents.

$$I_{da} = I_A = 15 \text{ A}$$

$$I_{db} = I_B = 5 \text{ A}$$

$$I_{dc} = I_C = 10 \text{ A}$$

$$\theta_{P-DA} = \cos^{-1} 0.8 = 36.86° \text{ (leading)},$$

$$\theta_{P-DB} = \cos^{-1} 1.0 = 0$$

$$\theta_{P-DC} = \cos^{-1} 0.9 = -25.84° \text{ (lagging)}$$

From Equation 1.7,

$$\theta_P = \theta_L - 30°$$

$$\theta_{L-A} = \theta_{P-DA} + 30° = 36.86° + 30° = +66.86°$$

$$\theta_{L-B} = \theta_{P-DB} + 30° = 0° + 30° = +30°$$

$$\theta_{L-C} = \theta_{P-DC} + 30° = -25.84° + 30° = +4.16°$$

The computations of both Examples 4.8 and 4.9 demonstrate that, given the phase currents, the line currents of a wye circuit are more readily determined than those of a delta circuit.

## 4.3 Combined Balanced or Unbalanced Three-Phase Circuits

As may often happen, a three-phase circuit will serve a mixture of delta and wye loads that could be both balanced and unbalanced loads. A common problem is to calculate line currents in the conductors of a three-phase circuit that serves a variety of loads.

In many instances, calculations are not required to determine line currents for some loads as the line currents may already be provided. For example, motor nameplates will state the motor's full load current (which, according to common practice, would be the line current). Often the full load power factor (which describes the phase current lead/lag with respect to phase voltage) is also included on the motor nameplate. In such cases, the motor data can easily be converted to lead/lag of currents with respect to line voltages.

In the previous section, it was shown how to calculate line currents for delta or wye loads whether the phases are balanced or unbalanced. Using the demonstrated techniques, currents in each of the three phases can readily be determined. Essentially, determination of the net current in a selected phase merely requires the addition of the "x" and "y" components of the contributing phasors to determine, respectively, the "x" and "y" components that determine the net currents in each of the common feeder phases. In the way of illustration, a three-phase feeder is shown in Figure 4.23 serving three loads that are designated as load 1, load 2, and load 3. The loads could be delta, wye, balanced, unbalanced, or any mixture thereof.

In Figure 4.23,

The current in conductor A is the vector sum of the currents in conductors E, H, and U.

The current in conductor B is the vector sum of the currents in conductors F, J, and V.

The current in conductor C is the vector sum of the currents in conductors G, K, and W.

The currents in conductors A, B, and C can be obtained by the vector additions of the currents in the respective branch conductors. To determine the line current in conductor B, the vectors representing the line currents in conductors F, J, and V are added as shown in Figure 4.24. (Current $I_B$ is determined first for the reason that the computations for $I_B$ are somewhat easier

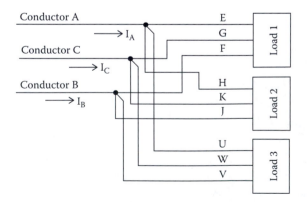

**FIGURE 4.23**
Three three-phase loads.

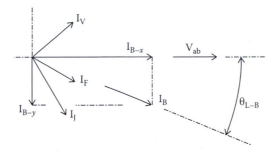

**FIGURE 4.24**
Three phasors determining $I_B$.

than those for currents $I_A$ and $I_C$.) In Figure 4.24, arbitrary phasor values are shown as representative of currents F, J, and V.

For the circuit configuration of Figure 4.23,

$$I_{B-x} = I_{F-x} + I_{J-x} + I_{Y-x}, \quad \text{and}$$

$$I_{B-y} = I_{F-y} + I_{J-y} + I_{Y-y},$$

where
  $I_{B-x}$ is the abscissa component of current $I_B$
  $I_{B-y}$ is the ordinate component of current $I_B$

$$I_{F-x} = I_F \cos \theta_{L-F}$$

$$I_{J-x} = I_J \cos \theta_{L-J}$$

$$I_{V-x} = I_Y \cos \theta_{L-Y}$$

$$I_{F-y} = I_F \sin \theta_{L-F}$$

$$I_{J-y} = I_J \sin \theta_{L-J}$$

$$I_{V-y} = I_Y \sin \theta_{L-Y}$$

$\theta_{L-F}$ is the lead/lag of line current $I_F$ with respect to line voltage $V_{ab}$
$\theta_{L-J}$ is the lead/lag of line current $I_J$ with respect to line voltage $V_{bc}$
$\theta_{L-V}$ is the lead/lag of line current $I_F$ with respect to line voltage $V_{ca}$

$$|I_B| = \{(I_{B-x})^2 + (I_{B-y})^2\}^{\frac{1}{2}}$$

$$\theta_{L-B} = \sin^{-1}(I_{B-y} \div I_B)$$

**Example 4.10**

**Problem**: For the configuration of Figure 4.23, assume

$$I_F = 12 \text{ A at } \theta_{L-F} = -30° \text{ (lagging with respect to } V_{ab})$$

$$I_J = 16 \text{ A at } \theta_{L-J} = -60° \text{ (lagging with respect to } V_{ab})$$

$$I_V = 13 \text{ A at } \theta_{L-Y} = +40° \text{ (leading with respect to } V_{ab})$$

Find current $I_B$ and its lead/lag with respect to voltage $V_{ab}$.

**Solution**

$$I_{F-x} = 12 \cos -30° = (12)(.866) = 10.392$$

$$I_{J-x} = 16 \cos -60° = (16)(.500) = 8.000$$

$$I_{V-x} = 13 \cos 40° = (13)(.766) = 9.958$$

$$I_{B-x} = I_{F-x} + I_{J-x} + I_{Y-x} = 10.392 + 8.000 + 9.958 = 28.350$$

$$I_{F-y} = (12 \sin -30°) = (12)(-.500) = -6.000$$

$$I_{J-y} = (16 \sin -60°) = (16)(-.866) = -13.856$$

$$I_{V-y} = (13 \sin 40°) = (13)(.642) = 8.356$$

$$I_{B-y} = I_{F-y} + I_{J-y} + I_{Y-y} = -6.000 - 13.856 + 8.356 = -11.500$$

$$|I_B| = \{(I_{B-x})^2 + (I_{B-y})^2\}^{\frac{1}{2}} = \{(28.350)^2 + (-11.500)^2\}^{\frac{1}{2}} = 30.593 \text{ A}$$

$$\theta_{L-B} = \sin^{-1}(I_{B-y} \div I_B) = \sin^{-1}(-11.500 \div 30.593)$$

$I_B$ is in Quadrant IV.

$$\sin^{-1}(-11.500 \div 30.593) = \sin^{-1}(-.376) = -22.080°$$

$$\theta_{L-B} = -22.080°$$

The computed values of $I_{B-x}$, $I_{B-y}$, $I_B$, and $\theta_{L-B}$ are shown, drawn to scale, in Figure 4.24. (The arc designating the value and position of $\theta_{L-B}$ in Figure 4.24 is shown with arrowheads at both ends to indicate it as either a negative value or a parameter without a positive direction indication.)

In order to arrive at the value of $I_A$ (Figure 4.23), the phasors determining $I_A$ are arranged as represented in Figure 4.25.

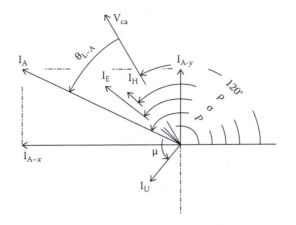

**FIGURE 4.25**
Three phasors determining $I_A$.

In Figure 4.25,

$$\sigma = 120° + \theta_{L-E}$$

$$\rho = 120° + \theta_{L-H}$$

$$\mu = 120° + \theta_{L-U}$$

$$I_{E-x} = I_E \cos \sigma$$

$$= I_E \cos (120° + \theta_{L-E})$$

$$= I_E [\cos 120° \cos \theta_{L-E} - \sin 120° \sin \theta_{L-E}]$$

$$\cos 120° = -1/2$$

$$\sin 120° = (\sqrt{3}/2)$$

$$I_{E-x} = I_E [(-1/2) \cos \theta_{L-E} - (\sqrt{3}/2) \sin \theta_{L-E}]$$

$$I_{E-x} = -I_E [(\sqrt{3}/2) \sin \theta_{L-E} + (1/2) \cos \theta_{L-E}], \quad \text{and}$$

$$I_{H-x} = -I_H [(\sqrt{3}/2) \sin \theta_{L-H} + (1/2) \cos \theta_{L-H}]$$

$$I_{U-x} = -I_X [(\sqrt{3}/2) \sin \theta_{L-X} + (1/2) \cos \theta_{L-X}]$$

$$I_{E-y} = I_E \sin \sigma = I_E \sin (120° + \theta_{L-E})$$

$$I_{E-y} = I_E [\sin 120° \cos \theta_{L-E} + \cos 120° \sin \theta_{L-E}]$$

$$= I_E [(\sqrt{3}/2) \cos \theta_{L-E} + (-1/2) \sin \theta_{L-E}]$$

$$I_{E-y} = I_E \left[ (\sqrt{3}/2) \cos \theta_{L-E} - (1/2) \sin \theta_{L-E} \right], \quad \text{and}$$

$$I_{H-y} = I_H \left[ (\sqrt{3}/2) \cos \theta_{L-H} - (1/2) \sin \theta_{L-H} \right]$$

$$I_{U-x} = I_X \left[ (\sqrt{3}/2) \cos \theta_{L-X} - (1/2) \sin \theta_{L-X} \right]$$

$$I_{A-x} = I_{E-x} + I_{H-x} + I_{X-x}$$

$$I_{A-y} = I_{E-y} + I_{H-y} + I_{X-y}$$

$$|I_A| = \{ (I_{A-x})^2 + (I_{A-y})^2 \}^{1/2}$$

Let,

$$\kappa = \sin^{-1} (I_{A-y} \div I_A)$$

**NOTE:** For suggestions on foolproof methods to calculate $\sin^{-1}$ values to determine $\lambda$, $\varphi$, $\kappa$, and $\zeta$, see Equation B.1.

$$\theta_{L-A} = \kappa - 120°$$

**Example 4.11**

**Problem:**
Let

$$I_E = 14 \text{ A at } +22° \text{ (leading with respect to } V_{ca})$$

$$I_H = 11 \text{ A at } +15° \text{ (leading with respect to } V_{ca})$$

$$I_U = 7 \text{ A at } +110° \text{ (leading with respect to } V_{ca})$$

Determine $I_A$.

**Solution**

$$I_{E-x} = -I_E \left[ (\sqrt{3}/2) \sin \theta_{L-E} + (1/2) \cos \theta_{L-E} \right]$$

$$= -(14) \left[ (\sqrt{3}/2) \sin (22°) + (1/2) \cos (22°) \right] = -11.030$$

$$I_{H-x} = -I_H \left[ (\sqrt{3}/2) \sin \theta_{L-H} + (1/2) \cos \theta_{L-H} \right]$$

$$= -(11) \left[ (\sqrt{3}/2) \sin (15°) + (1/2) \cos (15°) \right] = -7.777$$

$$I_{U-x} = -I_X \left[ (\sqrt{3}/2) \sin \theta_{L-X} + (1/2) \cos \theta_{L-X} \right]$$

$$= -(7) \left[ (\sqrt{3}/2) \sin (110°) + (1/2) \cos (110°) \right] = -4.499$$

$$I_{A-x} = I_{E-x} + I_{H-x} + I_{X-x}$$

$$I_{A-x} = -11.030 + (-7.777) + (-4.499) = -23.306$$

$$I_{E-y} = I_E [(\sqrt{3}/2) \cos \theta_{L-E} - (1/2) \sin \theta_{L-E}]$$

$$= (14) [(\sqrt{3}/2) \cos (22°) - (1/2) \sin (22°)] = 8.617$$

$$I_{H-y} = I_H [(\sqrt{3}/2) \cos \theta_{L-H} - (1/2) \sin \theta_{L-H}]$$

$$= (11) [(\sqrt{3}/2) \cos (15°) - (1/2) \sin (15°)] = 7.777$$

$$I_{U-y} = I_X [(\sqrt{3}/2) \cos \theta_{L-X} - (1/2) \sin \theta_{L-X}]$$

$$= (7) [(\sqrt{3}/2) \cos (110°) - (1/2) \sin (110°)] = -5.361$$

$$I_{A-y} = I_{E-y} + I_{H-y} + I_{X-y}$$

$$I_{A-y} = 8.617 + 7.777 + (-5.361) = 11.033$$

$$|I_A| = \{(I_{A-x})^2 + (I_{A-y})^2\}^{1/2}$$

$$|I_A| = \{(-23.306)^2 + (11.033)^2\}^{1/2} = 25.785$$

$$\kappa = \sin^{-1} (Y_A \div I_A) = \sin^{-1} (11.033 \div 25.785 = \sin^{-1} .4278$$

$I_A$ is in Quadrant II.

$$p = 154.66°$$

$$\theta_{L-A} = 120° - p = 120° - 154.66° = -34.66°$$

The computed values of currents $I_{A-x}$, $I_{A-y}$, $I_A$, and phase angle $\theta_{L-A}$ are shown, drawn to scale, in Figure 4.25.

In order to arrive at the value of $I_C$, the phasors determining $\underline{I_C}$ are arranged as represented in Figure 4.26.

In Figure 4.26,

$$q = 240° + \theta_{L-G}$$

$$r = 240° + \theta_{L-K}$$

$$s = 240° + \theta_{L-Z}$$

$$I_{G-x} = I_G \cos q$$

$$= I_G \cos (240° + \theta_{L-G})$$

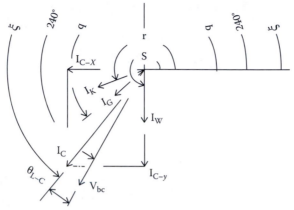

**FIGURE 4.26**
Three phasors determining $I_C$.

$$I_{G-x} = I_G [\cos 240° \cos \theta_{L-G} - \sin 240° \sin \theta_{L-G}]$$

$$\cos 240° = -\cos 60° = -(1/2)$$

$$\sin 240° = -\sin 60° = -(\sqrt{3}/2)$$

$$I_{G-x} = I_G [-(1/2) \cos \theta_{L-G} - (-\sqrt{3}/2) \sin \theta_{L-G}]$$

$$I_{G-x} = I_G [(\sqrt{3}/2) \sin \theta_{L-G} - (1/2) \cos \theta_{L-G}], \quad \text{and}$$

$$I_{K-x} = I_K [(\sqrt{3}/2) \sin \theta_{L-K} - (1/2) \cos \theta_{L-K}]$$

$$I_{W-x} = -I_Z [(\sqrt{3}/2) \sin \theta_{L-Z} - (1/2) \cos \theta_{L-Z}]$$

$$I_{G-y} = I_G \sin q = I_G \sin (240° + \theta_{L-G})$$

$$= I_G [\sin 240° \cos \theta_{L-G} + \cos 240° \sin \theta_{L-G}]$$

$$= I_G [-(\sqrt{3}/2) \cos \theta_{L-G} + (-1/2) \sin \theta_{L-G}]$$

$$I_{G-y} = -I_G [(\sqrt{3}/2) \cos \theta_{L-G} + (1/2) \sin \theta_{L-G}], \quad \text{and}$$

$$I_{K-y} = -I_K [(\sqrt{3}/2) \cos \theta_{L-K} + (1/2) \sin \theta_{L-K}]$$

$$I_{W-y} = -I_Z [(\sqrt{3}/2) \cos \theta_{L-Z} + (1/2) \sin \theta_{L-Z}]$$

$$I_{C-x} = I_{G-x} + I_{K-x} + I_{Z-x}$$

$$I_{C-y} = I_{G-y} + I_{K-y} + I_{Z-y}$$

$$I_C = \{(I_{C-x})^2 + (I_{C-y})^2\}^{1/2}$$

$$\zeta = \sin^{-1} (I_{C-y} \div I_C)$$

$$\theta_{L-C} = \zeta - 240°$$

**Example 4.12**

**Problem**:
Reference Figure 4.23. Let

$$I_G = 12 \text{ A at } -20° \text{ (lagging with respect to } V_{bc})$$

$$I_K = 15 \text{ A at } -40° \text{ (lagging with respect to } V_{bc})$$

$$I_W = 16 \text{ A at } +30° \text{ (leading with respect to } V_{bc})$$

Determine $I_C$.

**Solution**

$$I_{G-x} = I_G \left[ (\sqrt{3}/2) \sin \theta_{L-G} - (1/2) \cos \theta_{L-G} \right]$$

$$= (12) \left[ (\sqrt{3}/2) \sin -20° - (1/2) \cos -20° \right] = -9.1908$$

$$I_{K-x} = I_K \left[ (\sqrt{3}/2) \sin \theta_{L-K} - (1/2) \cos \theta_{L-K} \right]$$

$$= (15) \left[ (\sqrt{3}/2) \sin -40° - (1/2) \cos -40° \right] = -14.0925$$

$$I_{W-x} = I_Z \left[ (\sqrt{3}/2) \sin \theta_{L-z} - (1/2) \cos \theta_{L-z} \right]$$

$$= (16) \left[ (\sqrt{3}/2) \sin 30° - (1/2) \cos 30° \right] = 0$$

$$I_{C-x} = I_{G-x} + I_{K-x} + I_{W-x}$$

$$= -9.1908 - 14.0925 - 0 = -23.2833$$

$$I_{G-y} = -I_G \left[ (\sqrt{3}/2) \cos \theta_{L-G} + (1/2) \sin \theta_{L-G} \right]$$

$$= -(12) \left[ (\sqrt{3}/2) \cos -20° + (1/2) \sin -20° \right] = -7.7124$$

$$I_{K-y} = -I_K \left[ (\sqrt{3}/2) \cos \theta_{L-K} + (1/2) \sin \theta_{L-K} \right]$$

$$= -(15) \left[ (\sqrt{3}/2) \cos -40° + (1/2) \sin -40° \right] = -5.130$$

$$I_{W-y} = -I_Z \left[ (\sqrt{3}/2) \cos \theta_{L-W} + (1/2) \sin \theta_{L-W} \right]$$

$$= -(16) \left[ (\sqrt{3}/2) \cos 30° + (1/2) \sin 30° \right] = -16$$

$$I_{C-y} = I_{G-y} + I_{K-y} + I_{W-y}$$

$$= -7.7124 - 5.130 - 16.000 = -28.842$$

$$I_C = \{ (I_{C-x})^2 + (I_{C-y})^2 \}^{½}$$

$$= \{ (-23.2833)^2 + -28.842)^2 \}^{½} = 37.067 \text{ A}$$

$$\zeta = \sin^{-1} (I_{C-y} \div I_C) = \sin^{-1} (-28.842 \div 37.067) = \sin^{-1} (-0.778)$$

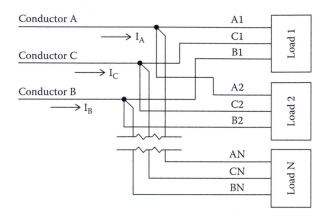

**FIGURE 4.27**
Two or more three-phase loads.

$I_C$ is in Quadrant III.

$$\zeta = 231.087°$$

$$\theta_{L-C} = \zeta - 240° = 231.087° - 240° = -8.913°$$

The computed values of $I_{C-x}$, $I_{C-y}$, $I_C$, and $\theta_{L-C}$ are shown, drawn to scale, in Figure 4.26.

It becomes apparent that when computing the net value of line currents in conductors that deliver power to a variety of three-phase loads that could be balanced or unbalanced, the parameters may be represented as shown in Equations 4.16 through 4.21. The parameters of the generalized Equations 4.16 through 4.21 are shown in Figure 4.27.

Given loads 1, 2, and 3, the applicable equations for determining the currents in conductors A, B, and C of the representation of Figure 4.27 then become

$$|I_A| = \left\{(I_{A-x})^2 + (I_{A-y})^2\right\}^{\frac{1}{2}} \tag{4.16}$$

and

$$\theta_{L-A} = (\kappa - 120°) \tag{4.17}$$

where
$\kappa = \sin^{-1}(I_{A-y} \div I_A)$
$I_{A-x} = \Sigma\,(I_{A1-x} + I_{A2-x} + I_{A3-x} \cdots + I_{AN-x})$
$I_{A-y} = \Sigma\,(I_{A1-y} + I_{A2-y} + I_{A3-y} \cdots + I_{AN-y})$
$I_{A1-x} = -I_{A1}\,(1/2)\,[(\sqrt{3})\sin\theta_{A1} + \cos\theta_{A1}]$

$I_{A2-x} = -I_{A2} (1/2) [(\sqrt{3}) \sin \theta_{A2} + \cos \theta_{A2}]$ ... to
$I_{AN-x} = -I_{AN} (1/2) [(\sqrt{3}) \sin \theta_{AN} + \cos \theta_{AN}]$
$I_{A1-y} = I_{A1} (1/2) [(\sqrt{3}) \cos \theta_{A1} - \sin \theta_{A1}]$ ... to
$I_{AN-y} = I_{A-N} (1/2) [(\sqrt{3}) \cos \theta_{AN} - \sin \theta_{AN}]$
$I_{A1}$ is the line current in branch of conductor A to load #1
$I_{A2}$ is the line current in branch of conductor A to load #2 ... to
$I_{AN}$ is the line current in branch of conductor A to load #N
$\theta_{A1}$ is the lead/lag of line current $I_{A1}$ with respect to line voltage $V_{ca}$
$\theta_{A2}$ is the lead/lag of line current $I_{A2}$ with respect to line voltage $V_{ca}$ ... to
$\theta_{AN}$ is the lead/lag of line current $I_{AN}$ with respect to line voltage $V_{ca}$

$$|I_B| = \left\{ (I_{B-x})^2 + (I_{B-y})^2 \right\}^{\frac{1}{2}} \tag{4.18}$$

and

$$\theta_{L-B} = \sin^{-1}(I_{B-y} \div I_B) \tag{4.19}$$

where

$I_{B-x} = \Sigma (I_{B1-x} + I_{B2-x} + I_{B3-x} \cdots + I_{BN-x})$
$I_{B-y} = \Sigma (I_{B1-y} + I_{B2-y} + I_{B3-y} \cdots + I_{BN-y})$
$I_{B1-x} = I_{B1} \cos \theta_{B1}$ ... to
$I_{BN-x} = I_{BN} \cos \theta_{BN}$
$I_{B1-y} = I_{B1} \sin \theta_{B1}$ ... to
$I_{BN-y} = I_{BN} \sin \theta_{BN}$
$I_{B1}$ is the line current in branch of conductor B to load #1
$I_{B2}$ is the line current in branch of conductor B to load #2 ... to
$I_{BN}$ is the line current in branch of conductor B to load #N
$\theta_{B1}$ is the lead/lag of line current $I_{B-1}$ with respect to line voltage $V_{ca}$
$\theta_{B2}$ is the lead/lag of line current $I_{A-2}$ with respect to line voltage $V_{ca}$ ... to
$\theta_{BN}$ is the lead/lag of line current $I_{A-N}$ with respect to line voltage $V_{ca}$
$I_{B1}$ is the line current in branch of conductor B to load #1
$I_{B2}$ is the line current in branch of conductor B to load #2 ... to
$I_{BN}$ is the line current in branch of conductor B to load #N
$\theta_{B1}$ is the lead/lag of line current $I_{B-1}$ with respect to line voltage $V_{ab}$
$\theta_{B2}$ is the lead/lag of line current $I_{B-2}$ with respect to line voltage $V_{ab}$ ... to
$\theta_{BN}$ is the lead/lag of line current $I_{B-N}$ with respect to line voltage $V_{ab}$

$$|I_C| = \left\{ (I_{C-x})^2 + (I_{C-y})^2 \right\}^{\frac{1}{2}} \tag{4.20}$$

and

$$\theta_{L-C} = \zeta - 240° \tag{4.21}$$

$$\zeta = \sin^{-1}(I_{C-y} \div I_C)$$

$$I_{C-x} = \Sigma (I_{C1-x} + I_{C2-x} + I_{C3-x} \cdots + I_{CN-x})$$

$$I_{C-y} = \Sigma (I_{C1-y} + I_{C2-y} + I_{C3-y} \cdots + I_{CN-y})$$

$I_{C1}$ is the current in branch of conductor C to load #1
$I_{C2}$ is the current in branch of conductor C to load #2 ... to
$I_{CN}$ is the current in branch of conductor C to load #N

$$I_{C1-x} = I_{C1} (1/2) [(\sqrt{3}) \sin \theta_{C1} - \cos \theta_{C1}]$$

$$I_{C2-x} = I_{C2} (1/2) [(\sqrt{3}) \sin \theta_{C2} - \cos \theta_{C2}] \ldots \text{ to}$$

$$I_{CN-x} = I_{CN} (1/2) [(\sqrt{3}) \sin \theta_{CN} - \cos \theta_{CN}]$$

$$I_{C1-y} = -I_{C1} (1/2) [(\sqrt{3}) \cos \theta_{C1} + \sin \theta_{C1}]$$

$$I_{C2-y} = -I_{C2} (1/2) [(\sqrt{3}) \cos \theta_{C2} + \sin \theta_{C2}] \ldots \text{ to}$$

$$I_{CN-y} = -I_{C-N} (1/2) [(\sqrt{3}) \cos \theta_{C-N} + \sin \theta_{C-N}]$$

$\theta_{C1}$ is the lead/lag of line current $I_{C1}$ with respect to line voltage $V_{bc}$
$\theta_{C2}$ is the lead/lag of line current $I_{C2}$ with respect to line voltage $V_{bc}$ ... to
$\theta_{CN}$ is the lead/lag of line current $I_{CN}$ with respect to line voltage $V_{bc}$

## Problems

1. A balanced wye circuit has phase currents of 20 A and a lagging PF of 0.90. What is the line current and lead/lag with respect to line voltage?
2. A balanced three-phase delta circuit has phase currents of 10 A and a lagging power factor of 0.8. What are the line currents?
3. Assume the phasor for current $I_1$ has abscissa component $I_{1-x}$ and ordinate component $I_{1-y}$. The phasor for $I_2$ has $I_{2-x}$ and ordinate component $I_{2-y}$. If $I_3 = I_2 - I_1$, then what is the abscissa component of $I_3$ and the ordinate component of $I_3$?

4. The abscissa vector components determining the phasor of the current in conductor A of a three-phase circuit are $12 + 7 - 9 + 3$. The ordinate components of the phasor determining the current in conductor A are $9 - 14 + 3 - 2$. What is the magnitude of the current in conductor A?

5. A 480-3-60 VAC delta circuit has a 10 kW heater between phase A and phase B, a 10 kW heater between phase B and phase C, and a 5 kW heater between phase C and phase A. Determine the line currents in, respectively, conductors A, B, and C.

6. A branch circuit of a four-wire 480/3/60 service is used for 277/1/60 VAC building lighting. The electrical conductors are designated A, B, C, and D, with conductor D being the neutral conductor. The largest current expected in circuits A–D, B–D, or C–D is 100 A each. What is the largest possible current that can be expected in conductor D?

7. A four-wire wye circuit that has conductors A, B, C, and D, with conductor D as the ground conductor, has the following loading: phase a–d—12 A at lagging PF of 0.8; phase b–d—17 A with lagging power factor of 0.7; and phase c–d—15 A with lagging power factor of 0.75. What is the current in conductor A?

8. Three equally sized, single-phase heaters are used to warm the crankcase of a marine diesel engine. The heaters are energized from a 208/3/60 electrical source and wired in a delta configuration. Each heater is stamped 10 A and 208 VAC. Determine line currents to the heaters.

9. A 240/1/60 line serves two separate loads that are connected in parallel. The first load is capacitive with 10 A leading line voltage at 90°. The second load is 30 A lagging at 30°. What are the current and the lead/lag in the common 240/1/60 line?

10. A three-phase circuit with conductors A, B, and C delivers voltage to a number of three-phase loads. The voltage sequence is A–B, B–C, and C–A. The sum of the abscissa phasors for current in conductor A is –10, and the sum of the ordinate phasors is +10. Assuming that voltage C–A leads voltage A–B by 120.0°, determine the lead or lag of the current in conductor A with respect to voltage C–A.

11. A three-phase four-wire circuit with conductors A, B, C, and neutral ground conductor D delivers power to both an induction motor and building lighting. The nameplate on the motor states that the FLA is 12 A and the PF is 0.6. The lighting circuits are connected in a wye configuration, and all lighting is at PF equal to 0.7, lagging. Current in the lighting circuit common to conductor A is 10 A. Current in the lighting circuit connected to conductor B is 15 A. Current in the lighting circuit connected to conductor C is 20 A. Determine the currents in, respectively, conductors A and B.

12. A 480/3/60 VAC source with conductors denoted as A, B, and C delivers power to two induction motors. One of the motors has nameplate data, "FLA of 20 A and PF 0.70." The second motor has nameplate data, "FLA 30 A and PF 0.60." Determine the full load, the currents, and power factor of source conductors A, B, and C.

13. A three-phase delta circuit has the following loading: phase a–b is 7 A at power factor 0.7 lagging, phase b–c is 8 A at power factor 0.8 lagging, and phase c–a is 9 A at power factor 0.9 leading. Determine the current in conductor A.

14. A common three-phase line serves two induction motors. The name plate of one motor states, "wye… FLA: 7 A, PF: .9." The nameplate of the second motor states, "delta…FLA: 8A, PF: .6." Determine the full load current and power factor in the common three-phase feeder.

15. A three-phase feeder, with conductors denoted A, B, and C, delivers power to two separate three phase users designated as user #1 and user #2. Conductor A has branch conductor P to user #1 and conductor S to user #2. Conductor B has branch conductor Q to user #1 and conductor T to user #2. Conductor C has branch conductor R to user #1 and conductor U to user #2. The currents in the branch conductors are P: 50 A leading line voltage by 20°; S: 15 A leading line voltage by 25°; Q: 10 A leading line voltage by 20°; T: 40 A leading line voltage by 10°; R: 20 A lagging line voltage by 10°; and U: 60 A leading line voltage by 1°. Determine the current in conductor A.

16. A building has several large air-conditioning chillers that are served by a 13.8 kV supply. All of the chiller motors have nameplate power factors of 0.8 and the sum of the nameplate FLAs is 150 A. The a–b phase, and only the a–b phase, has an added resistive load of 50 A. Determine the currents in conductors A, B, and C.

# 5

## Calculating Three-Phase Power

Very often the need will arise for the calculation of the power consumption of an existing three-phase circuit. Similarly, the power of a proposed circuit may require determination. Power consumption calculations of a single, balanced three-phase circuit can be performed with relative ease. In contrast, the power of a three-phase circuit that, say, feeds one or more unbalanced three-phase circuits can present a greater challenge. This section deals with the computations that will allow a person to complete the power calculations of any type of three-phase circuit, balanced or unbalanced. Reviewing the calculations of single-phase circuits is a preferred entry into the subject as those calculations are the easiest to perform. After all, three-phase power consists of three single phases.

## 5.1 Units

A few words on the subject of power and energy! By definition, power is the rate of consumption of energy per unit of time. The basic unit of electrical power is the watt, which is a unit of the internationally recognized SI (Système International) system of units. The joule is the SI unit of energy, and 1 W is the consumption rate of 1 J/s. A more common unit of energy usage in the industry is the watt-hour, which would be the energy consumption resulting from a power usage of 1 W for 1 h.

By far the largest usage of three-phase electrical power can be attributed to three categories: lighting, electric motors, and heating. The units used to quantify these usages of electrical power have changed over the years, and it can be anticipated there will be more changes in the future. The following is a brief review of the units used.

Electrical lighting, and especially incandescent lighting, has traditionally been defined by the amount of power to activate a light, that is, by the wattage. This practice is changing as more efficient light sources have become available. Gradually, the various types of lights are being defined by the amount of light produced rather than by the amount of electrical power the light requires. One unit that seems to be gaining in acceptance is the "lumen," which is a metric unit. There are of course other popular units of measure for defining the amount of light a device produces. In any event, the unit of the electrical power required to activate a light will most certainly remain the "watt."

Another big change occurring in North America is in the way electrical power of motors is defined. Traditionally, the size of a motor has been defined by "horsepower." Because of globalization and increased international trade, the term horsepower is gradually being replaced or augmented by "watt." In electrical units, 1 horsepower = 746 W. For example, a 1 kW motor with an efficiency of 74.6% would require 1 kW (1000 W) of electrical energy to operate and would produce 1 kW × 0.746 kW of output power (brake power), or 746 W of output, which would be the equivalent to 1 horsepower of output. A 1 kW motor with an efficiency of 80% would produce 0.80 × 1 kW of output power or (800/746) = 1.07 HP.

When describing heater output, the BTU (British Thermal Unit) has long been used in North America and in Commonwealth Countries. In the United States, the BTU will probably continue its use for the foreseeable future, although overseas the BTU is being replaced by the "calorie" as well as other units. Nevertheless, electric resistance heating is mostly classified by its input power (wattage) rating, which, of course, is the same as its output power.

## 5.2 Calculating Power of a Single-Phase Circuit

Instantaneous power in a single-phase circuit is given by the expression

$$P^i = v^i \cdot i^i \, (\text{W})$$

From Equation 1.1,

$$v^i = (v_{PK}) \sin \omega t$$

and from Equation 1.2,

$$i^i = i_{PK} \sin(\omega t + \theta_{SP})$$

Therefore,

$$P^i = [(v_{PK}) \sin \omega t][i_{PK} \sin(\omega t + \theta_{SP})], \quad \text{or}$$
$$P^i = (v_{PK})(i_{PK})(\sin \omega t)[\sin(\omega t + \theta_{SP})]$$

Since, $\sin(\omega t + \theta_{SP}) = \sin \omega t \cos \theta_{SP} + \cos \omega t \sin \theta_{SP}$,

$$P^i = (v_{PK})(i_{PK})(\sin \omega t)[\sin \omega t \cos \theta_{SP} + \cos \omega t \sin \theta_{SP}], \quad \text{or}$$
$$P^i = (v_{PK})(i_{PK})(\cos \theta_{SP})(\sin \omega t)^2 + (v_{PK})(i_{PK})(\sin \theta_{SP})(\sin \omega t)(\cos \omega t)$$

(5.1)

The equation for instantaneous power ($P^i$) is often presented in a different form. One common version is

$$P^i = [(v_{PK})(i_{PK})/2]\cos \theta_{SP} - [(v_{PK})(i_{PK})/2]\cos(2\omega t + \theta_{SP})$$

(5.2)

(Reference 4.1) As demonstrated in Section A.1, Equations 5.1 and 5.2 are equivalent equations, although each is stated in a different form.

### Example 5.1

It will be informative to plot the computed values of Equation 5.1 for one complete cycle ($\omega t = 0°$ to $\omega t = 360°$) of the variable $\omega t$ for a typical single-phase application. The plot provides a pictorial view of the instantaneous value of power as a function of time. In the way of illustration, consider a specific case having the following parameters:

$$V_{rms} = 480 \text{ VAC, single phase}$$
$$v_{PK} = 480\sqrt{2} = 678.82 \text{ VAC}$$
$$I_{rms} = 10 \text{ A}$$
$$i_{PK} = 10\sqrt{2} = 14.14 \text{ A}$$

Power Factor = 0.70, lagging and $\cos \theta = 0.70$, and $\theta = -45.57°$
Equation 5.1 states,

$$P^i = (v_{PK})(i_{PK})(\cos \theta_{SP})(\sin^2 \omega t) + (v_{PK})(i_{PK})(\sin \theta_{SP})(\sin \omega t)(\cos \omega t)$$

Let

$$A = (v_{PK})(i_{PK})\cos \theta \sin^2 \omega t$$
$$B = (v_{PK})(i_{PK})\sin \theta \sin \omega t \cos \omega t$$
$$P^i = A + B$$
$$A = (480\sqrt{2})(10\sqrt{2})(0.7)\sin^2 \omega t$$
$$A = (6720.00)\sin^2 \omega t, \text{ and}$$
$$B = (480\sqrt{2})(10\sqrt{2})(\sin - 45.57°)\sin \omega t \cos \omega t$$
$$B = (9600)(-.7141)\sin \omega t \cos \omega t$$
$$B = (-6855.36)\sin \omega t \cos \omega t$$

The computed values of Function A and Function B in this example are contained in Section A.2. A scaled plot of the values of Function A is

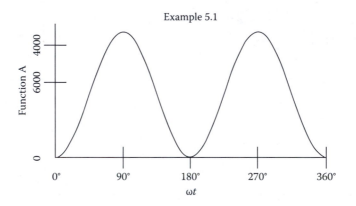

**FIGURE 5.1**
Function "A" vs. ω*t*.

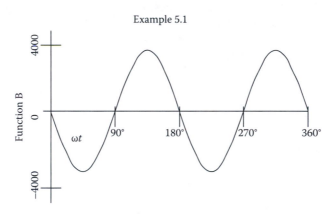

**FIGURE 5.2**
Function "B" vs. ω*t*.

shown in Figure 5.1. A scaled plot of the values of Function B is shown in Figure 5.2, and a plot of Function A plus Function B is shown in Figure 5.3.

It is pertinent to note that in Figure 5.3, which describes power flow in a typical single-phase circuit with a power factor of 0.70, power flow has two large peaks in the positive direction and two smaller valleys in the negative direction. This power flow will be compared to power flow in a three-phase circuit that, as demonstrated later, is drastically different.

According to Equation 5.1, the instantaneous power in a single-phase circuit is

$$P_i = (v_{PK})(i_{PK})(\cos \theta_{SP})(\sin^2 \omega t) + (v_{PK})(i_{PK})(\sin \theta_{SP})(\sin \omega t)(\cos \omega t)$$

Generally speaking, the average power is of the most interest. To determine the average power, a single cycle of ω*t* is considered, that is, from

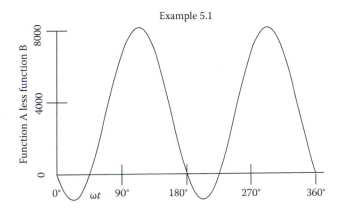

**FIGURE 5.3**
(Function A less Function B) vs. $\omega t$.

the value of $\omega t = 0$ to $\omega t = 360°$. To determine average power, the net power over a single cycle is determined and then divided by the time of one period.

Let T = the time for a single cycle (from $\omega t = 0°$ to $\omega t = 360°$). The net power in the sample period is determined from integration of the equation for instantaneous power (Equation 5.1) with respect to $t$.

Let

$$P^i = A(t) + B(t), \text{ whereby}$$
$$A(t) = (v_{PK})(i_{PK})\cos\theta\sin^2\omega t, \text{ and}$$
$$B(t) = (v_{PK})(i_{PK})\sin\theta\sin\omega t\cos\omega t$$
$$\int A(t) = \int (v_{PK})(i_{PK})\cos\theta\sin^2\omega t \, dt = (v_{PK})(i_{PK})\cos\theta \int \sin^2\omega t \, dt$$
$$\int A(t) = (v_{PK})(i_{PK})\cos\theta \, [t/2 - (\sin 2\omega t)/4\omega], \text{ evaluated from } t = 0 \text{ to } t = T.$$

The integral is evaluated from $t = 0$ to $t = T$ (with value "T" occurring at $\omega t = 360°$)

$$\int A(t) = [(v_{PK})(i_{PK})\cos\theta]\{[T/2 - (\sin 2\omega t)/4\omega] - [0/2 - (\sin 0)/4\omega]\}$$
$$\int A(t) = [(v_{PK})(i_{PK})\cos\theta](T/2)$$

The average value of $\int A(t)$ throughout the period is

$$P = [(v_{PK})(i_{PK})\cos\theta][T/2]/T = [(v_{PK})(i_{PK})\cos\theta]/2$$

It will be apparent that the function B($t$) throughout the period $t = 0$ to $t = T$ provides no net addition to the value of function P. This is the case

since half of the function is positive during the period and half is negative, with the result that there is no net contribution of the function to the
value of power throughout the period. Consequently, the function A($t$)
solely determines the value of power throughout one cycle.

Voltages and currents are commonly used in rms (root mean squared)
values. By definition,

$$V = \frac{v_{PK}}{\sqrt{2}}$$

and

$$I = \frac{i_{PK}}{\sqrt{2}}$$

and

$$P = \frac{(v_{PK})(i_{PK})\cos\theta}{2}$$

Substituting rms values for peak values of voltage and current gives

$$P = VI\cos\theta \qquad\qquad (5.3)$$

where
   P is the power (watts)
   V is the potential (rms voltage)
   I is the current (rms amperage)
   cos θ is the power factor

(The valid range of θ for a single-phase circuit is from −90° + 90°,
and the power factor throughout the period will always be between 0
and +1.0.)

Equation 5.3 is the commonly recognized equation for calculating
power in a single-phase circuit.

## 5.3 Calculating Power in Balanced Three-Phase Circuits

By definition, a three-phase circuit consists of three separate, single-phase
circuits. A typical trace of the three voltages with respect to time would be
similar to that represented in Figure 1.3. The phases are commonly identified
as phase A, phase B, and phase C, and the common sequence is A–B–C. Each
of the phases is 120° from one another in a time plot.

There are two basic types of three-phase circuits: delta and wye. A typical delta circuit is shown in Figure 1.14, and a typical wye circuit is shown

in Figure 1.15. In a delta circuit, phase voltage equals line voltage, but phase current is not equal to line current. In a wye circuit, line current equals phase current but phase voltage is not equal to line voltage. So, in order to perform correct power calculations, one must make the distinction between line parameters and phase parameters and then apply the appropriate calculation.

A wye circuit could be the three-wire type or the four-wire type. If a wye circuit is the four-wire type, a fourth conductor is extended from the wye neutral point to ground. A three-wire wye circuit is absent a conductor extending from the neutral point to ground.

To calculate the power of a three-phase circuit, specific parameters must be known. Voltage must be known. Also, the values of currents and the leads or lags of currents with respect to voltages are necessary.

Power factor in a linear, single-phase circuit is a clearly understood condition. More specifically, power factor in a single-phase circuit is the cosine of the angle between current and voltage. A leading current with a lead angle of, say, 30° has a power factor equal to the cosine of 30° or 0.866. A lagging current of 30° would likewise have a power factor of 0.866. In three-phase circuits, the matter of power factor is somewhat more complicated.

A motor with a nameplate that bears the motor power factor is stating the power factor of the phases. In other words, a motor power factor is the cosine of the angle between the phase current and the phase voltage whether the motor is a wye wound motor or a delta wound motor. The most common type of three-phase motor is the induction motor that always has a lagging power factor. Equation 1.7 gives the relationship between phase lead/lag and line lead/lag for balanced wye circuits and balanced delta circuits. Equation 1.7 can be helpful when a person needs to determine line currents and the associated lead/lag of the line currents in order to perform power calculations.

In a wye circuit, line currents and line voltages of an existing installation can generally be measured with ease, but phase voltages for a three-wire wye circuit may be difficult to read. In a delta circuit line, voltages, which are the same as phase voltages, can be conveniently read but phase currents may be difficult to read. For example, to measure the phase currents of a delta wound motor would require removing the terminal box cover and moving existing connections. Particularly with high-voltage motors an effort of this type may be especially hazardous and undesirable.

By definition, a balanced circuit has line voltages, line currents, phase voltages, phase currents, and the lead/lag of the currents that are all identical. According to Equation 5.3, the power of a single-phase circuit is described by the relationship

$$P = VI\cos\theta,$$ where the parameters are the single-phase parameters.

It follows that the power for a balanced three-phase circuit is

$$P = 3V_P I_P \cos\theta_P \qquad (5.4)$$

where
  P is the power (watts)
  $V_P$ is the phase voltage (rms)
  $I_P$ is the phase current (rms)
  $\cos\theta_P$ is the power factor of phase

$\theta_P$ is the angle of lead or angle of lag (radians or degrees) (phase current with respect to phase voltage) (the subscript "P" designates "phase")

for a lagging power factor $\theta_P < 0$
for a leading power factor $\theta_P > 0$

As in a single-phase circuit, instantaneous power of each of the phases of a three phase is given by Equation 5.1.

### Example 5.2

In Example 5.1, the values of instantaneous power are plotted for one complete cycle in a typical single-phase application. As will be demonstrated, a plot of three-phase power throughout the same period has a very different appearance. Consider a case having the following parameters applicable to a three-phase motor:

$$V_{rms} = 480 \text{ VAC, three phase}$$
$$V_{PK} = 480\sqrt{2}\text{VAC}$$
$$I_{rms} = 10 \text{ A}\left(\text{phase current}\right)$$
$$I_{PK} = 10\sqrt{2} \text{ A}$$

Power Factor = 0.70, lagging
Therefore,

$$\cos\theta = 0.70, \quad \text{and}$$
$$\theta = -45.57°$$

Equation 5.1 states that for each of the three phases

$$P^i = V_{PK}I_{PK}\cos\theta\sin^2\omega t + V_{PK}I_{PK}\sin\theta\sin\omega t\cos\omega t$$

It may be seen that the power factor is the same as that assumed in Example 5.1. In Example 5.2, the user is a three-phase motor.

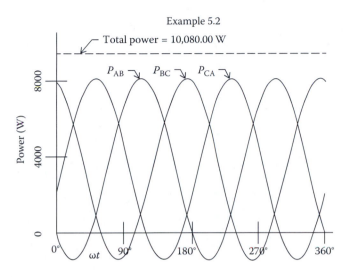

**FIGURE 5.4**
Three-phase power vs. ω*t*.

A summary of the power computations of instantaneous power vs. ω*t* for Example 5.2 is tabulated in Section A.5. A plot of the values of power for the assumed parameters for one complete cycle is shown in Figure 5.4. It is significant to note that the value of total instantaneous power throughout the cycle is constant. In other words, the sum of the powers of the three phases is calculated to be exactly a constant value; in this instance, 10,080.00 W. The computations of Examples 5.1 and 5.2 demonstrate one of the significant differences between a single-phase motor and a three-phase motor. The power of the single-phase motor in a single cycle has two sharp peaks and two sharp valleys per cycle, whereas the power of the three-phase motor is exactly constant for all values of ω*t*.

As mentioned, the nameplate on a three-phase motor will bear the power factor that is applicable to the phase currents and phase voltages. The nameplate will also state the full load amperage, as "FLA," which is the magnitude of the line current. Although seemingly inconsistent, this is nonetheless standard practice: motor nameplates state the power factor of the phase currents not the line currents.

### 5.3.1 Calculating Power in Balanced Three-Phase Wye Circuits

In a balanced three-phase wye circuit, the phase currents of all three phases are of the same magnitude as the line currents, and all three currents are at the same lead/lag angle. The phase voltages are at a magnitude that is a fixed ratio of the line voltages.

In the typical three-phase wye circuit as represented in Figure 1.15, the line current in conductor A is the current of phase A–D. The line current in conductor B is the current of phase B–D, and the line current in conductor

C is the current of phase B–D. For a wye circuit, the magnitude of the phase voltages are a fixed ratio of the line voltages and are given by the expression

$$V_L = \left(\sqrt{3}\right)V_P \tag{5.5}$$

or

$$V_P = \left[\frac{1}{\sqrt{3}}\right]V_L$$

where
  $V_L$ is the line voltage (rms)
  $V_P$ is the phase voltage (rms)
  The lead/lag of the line currents in a wye circuit are related to the lead/lag
    of the phase currents by the general expression
  $\theta_P = \theta_L - 30°$ (according to Equation 1.7)

where
  $\theta_P$ is the lead/lag of phase current with respect to phase voltage
  $\theta_L$ is the lead/lag of line current with respect to line voltage

More specifically,

$$\theta_{P-A/AD} = \theta_{L-A/CA} - 30°$$
$$\theta_{P-B/BD} = \theta_{L-B/AB} - 30°$$
$$\theta_{P-C/CD} = \theta_{L-C/BC} - 30°$$

### 5.3.1.1 Calculating Power in Balanced Three-Phase Wye Circuits: Resistive Loads

#### 5.3.1.1.1 Using Phase Parameters

$$P = 3V_P I_P \cos\theta_P$$

In a resistive circuit, the power factor is unity, or $\cos\theta_P = 1.0$. Thus,

$$P = 3V_P I_P \tag{5.6}$$

### 5.3.1.1.2 Using Line Parameters

From Equation 5.4,

$$P = 3V_P I_P \cos\theta_P$$

From Equation 5.4,

$$V_L = \left(\sqrt{3}\right)V_P, \quad \text{or}$$
$$V_P = \left[1/\sqrt{3}\right]V_L$$

In a balanced wye circuit, the magnitude of the line current is identical to the magnitude of the phase current (although the lead/lag of the line current with respect to phase voltage is different from the lead/lag of the current with respect to line voltage).

$$I_P = I_L$$

For a balanced or unbalanced wye circuit, the current lead/lag are related by the expression

$$\theta_P = \theta_L - 30° \tag{1.7}$$

Since $\cos\theta_P = 1.0$ for resistive loads, $\theta_P = 0°$

$$\theta_P = 0° = \theta_L - 30°$$
$$\theta_L = +30°$$

Thus,
The net power for all three phases becomes

$$P = 3V_P I_P \cos\theta_P = 3\left\{\left[1/(\sqrt{3})\right]V_L\right\}I_L \cos(\theta_L - 30°)$$

$$P = \sqrt{3}V_L I_L \cos(\theta_L - 30°) = \sqrt{3}V_L I_L \cos(30° - 30°)$$

$$= \sqrt{3}V_L I_L \cos(0°) = \sqrt{3}V_L I_L(1)$$

$$P = \sqrt{3}V_L I_L \tag{5.7}$$

### 5.3.1.2 Calculating Power in Balanced Three-Phase Wye Circuits: Inductive or Capacitive Loads

#### 5.3.1.2.1 Using Phase Parameters

From Equation 5.4,

$$P = 3V_P I_P \cos\theta_P \tag{5.8}$$

#### 5.3.1.2.2 Using Line Parameters

From Equation 5.4,

$$P = 3V_P I_P \cos\theta_P$$

From Equation 1.7, $\theta_P = \theta_L - 30°$

$$V_P = \left[1/\sqrt{3}\right]V_L$$

In a balanced wye circuit, the magnitude of the line current is identical to the magnitude of the phase current.

$$I_P = I_L$$
$$P = 3V_P I_P \cos\theta_P$$

From Equation 1.7,

$$\theta_P = \theta_L - 30°$$
$$P = 3\left[1/\sqrt{3}\right]V_L I_L \cos(\theta_L - 30°)$$
$$P = \left(\sqrt{3}\right)V_L I_L \cos(\theta_L - 30°) \tag{5.9}$$

NOTE: Equation 5.9 is applicable to any balanced three-phase load whether the user is wye, delta, or a mix. The reasons for this condition are explained later.

### 5.3.2 Calculating Power in Balanced Delta Three-Phase Circuits

In a balanced three-phase delta circuit, the phase voltages are the same as the line voltages and the line currents are a fixed ratio of the phase currents.

In the representation of a typical delta circuit in Figure 1.14, line voltage A–B is the voltage of phase A–B. Line voltage B–C is the voltage of phase B–C, and line voltage B–C is the voltage of phase B–C.

Since the phase voltages are of the same magnitude as the line voltages,

$$V_L = V_P$$

Phase currents are related to line currents by the expression,

$$I_L = \left(\sqrt{3}\right)I_P$$

As in a balanced wye circuits, the lead/lag of the line currents in a balanced delta circuit are related to the lead/lag of the phase currents by the expression $\theta_P = \theta_L - 30°$ (according to Equation 1.7)

$$\theta_{P-A/AD} = \theta_{L-A/CA} - 30°$$
$$\theta_{P-B/BD} = \theta_{L-B/AB} - 30°$$
$$\theta_{P-C/CD} = \theta_{L-C/BC} - 30°$$

### 5.3.2.1 Calculating Power in Balanced Three-Phase Delta Circuits: Resistive Loads

#### 5.3.2.1.1 Using Phase Parameters
From Equation 5.4,

$$P = 3V_P I_P \cos\theta_P$$

Since the power factor in a resistive load is unity,

$$P = 3V_P I_P \qquad (5.10)$$

#### 5.3.2.1.2 Using Line Parameters

$$P = 3V_P I_P \cos\theta_P$$
$$V_L = V_P$$
$$I_L = \left(\sqrt{3}\right)I_P, \quad \text{or} \quad I_P = \left[1/\left(\sqrt{3}\right)\right]I_L$$
$$\theta_P = \theta_L - 30°$$
$$\theta_P = 0° = \theta_L - 30°$$
$$\theta_L = +30°$$
$$P = 3V_P I_P \cos\theta_P = 3V_L\left\{\left[1/\sqrt{3}\right]I_L\right\}\cos(\theta_L - 30°)$$
$$P = 3V_L\left\{\left[1/\sqrt{3}\right]I_L\right\}\cos(+30° - 30°) = P = 3V_L\left\{\left[1/\sqrt{3}\right]I_L\right\}\cos(0°)$$
$$P = 3V_L\left\{\left[1/\left(\sqrt{3}\right)\right]I_L\right\} = \left(\sqrt{3}\right)V_L I_L$$
$$P = \left(\sqrt{3}\right)V_L I_L \qquad (5.11)$$

### 5.3.2.2 Calculating Power in Balanced Three-Phase Delta Circuits: Inductive or Capacitive

*5.3.2.2.1 Using Phase Parameters*
From Equation 5.4,

$$P = 3V_P I_P \cos\theta_P \tag{5.12}$$

*5.3.2.2.2 Using Line Parameters*

$$P = 3V_P I_P \cos\theta_P$$

$$I_L = \left(\sqrt{3}\right)I_P$$

$$I_P = \left[1/\left(\sqrt{3}\right)\right]I_L$$

$$V_P = V_L \tag{5.13}$$

$$\theta_P = \theta_L - 30°$$

$$P = 3V_P I_P \cos\theta_P = 3V_L\left[1/\left(\sqrt{3}\right)\right]I_L \cos(\theta_L - 30°)$$

$$P = \sqrt{3}V_L I_L \cos(\theta_L - 30°)$$

## 5.4 Calculating Power in Unbalanced Three-Phase Circuits

Unlike power computations for unbalanced wye circuits, the computations of power for unbalanced delta circuits cannot be performed easily. This is in large part due to the fact that in a delta circuit, line currents are the product of two phase currents. If a delta circuit is a balanced circuit, the computations are relatively straightforward. On the other hand, unbalanced delta circuits require more attention to detail and more time overall.

### 5.4.1 Calculating Power in Unbalanced Three-Phase Wye Circuits

A viable method to determine the total power of an unbalanced three-phase wye circuit is to treat the circuit as a combination of three single-phase circuits. The total power then becomes the sum of the three individual power determinations. The computations are made especially easy since phase current is equal to line current and phase voltage is a fixed ratio of line voltage.

### 5.4.1.1 Calculating Power in Unbalanced Three-Phase Wye Circuits: Resistive Loads

*5.4.1.1.1 Using Phase Parameters*

For phase A–D,

$$P_A = V_{A-D} I_A \cos \theta_{P-AD}$$

For phase B–D,

$$P_B = V_{B-D} I_B \cos \theta_{P-BD}$$

For phase C–D,

$$P_C = V_{C-D} I_C \cos \theta_{P-CD}$$

where
  $P_A$ is the power attributed to phase A
  $P_B$ is the power attributed to phase B
  $P_C$ is the power attributed to phase C
  $I_A$ is the current in conductor A and phase A–D
  $I_B$ is the current in conductor B and phase B–D
  $I_A$ is the current in conductor A and phase C–D
  $\theta_{P-AD}$ is the lead/lag of current in phase A with respect to voltage $V_{A-D}$
  $\theta_{P-BD}$ is the lead/lag of current in phase B with respect to voltage $V_{B-D}$
  $\theta_{P-CD}$ is the lead/lag of current in phase C with respect to voltage $V_{C-D}$

For resistive loads, $\cos \theta_{P-AD} = \cos \theta_{P-BD} = \cos \theta_{P-CD} = 1.0$

$$P_A = V_{A-D} I_A$$
$$P_B = V_{B-D} I_B$$
$$P_C = V_{C-D} I_C$$
$$P_T = P_A + P_B + P_C$$
$$P_T = V_{A-D} I_A + V_{B-D} I_B + V_{C-D} IC$$

Since $V_{A-D} = V_{B-D} = V_{C-D} = V_P$

$$P_T = V_P [I_A + I_B + I_C] \tag{5.14}$$

**Example 5.3**

Consider a wye circuit the equivalent to that of Figure 1.15. Assume the following conditions:

Voltage: 480/3/60

Currents:

$$I_A = 7 \text{ A}$$

$$I_B = 11 \text{ A}$$

$$I_C = 15 \text{ A}$$

Calculate total power.

$$P_T = V_P[I_A + I_B + I_C]$$
$$V_P = 480/\sqrt{3} = 277.128 \text{ V}$$

According to Equation 6.14,

$$P_T = V_P[I_A + I_B + I_C] = 277.128 \ [7 + 11 + 15] = 9145.228 \text{ W}$$

### 5.4.1.1.2 Using Line Parameters
Phase currents equal line currents.
  For phase A–D,

$$P_A = V_{DA}I_A \cos\theta_{P-A/AD}$$

$$I_{P-A} = I_{L-A} = I_A$$

$$V_{P-AD} = \left[1/\sqrt{3}\right]V_{L-CA}$$

$$\theta_{P-A/AD} = \theta_{L-A/CA} - 30°$$

$$P_A = V_{DA}I_A \cos\theta_{P-AD} = \left\{\left[1/\sqrt{3}\right]V_{L-CA}\right\}[I_A]\cos(\theta_{L-A/CA} - 30°)$$

$$P_T = P_A + P_B + P_C$$

The general form of the applicable equation is

$$P_T = V_L\left(1/\sqrt{3}\right)[I_{L-A}\cos(\theta_{L-A/CA} - 30°) + I_{L-B}\cos(\theta_{L-B/AB} - 30)°$$

$$+ I_{L-C}\cos(\theta_{L-C/BC} - 30°) \tag{5.15}$$

It is noted that for resistive loads, $\theta_{P-A} = 0°$

$$\theta_{P-A} = \theta_{L-A} = 30° = 0°$$
$$\theta_{L-A} = +30°$$
$$\cos(\theta_{L-A} - 30'') = \cos(+30° - 30°) = \cos 0° = 1.0$$
$$P_{DA} = \left\{[1/\sqrt{3}]V_{L-CA}\right\}[I_A](1.0) = (1/\sqrt{3})V_{L-CA}\,I_{L-A}$$

For phase B–D,

$$P_B = \left(1/\sqrt{3}\right)V_{L-AB}\,I_{L-B}$$

For phase C–D,

$$P_C = \left(1/\sqrt{3}\right)V_{L-CA}\,I_{L-C}$$

The net power then is

$$P_T = P_A + P_B + P_C$$

Since $V_{L-CA} = V_{L-AB} = V_{L-CA} = V_L$

$$P_T = \left(1/\sqrt{3}\right)V_L I_{L-A} + \left(1/\sqrt{3}\right)V_L I_{L-B} + \left(1/\sqrt{3}\right)V_L I_{L-C}$$

$$P_T = \left(1/\sqrt{3}\right)V_L[I_{L-A} + I_{L-B} + I_{L-C}]$$

**Example 5.4**

Assume the parameters of Example 5.3 and calculate total power using line parameters.

Voltage: 480/3/60

Currents:

$$I_A = 7\text{ A}$$
$$I_B = 11\text{ A}$$
$$I_C = 15\text{ A}$$

Calculate total power.
In wye circuits, phase currents equal line currents.

$$P_T = V_P[I_A + I_B + I_C]$$
$$V_P = 480/\sqrt{3} = 277.128\text{ V}$$

According to Equation 6.15,

$$P_T = V_P[I_A + I_B + I_C] = 277.128\,[7 + 11 + 15] = 9{,}145.228\text{ W}$$

### 5.4.1.2 Calculating Power in Unbalanced Three-Phase Wye Circuits: Inductive or Capacitive Loads

#### 5.4.1.2.1 Using Phase Parameters

$$P_A = V_{A-D}I_A \cos\theta_{P-AD}$$
$$P_B = V_{B-D}I_B \cos\theta_{P-BD}$$
$$P_C = V_{C-D}I_C \cos\theta_{P-CD}$$

The net power is

$$P_T = P_A + P_B + P_C$$
$$P_T = V_{A-D}I_A \cos\theta_{P-AD} + V_{B-D}I_B \cos\theta_{P-BD} + V_{C-D}I_C \cos\theta_{P-CD}$$

Since $V_{A-D} = V_{B-D} = V_{C-D} = V_P$

$$P_T = V_P[I_A\cos\theta_{P-AD} + I_B\cos\theta_{P-BD} + I_C\cos\theta_{P-CD}] \tag{5.16}$$

**Example 5.5**

Assume the following conditions for an unbalanced wye circuit:

Voltage: 480/3/60

Currents:

$$I_A = 8\text{ A}, \quad PF_A = 0.60$$
$$I_B = 12\text{ A}, \quad PF_B = 0.70$$
$$I_C = 16\text{ A}, \quad PF_C = 0.80$$

Calculate total power using phase parameters.

Power factors are descriptive of the lag/lead of phase current with respect to phase voltage. Therefore,

$$\cos\theta_{P-AD} = 0.60, \text{lagging}$$
$$\cos\theta_{P-BD} = 0.70, \text{lagging}$$
$$\cos\theta_{P-CD} = 0.80, \text{lagging}$$
$$V_P = V_L/\sqrt{3}$$
$$P_T = V_P[I_A\cos\theta_{P-AD} + I_B\cos\theta_{P-BD} + I_C\cos\theta_{P-CD}]$$
$$P_T = \left[480/\sqrt{3}\right][(8)(.60) + (12)(.70) + (16)(.80)]$$
$$P_T = (277.128)[4.8 + 8.4 + 12.8] = (277.18)[26] = 7,205.33\text{ W}$$

### 5.4.1.2.2 Using Line Parameters

$$P_A = V_{AD} I_{P-A} \cos \theta_{P-AD}$$

$$V_{A-D} = \left[ 1/\sqrt{3} \right] V_{L-CA}$$

$$I_{P-A} = I_A$$

$$\theta_{P-A/AD} = \theta_{L-A/CA} - 30°$$

$$P_A = V_{AD} I_{P-A} \cos \theta_{P-AD} = \left[ 1/\sqrt{3} \right] V_{L-CA} I_A \cos (\theta_{L-A/CA} - 30°)$$

Similarly,

$$P_B = \left[ 1/\sqrt{3} \right] V_{L-AB} I_B \cos (\theta_{L-B/AB} - 30°)$$

$$P_C = \left[ 1/\sqrt{3} \right] V_{L-BC} I_C \cos (\theta_{L-C/BC} - 30°)$$

Since $V_{L-CA} = V_{L-AB} = V_{L-BC} = V_L$

$$P_T = P_A + P_B + P_C$$

$$P_T = (1/\sqrt{3}) V_L [I_A \cos(\theta_{L-A/CA} - 30°) + I_B \cos (\theta_{L-B/AB} - 30°)$$

$$+ I_C \cos(\theta_{L-C/BC} - 30°)] \tag{5.17}$$

**Example 5.6**

Assume the same circuit parameters used in Example 5.3 for a wye circuit and calculate total circuit power using line parameters.

Voltage: 480/3/60

Circuit: wye

Currents:

$$I_A = 8 \text{ A}, \quad PF_A = .60, \text{ lagging}$$

$$I_B = 12 \text{ A}, \quad PF_B = .70, \text{ lagging}$$

$$I_C = 16 \text{ A}, \quad PF_C = .80, \text{ lagging}$$

Calculate total power using line values.

Power factors are descriptive of the lead/lag of phase current with respect to phase voltage. Therefore,

$$\cos\theta_{P-AD} = 0.60$$
$$\theta_{P-AD} = -53.130°$$
$$\cos\theta_{P-BD} = 0.70$$
$$\theta_{P-BD} = -45.572°$$
$$\cos\theta_{P-CD} = 0.80$$
$$\theta_{P-CD} = -36.869°$$

According to Equation 5.17,

$$P_T = \left(1/\sqrt{3}\right)V_L[I_A\cos(\theta_{L-A/CA} - 30°) + I_B\cos(\theta_{L-B/AB} - 30°)$$
$$+ I_C\cos(\theta_{L-C/BC} - 30°)]$$
$$\theta_{P-A/AD} = \theta_{L-A/CA} - 30° = -53.130°$$
$$\theta_{L-A/CA} = -53.130° + 30° = -23.130°$$
$$\theta_{P-B/BD} = \theta_{L-B/AB} - 30° = -45.572°$$
$$\theta_{L-B/AB} = -45.572° + 30° = -15.572°$$
$$\theta_{P-C/CD} = \theta_{L-C/BC} - 30° = -36.869°$$
$$\theta_{L-C/CD} = -36.869° + 30° = -6.869°$$
$$P_T = \left(1/\sqrt{3}\right)V_L[I_A\cos(\theta_{L-A/CA} - 30°) + I_B\cos(\theta_{L-B/AB} - 30°)$$
$$+ I_C\cos(\theta_{L-C/BC} - 30°)]$$
$$P_T = \left(1/\sqrt{3}\right)(480)[(8)\cos(-23.130° - 30°)$$
$$+ (12)\cos(-15.572° - 30°) + (16)\cos(-6.869° - 30°)]$$
$$P_T = \left(1/\sqrt{3}\right)(480)[(8)(.6) + (12)(.7) + (16)(.8)]$$
$$P_T = 7,205.33 \text{ W}$$

This computed value of power is in agreement with the calculation of total circuit power using phase parameters.

## 5.4.2 Calculating Power in Unbalanced Three-Phase Delta Circuits

Much as in a wye circuit, total power in an unbalanced three-phase delta circuit can be determined by treating the circuit as a combination of three single-phase circuits. The power of each of the three phases is separately determined and the total of the three becomes the three-phase power of the circuit. This method, of course, assumes that the values of the currents and the respective leads/lags of the phases are available. The alternate method is to determine circuit power from the properties of the line currents.

### 5.4.2.1 Calculating Power in Unbalanced Three-Phase Delta Circuits: Resistive Loads

*5.4.2.1.1 Using Phase Parameters*

$$P_{AB} = V_{A-B}I_{P-AB}\cos\theta_{P-AB}, P_{BC} = V_{B-C}I_{P-BC}\cos\theta_{P-BC} \text{ and}$$

$$P_{CA} = V_{C-A}I_{P-CA}\cos\theta_{P-CA}$$

For a circuit with all loads resistive,

$$\cos\theta_{P-AB} = 1.0, \cos\theta_{P-BC} = 1.0 \quad \text{and} \quad \cos\theta_{P-CA} = 1.0$$

$$P_{AB} = V_{L-AB}I_{P-AB}$$

$$P_{BC} = V_{L-BC}I_{P-BC}$$

$$P_{CA} = V_{L-CA}I_{P-CA}$$

$$P_T = P_{AB} + P_{BC} + P_{CA}, \quad \text{or}$$

$$P_T = V_{L-AB}I_{P-AB} + V_{L-BC}I_{P-BC} + V_{L-CA}I_{P-CA}$$

Since $V_{L-AB} = V_{L-BC} = V_{L-CA} = V_L$,

$$P_T = V_L I_{P-AB} + V_L I_{P-BC} + V_L I_{P-CA} = V_L[I_{P-AB} + I_{P-BC} + I_{P-CA}]$$

$$P_T = V_L[I_{P-AB} + I_{P-BC} + I_{P-CA}]$$

(5.18)

*5.4.2.1.2 Using Line Parameters*

Line parameters are identified as

Voltages: $V_{A-B}$, $V_{B-C}$ and $V_{C-A}$

Currents:

$$I_A - \text{Current in conductor A}$$
$$I_B - \text{Current in conductor B}$$
$$I_C - \text{Current in conductor C}$$

Lead/lag (currents with respect to line voltages):

$\theta_{L-A/CA}$ – lead/lag of line current A with respect to line voltage C–A
$\theta_{L-B/AB}$ – lead/lag of line current B with respect to line voltage A–B
$\theta_{L-C/BC}$ – lead/lag of line current C with respect to line voltage B–C

The applicable equation for circuit power as a function of line parameters with all resistance loads would be the same as that for inductive or capacitive loads as treated in Section 5.2.2.2.2

$$P_T = V_L\left(1/\sqrt{3}\right)[I_{L-A}\cos(\theta_{L-A/CA} - 30°) + I_{L-B}\cos(\theta_{L-B/AB} - 30°)$$

$$+ I_{L-C}\cos(\theta_{L-C/BC} - 30°)] \tag{5.19}$$

### 5.4.2.2 Calculating Power in Unbalanced Three-Phase Delta Circuits: Inductive or Capacitive

#### 5.4.2.2.1 Using Phase Parameters

$$P_{AB} = V_{A-B}I_{A-B}\cos\theta_{P-AB}, \quad P_{BC} = V_{B-C}I_{B-C}\cos\theta_{P-BC} \quad \text{and}$$

$$P_{CA} = V_{C-A}I_{C-A}\cos\theta_{P-CA}$$

$$P_T = P_{AB} + P_{BC} + P_{CA}, \text{ or}$$

$$P_T = V_{A-B}I_{A-B}\cos\theta_{P-AB} + V_{B-C}I_{B-C}\cos\theta_{P-BC} + V_{C-A}I_{C-A}\cos\theta_{P-CA}$$

Since $V_{A-B} = V_{B-C} = V_{C-A} = V_L = V_P$

$$P_T = V_P I_{A-B}\cos\theta_{P-AB} + V_P I_{B-C}\cos\theta_{P-BC} + V_P I_{C-A}\cos\theta_{P-CA}$$

$$P_T = V_P[I_{A-B}\cos\theta_{P-AB} + I_{B-C}\cos\theta_{P-BC} + I_{C-A}\cos\theta_{P-CA}] \tag{5.20}$$

**Example 5.7**

Assume the following conditions for an unbalanced delta circuit:

Line potential: 480/3/60 V

$$I_{ab} = 5 \text{ A at PF} = 1.00$$
$$I_{bc} = 10 \text{ A at PF} = 0.90 \text{ lagging}$$
$$I_{ca} = 15 \text{ A at PF} = 0.80 \text{ leading}$$

Find line currents in conductors A, B, and C.

**Solution**

$$\theta_{P-AB} = \cos^{-1}1.00 = 0$$

$$\theta_{P-BC} = \cos^{-1}0.90 = -25.84193°$$

$$\theta_{P-CA} = -\cos^{-1}0.80 = +36.86989°$$

According to Equation 5.20,

$$P_T = V_P[I_{A-B}\cos\theta_{P-AB} + I_{B-C}\cos\theta_{P-BC} + I_{C-A}\cos\theta_{P-CA}]$$

$$P_T = (480)[(5)(1) + (10)(.90) + (15)(.80)]$$

$$P_T = (480)[5 + 9 + 12] = (480)[26] = 12480 \text{ W}$$

*5.4.2.2.2 Using Line Parameters*
Line parameters are identified as

$$\text{Voltages}: V_{A-B}, \quad V_{B-C}, \quad \text{and} \quad V_{C-A}$$

Currents:

$$I_A - \text{Current in conductor A}$$
$$I_B - \text{Current in conductor B}$$
$$I_C - \text{Current in conductor C}$$

Lead/lag (currents with respect to line voltages):

$\theta_{L-A/CA}$ – lead/lag of line current A with respect to line voltage C–A
$\theta_{L-B/AB}$ – lead/lag of line current B with respect to line voltage A–B
$\theta_{L-C/BC}$ – lead/lag of line current C with respect to line voltage B–C

Total circuit power can be determined by assuming that there is a single user and that user is a wye circuit. This method is explained in detail in Section A.3.

$$P_{A-D} = V_{P-AD}I_{P-A}\cos\theta_{P-A/AD}$$

$$V_{P-AD} = \left(1/\sqrt{3}\right)V_{CA}$$

$$I_{P-A} = I_{L-A}$$

$$\theta_{P-A/AD} = \theta_{L-A/CA} - 30°$$

$$P_{A-D} = \left(1/\sqrt{3}\right)V_{CA}I_{L-A}\cos(\theta_{L-A/CA} - 30°)$$

Similarly,

$$P_{B-D} = \left(1/\sqrt{3}\right) V_{AB} I_{L-B} \cos\left(\theta_{L-B/AB} - 30°\right), \quad \text{and}$$

$$P_{C-D} = \left(1/\sqrt{3}\right) V_{BC} I_{L-C} \cos\left(\theta_{L-C/BC} - 30°\right)$$

$$P_T = P_{A-D} + P_{B-D} + P_{C-D}, \quad \text{or}$$

$$P_T = \left(1/\sqrt{3}\right) V_{CA} I_{L-A} \cos\left(\theta_{L-A/CA} - 30°\right) + \left(1/\sqrt{3}\right) V_{AB} I_{L-B} \cos\left(\theta_{L-B/AB} - 30°\right)$$

$$+ \left(1/\sqrt{3}\right) V_{BC} I_{L-C} \cos\left(\theta_{L-C/BC} - 30°\right)$$

Since $V_{L-AB} = V_{L-BC} = V_{L-CA} = V_L$

$$P_T = V_L\left(1/\sqrt{3}\right)[I_A \cos\left(\theta_{L-A/CA} - 30°\right) + I_B \cos\left(\theta_{L-B/AB} - 30°\right)$$

$$+ I_C \cos\left(\theta_{L-C/BC} - 30°\right)] \tag{5.21}$$

### Example 5.8

Assume the same delta phase values that were used in Example 5.7. Compute the associated line parameters and then compute total power using the values of line parameters.

Line potential: 480-3-60 V

$$I_{ab} = 5 \text{ A at PF} = 1.00$$
$$I_{bc} = 10 \text{ A at PF} = 0.90 \text{ lagging}$$
$$I_{ca} = 15 \text{ A at PF} = 0.80 \text{ leading}$$

Find line currents in conductors A, B, and C.

**Solution**

$$\theta_{P-AB} = \cos^{-1} 1.00 = 0$$

$$\theta_{P-BC} = \cos^{-1} 0.90 = -25.84193°$$

$$\theta_{P-CA} = -\cos^{-1} 0.80 = +36.86989°$$

Next, the associated line parameters are determined.

From Equations 4.4 and 4.5 (contained in Section A.5),

$$I_A = \left\{ (I_{A-x})^2 + (I_{A-y})^2 \right\}^{\frac{1}{2}}$$

$$\theta_{L-A} = (120° - \lambda)$$

where

$$I_{ba-x} = -I_{ab} \cos \theta_{P-AB} = -(5) \cos 0 = -5$$

$$I_{ba-y} = -I_{ab} \sin \theta_{P-AB} = -I_{ba}(0) = 0$$

$$I_{ca-x} = -I_{ca}(1/2) \left[ \left( \sqrt{3} \right) \sin \theta_{P-CA} + \cos \theta_{P-CA} \right]$$

$$I_{ca-x} = -(15)(1/2) \left[ \left( \sqrt{3} \right) \sin \theta_{P-CA} + \cos \theta_{P-CA} \right]$$

$$= -(7.5) \left[ \left( \sqrt{3} \right) \sin (36.86989°) + \cos (36.86989°) \right]$$

$$= -(7.5) \left[ \left( \sqrt{3} \right) (0.6) + (0.8) \right] = -(7.5)[1.039230 + 0.8] = -13.794228$$

$$I_{ca-y} = I_{ca}(1/2) \left[ \left( \sqrt{3} \right) \cos \theta_{P-CA} - \sin \theta_{P-CA} \right]$$

$$I_{ca-y} = I_{ca}(1/2) \left[ \left( \sqrt{3} \right) \cos (36.86989°) - \sin (36.86989°) \right]$$

$$= (7.5) \left[ \left( \sqrt{3} \right) (.80) - (-0.6) \right] = 5.892304$$

$$I_A = I_{ba-x} + I_{ca-x} = -5 + (-13.794228) = -18.794228$$

$$I_{A-y} = I_{ba-y} + I_{ca-y} = 0 + 5.892304 = 5.892304$$

$$I_A = \left\{ (I_{A-x})^2 + (I_{A-y})^2 \right\}^{\frac{1}{2}} = \left\{ (-18.794229)^2 + (5.892304)^2 \right\}^{\frac{1}{2}}$$

$$= 19.696250 \text{ A}$$

$$\lambda = \sin^{-1}(I_{A-y} \div I_A) = \sin^1[5.892304 \div 19.696250] = \sin^{-1}.2991586$$

$I_A$ is in Quadrant II

$$\lambda = \sin^{-1}.299158 = 162.5929°$$

$$\theta_{L-A} = (\lambda - 120°) = (162.5929° - 120°) = 42.5929°$$

From Equations 4.6 and 4.7,

$$I_B = \left\{ (I_{B-x})^2 + (I_{B-y})^2 \right\}^{\frac{1}{2}}$$

$$\theta_{L-B} = \sin^{-1}\left( I_{B-y} \div I_B \right)$$

where

$$I_B = I_{ab-x} + I_{cb-x}$$

$$I_B = I_{ab-y} + I_{cb-y}$$

$$I_{ab-x} = I_{ab}\cos\theta_{P-AB} = (5)\cos 0 = 5$$

$$I_{ab-y} = I_{ab}\sin\theta_{P-AB} = (5)\sin 0 = 0$$

$$I_{cb-x} = -I_{bc}(1/2)\left[ \left(\sqrt{3}\right)\sin\theta_{P-BC} - \cos\theta_{P-BC} \right]$$

$$= -(10)(1/2)\left[ \left(\sqrt{3}\right)\sin -25.84193° - \cos -25.84193° \right]$$

$$= -(5)\left[ \left(\sqrt{3}\right)(-0.4358898) - (.9) \right] = -(5)\left[ (-.754983) - (.9) \right]$$

$$= 8.274917$$

$$I_{cb-y} = I_{cb}(1/2)\left[ \left(\sqrt{3}\right)\cos\theta_{P-BC} + \sin\theta_{P-BC} \right]$$

$$= (10)(1/2)\left[ \left(\sqrt{3}\right)\cos -25.84193° + \sin -25.84193° \right]$$

$$= (5)\left[ \left(\sqrt{3}\right) + (-0.435889) \right] = (5)\left[ 1.55884 - 0.435889 \right]$$

$$= 5.61475$$

$$I_{B-x} = I_{ab-x} + I_{cb-x} = 5 + (8.274917) = 13.274917$$

$$I_{B-y} = I_{ab-y} + I_{cb-y} = 0 + 5.61475 = 5.61475$$

$$I_B = \left\{ (I_{B-x})^2 + (I_{B-y})^2 \right\}^{\frac{1}{2}} = \left\{ (13.274917)^2 + (5.61475)^2 \right\}^{\frac{1}{2}}$$

$$= 14.413495 \text{ A}$$

$$\theta_{L-B} = -\sin^{-1}(I_{B-y} \div I_B) = \sin^{-1}(5.61475 \div 14.41349) = 22.9263°$$

From Equations 4.8 and 4.9,

$$I_C = \left\{ (I_{C-x})^2 + (I_{C-y})^2 \right\}^{\frac{1}{2}}$$

where

$$I_{bc-x} = I_{bc}(1/2)\left[\left(\sqrt{3}\right)\sin\theta_{P-BC} - \cos\theta_{P-BC}\right]$$

$$= (10)(1/2)\left[\left(\sqrt{3}\right)\sin-25.84193° - \cos-25.84193°\right]$$

$$= (5)\left[\left(\sqrt{3}\right)(-0.435889) - (0.9)\right] = (5)[-.754983 - 0.9] = -8.27491$$

$$I_{bc-y} = -I_{bc}(1/2)\left[\left(\sqrt{3}\right)\cos\theta_{P-BC} + \sin\theta_{P-BC}\right]$$

$$= -(10)(1/2)\left[\left(\sqrt{3}\right)\cos-25.84193° + \sin-25.84193°\right]$$

$$= -(5)\left[\left(\sqrt{3}\right)(.9) + (-.435889)\right] = -(5)[1.55884 - .435889] = -5.61478$$

$$I_{ac-x} = I_{ca}(1/2)\left[\left(\sqrt{3}\right)\sin\theta_{P-CA} + \cos\theta_{P-CA}\right]$$

$$= (15)(1/2)\left[\left(\sqrt{3}\right)\sin36.86989° + \cos36.86989°\right]$$

$$= (7.5)\left[\left(\sqrt{3}\right) + (.8)\right] = (7.5)[1.039230 + .8] = 13.794228$$

$$I_{ac-y} = -I_{ca}(1/2)\left[\left(\sqrt{3}\right)\cos\theta_{P-CA} - \sin\theta_{P-CA}\right]$$

$$= -(15)(1/2)\left[\left(\sqrt{3}\right)\cos36.86989° - \sin36.86989°\right]$$

$$= -(7.5)\left[\left(\sqrt{3}\right)(.8) - (0.6)\right] = -(7.5)[1.385640 - 0.6] = -5.89230$$

$$I_{C-x} = I_{bc-x} + I_{ac-x} = -8.27491 + 13.794228 = 5.519318$$

$$I_{C-y} = I_{bc-y} + I_{ac-y} = -5.61478 + (-5.89230) = -11.50708$$

$$I_C = \left\{(I_{C-x})^2 + (I_{C-y})^2\right\}^{\frac{1}{2}} = \left\{(5.519318)^2 + (-11.50708)^2\right\}^{\frac{1}{2}}$$

$$= 12.762278 \text{ A}$$

$$\theta_{L-C} = (\varphi - 240°)$$

$I_C$ is in Quadrant IV

$$\varphi = \sin^{-1}\left(I_{C-y} \div I_C\right) = \sin^{-1}(-11.50708 \div 12.76227) = 295.6244°$$

$$\theta_{L-C} = (\varphi - 240°) = 295.6244° - 240° = 55.6244°$$

Summary of calculated values for the delta circuit (Table 5.1)

**TABLE 5.1**

Calculated Values for Delta Circuit (Example 5.8)

| Line Current | Lead/Lag of Line Current |
|---|---|
| $I_A = 19.6962506$ A | $\theta_{L-A/CA} = +42.5929°$ |
| $I_B = 14.413495$ A | $\theta_{L-B/AB} = +22.9263°$ |
| $I_C = 12.762278$ A | $\theta_{L-C/BC} = +55.6244°$ |

According to Equation 5.17,

$$P_T = V_L \left(1/\sqrt{3}\right) [I_{L-A} \cos(\theta_{L-A/CA} - 30°) + I_{L-B} \cos(\theta_{L-B/AB} - 30°)$$

$$+ I_{L-C} \cos(\theta_{L-C/BC} - 30°)]$$

$$P_T = \left(480\right)\left(1/\sqrt{3}\right)[(19.6962506) \cos(42.5929° - 30°)$$

$$+ (14.413495) \cos(22.9263° - 30°) + (12.762278) \cos(55.6244° - 30°]$$

$$P_T = (277.12812) \ [(19.6962506) \cos(12.5929°)$$

$$+ (14.413495) \cos(-7.0737°) + (12.762278) \cos(25.6244°)]$$

$$P_T = (277.12812) [19.22243 + 14.30378 + 11.50719]$$

$$P_T = (277.12812)[45.0334] = 12480.02 \text{ W} = 12480 \text{ W}$$

Thus, it is seen that the computation using line parameters yields the same result that was obtained using phase parameters as demonstrated in Example 5.7.

## 5.5 Calculating Power in a Three-Phase Circuit with Mixed Wye and Delta Loads

It is common to find a three-phase feeder that serves a mix of both wye and delta three-phase circuits. There could also be single-phase loads taken from any two of the phases. If data were available for all of the users, then the total power consumption is the sum of the power of all of the users. This total could be determined from either phase parameters or line parameters of the users. It can also be determined if adequate line data for the common feeder is available.

### 5.5.1 Using Phase Parameters

For wye loads, the power can be determined by one of the applicable equations presented earlier for wye loads using phase data. Similarly, the individual loads for delta circuits can be determined with an applicable delta equation. This would also be true of single-phase loads. The total power is then the sum of the individual loads.

## 5.5.2 Using Line Parameters

Earlier, the various equations are treated for both balanced and unbalanced circuits including both wye and delta circuits. If the type of load is known, then the applicable equation may be selected. What if, on the other hand, the type of load is unknown? Or, what if the load is a mix of wye, delta, and single phase? The answer is relatively simple!

It may be noted that of all the equations of power those that define power from line parameters are identical for both wye and delta circuits. The general form for power as determined from line parameters is that which is defined by Equation 5.17 or Equation 5.21, which are identical:

$$P_T = \left(1/\sqrt{3}\right) V_L \left[ I_A \cos(\theta_{L-A/CA} - 30°) + I_B \cos(\theta_{L-B/AB} - 30°) + I_C \cos(\theta_{L-C/BC} - 30°) \right]$$

The same general rule is applicable to the aforementioned equations that use phase values. Of course, the numerical value of phase potential in a particular circuit will be different from the numerical value of line potential.

A summary of the equations developed in Chapter 5 is presented in Section B.3.

## 5.6 Nonlinear Electrical Circuits

A linear electrical parameter, as voltage or current, is one that has values that with time closely approximates a sine wave function. A nonlinear electrical parameter is one that does not approximate a sine wave. If a system has a "stiff" voltage source, the voltage would be less likely to be effected by a nonlinear current. A stiff voltage source, for example, would be a transformer with a relatively high VA rating. If the voltage source has a relatively low VA rating, it very well could be that a nonlinear current could cause the voltage to deviate from the shape of a sine wave. In short, there are both nonlinear voltages and nonlinear currents.

Most electrical circuits can be characterized as linear. The increasingly greater use of thyristors has been the primary cause of the appearance of nonlinear circuits, although a smaller number of nonlinear circuits were present before the prevalence of thyristors. Some of the more common, contemporary causes of nonlinear circuits are

1. Electric arc furnaces
2. Fluorescent lighting
3. Switch mode power supplies for personal computers
4. Battery chargers
5. Variable frequency drives
6. Solid state inverters
7. Electric welders
8. Uninterruptible power supplies
9. Untuned capacitor banks

Although there is a variety of causes of nonlinearity in circuits, in most instances there are no resulting problems. Generally, nonlinear currents represent a small percentage of the net current consumption at a facility and most often corrective measures are unwarranted. On the other hand, some circumstances might merit ameliorating measures. For example, it very well could be that a nonlinear circuit is causing a watt-hour meter to read high. If significant savings can be realized, then a user of electricity may wish to take corrective action.

Nonlinear circuits may appear in a variety of forms. Some typical nonlinear currents are depicted in Figures 5.5 and 5.6. In both of these illustrations, a "stiff" voltage source is assumed. In other words, the voltages are assumed to be linear and essentially unaffected by nonlinear currents.

The number of different shapes of possible nonlinear circuits is unlimited as there is a large number of possible causes of nonlinearity. The nonlinear current of Figure 5.5 is called a "spiked" current and is typical of a nonlinear current trace that might result when a capacitor is in the circuit. The representation of Figure 5.6 shows a "flattened" current and is typical of a 60 Hz circuit that is distorted by the presence of third and fifth harmonic currents.

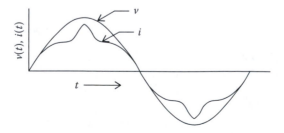

**FIGURE 5.5**
Spiked nonlinear current.

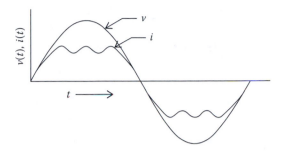

**FIGURE 5.6**
Flattened nonlinear current.

### 5.6.1 Linear Circuits and Power Factor

A linear, instantaneous voltage can very nearly be described by the expression

$$v^i = (V_{PK})\sin \omega t$$

A linear instantaneous current can be described by the expression

$$i^i = (I_{PK})\sin (\omega t + \theta_{SP})$$

Instantaneous power, linear or otherwise, is defined as

$$P^i = v^i \cdot i^i$$

The more common expressions of voltage and current use rms voltage and rms current. According to the accepted and proven relationship,

$$V(rms) = (V_{PK}) \div \sqrt{2}, \quad \text{and}$$
$$I(rms) = (V_{PK}) \div \sqrt{2}$$

The reason for the use of the root mean square (rms) values in the expressions for linear electrical systems is largely due to the analogy between ac and dc systems. In both dc and ac systems, power in a purely resistive circuit is defined by the expression

$$P = VI, \text{where}$$
$$V = \text{potential}(\text{rms voltage})$$
$$I = \text{current } (\text{rms amps})$$
$$P = \text{power } (\text{watts})$$

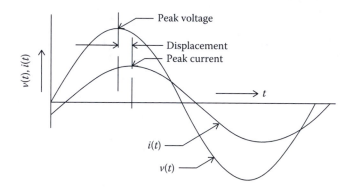

**FIGURE 5.7**
Displaced power factor.

If the current leads or lags voltage in an ac circuit, then a power factor is needed in the computation of power, namely, as

$$P = VI\cos\theta, \text{where}$$
$$\theta = \text{lead/lag of current with respect to voltage}$$

The power factor of a linear circuit is mostly known simply as "power factor," but it is also commonly known as a "displacement power factor." In a linear circuit that has a displacement power factor, the peak of the current trace is displaced to some extent from the peak of the voltage trace. The peak of the current trace represented in Figure 5.7 can be described as being displaced from the peak of the voltage trace. Actually, a displacement power factor is a special case of the more generic definition of power factor. A more encompassing definition of power factor defines power factor as the ratio of real power to total power. This later mention definition of power factor is applicable to both linear and nonlinear circuits. The terms applicable to the generic definition of power factor are

Real power, also called active power, P (W)
Reactive power, volt-ampere, Q (VAR)
Total power, or complex power, volt-amperes, S
Apparent power, |S| volt-amperes

In a linear, single-phase circuit, the angle between real power and total power is equivalent to $\theta_{SP}$ and the power factor is PF = $\cos\theta_{SP}$, which is also equal to the ratio of real power to total power, or
PF = $|P| \div |S|$. The values of P, Q, and S are typically represented as shown in Figure 5.8.

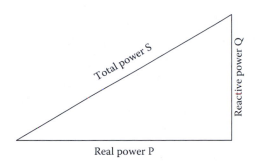

**FIGURE 5.8**
S, P, and Q.

### 5.6.2 Root Mean Square

The values of root mean square current and root mean square voltage are helpful in determine the total power, S of linear systems. For any function that can be described by an equation, the root mean square is defined by the relationship

$$f(t)\,\text{rms} = \left\{ 1/(T_2 - T_1) \int_{T1}^{T2} [f(t)]^2 \, dt \right\}^{\frac{1}{2}} \tag{5.22}$$

The common equations for voltage and current are normally given as rms values. In the way of illustration, consider a voltage is described by the expression $v^i = (V_{PK}) \sin \omega t$. If $(V_{PK}) = 1$ then,

$$V_{rms} = \left\{ 1/(T_2 - T_1) \int_{T1}^{T2} [f(\sin t)]^2 \, dt \right\}^{\frac{1}{2}}$$

$$= \left\{ [1/(T_2 - T_1)] [T_2/2 - (\sin T_2)/4] \; [T_1/2 - (\sin T_1)/4] \right\}^{\frac{1}{2}}$$

Evaluated from $\omega t = 0$ to $\omega t = 180°$ (in radians, $\omega t = 0$ to $\omega t = \pi$),

$$V_{rms} = \left\{ 1/(\pi - 0) ][\pi/2 - (\sin \pi)/4] - [0/2 - (\sin 0)/4] \right\}^{\frac{1}{2}} = 1/\sqrt{2}$$

So, for a sine wave voltage,

$$\text{Vrms} = (V_{PK})/\sqrt{2} = (V_{PK})(0.70710)$$

If a variable is described by a curve but the curve cannot readily be described by an equation, the root mean square may be closely approximated by the use of evenly spaced values of the function. More specifically,

$$f(X)_{rms} = \left\{ [1/n]\left[ X_1^2 + X_2^2 + \cdots X_n^2 \right] \right\}^{\frac{1}{2}} \tag{5.23}$$

$X_1, X_2, X_n$ are evenly spaced values within the region of the variable $X$ under consideration.

### Example 5.9

In this example, the root mean square of a sine function is to be determined by using separate values of the sine function and the expression

$$Xrms = \left\{ [1/n]\left[ X_1^2 + X_2^2 + \cdots X_n^2 \right] \right\}^{\frac{1}{2}}$$

Values of the function $v^i = (V_{PK}) \sin \omega t$, for $(V_{PK}) = 1.0$, have been calculated for values of $\omega t$ every 5°, starting at $\omega t = 0°$ and ending at $\omega t = 90°$. Because of the symmetry of the sine function, it is not necessary to compute the value of rms from $\omega t = 0°$ and ending at $\omega t = 180°$. The calculated values are tabulated in Section A.6. The corresponding values of $[v^i]^2$ are also tabulated in the table. As noted in Section A.6, the net value of $\left[ (\omega t_1)^2 + (\omega t_2)^2 + \cdots (\omega t)_n^2 \right]$ was determined to be 9.4993. There were 19 values of $\omega t$. So, the root mean square voltage is computed to be

$$Vrms = \left[ (1/19)(9.4993) \right]^{\frac{1}{2}} = 0.70708$$

The computed value of 0.70708 compares very well with the mathematically precise value of 0.70710. It may be concluded that the described procedure is a valid method of estimating the rms values of functions.

While RMS values of voltage and current may be used to calculate the power of linear systems, the use of RMS values of nonlinear systems cannot legitimately be used to compute the power of nonlinear systems. For nonlinear systems, a different method is needed.

## 5.6.3 Nonlinear Circuits and Power Factor

A common method of determining the power of a nonlinear system involves plotting the product of voltage and current and then determining the average throughout one cycle. In the way of illustration, consider a linear case for which the equations of voltage, current, and power are readily available.

### Example 5.10

In this example, the circuit power, P, of a linear system is determined graphically and compared to the value of P as computed by the conventional, and commonly accepted, expression for power.

Assume a circuit for which

$$V = 240\,VAC(rms)$$
$$I = 10\ A(rms)$$
$$PF = 0.70, \text{leading}$$

Determine the value of power from a plot of instantaneous power with time.

$$v^i = (V_{PK})\sin\omega t = (240)(\sqrt{2})\,(\sin\omega t = 339.4112\sin\omega t$$
$$i^i = I_{PK}\sin(\omega t + \theta_{SP}) = (10)(\sqrt{2})\,(\sin(\omega t + \theta_{SP}))$$
$$\theta_{SP} = \cos^{-1}(.70) = +45.57°$$
$$i^i = I_{PK}\sin(\omega t + \theta_{SP}) = (10)(\sqrt{2})\,(\sin(\omega t + 45.57) = 14.12\sin(\omega t + 45.57)$$
$$P^i = v^i \cdot i^i$$

Values of $v^i$ and $i^i$, and $P^i$ are computed and tabulated in Section A.7. A plot of instantaneous power, $P^i$, is shown in Figure 5.9. The computed values are used to estimate the area under the curve of $P^i$. The area under the plot of $P^i$ is determined in the following manner. Only the first half (0°–180°) of a full cycle is used since the second half of the cycle would be a mirror image of the first half and the average power would be the same as that of the first half cycle. (A full cycle is considered as extending from $\omega t = 0°$ to $\omega t = 360°$.) The method used here is known as the midpoint approximation. The method assumes that the area under the curve may be approximated by a number of narrow rectangles each of which have the midpoint of the tops of the rectangles bisected by the curve. There are several methods of approximating the area under a curve but the midpoint approximation method is the most intuitive. In this example, separate sections of a width equal to $\omega t = 5°$ are considered throughout the period from $\omega t = 0°$ to $\omega t = 180°$ and values of $P^i$ are computed at every 5° starting at the value of $\omega t$ centered at $\omega t = 2.5°$. Consequently, there is a total of 36 points. It is assumed that at each point there is an area that is of a width of 5° and of a height equal to the computed value of $P^i$. The sum of the areas was determined to be equal to (60443.20) (5°). The average area is

$$P = (60443.20)\,(5°) \div 36 = (1678.97)(5°).$$

The average value of $P^i$ throughout the period from $\omega t = 0°$ to $\omega t = 180°$ is (1678.97) (5°) ÷ (5°) or 1678.97 W.

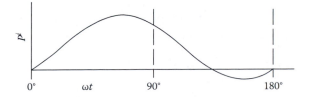

**FIGURE 5.9**
Instantaneous power.

If the power is computed using the commonly accepted expression with known rms values, the power computes to

$$P = VI(PF) = (240)(10)(0.7) = 1680 \text{ W}$$

## 5.7 Conclusion

In this exercise, the value of power computed by estimating the area under the instantaneous power curve was within 0.060% of the value computed using the mathematically precise values of rms voltage, rms current, and power factor. The degree of accuracy obtained in this example would generally be considered as acceptable for most practical applications. Nevertheless, better accuracy could generally be obtained by assuming areas of a smaller width. It may be concluded that this approach offers a viable method of determining the average power of nonlinear functions.

With the described approach, the value of power may be determined for nonlinear systems. The following is an example of the method as used with a nonlinear current.

### Example 5.11

In this example, the nonlinear plot of current and the linear plot of voltage in Figure 5.6 are considered. The power, volt-amps, and power factor are calculated.

The following conditions are assumed:

Current peak of $i^i = (10)(\sqrt{2})$. (This is the same peak current that would be found if the current were a sine wave and 10 A rms.) The voltage is assumed to be voltage V = 240 (rms) with a peak of $v^i = (240)(\sqrt{2})$. To calculate the area under the current curve, values of $i^i$ are taken from Figure 5.6 for every 5° value of $\omega t$ starting at $\omega t = 2.5°$. Values of $v^i$ are calculated at the same points. The computed values are tabulated in Section A.8 along with the calculated values of $P^i$. Because of symmetry, it is necessary to only determine average $P^i$ from $\omega t = 0°$ to $\omega t = 90°$. As shown in Section A.8, the net area under the plot of $P^i$ computes to Area = (36072.73) (5°). The average area is P = (36072.73) (5°) ÷ 18 = (2004.04)(5°). So, the average value of $P^i$ throughout the period from $\omega t = 0°$ to $\omega t = 90°$ is (2004.04) (5°) ÷ (5°) or 2004.04 W. Because of the symmetry, the average power would likewise be 2004.04 from $\omega t = 0°$ to $\omega t = 180°$ or from $\omega t = 0°$ to $\omega t = 360°$.

Using the generic definition of power factor,

$$PF = (\text{real power} \div \text{total power}) = 2004.04 \div 2040.96 = 0.982$$

In Example 5.11, it was determined that for the nonlinear current profile of Figure 5.5 with a peak current of 14.142 A the real power computes to 2004.04 W and the power factor to 0.982. It is noted that if the current were described by a sine wave function of 10 A rms with a peak of 14.142 the power would be 2400 W with a power factor of 1.00.

## Problems

1. What is the current draw of a 1 kW motor with an efficiency of 85% and a power factor of 0.80 on a 240-1-60 electrical line draws?

2. What is the equivalent of 1 W in SI units?

3. Describe the equation of instantaneous power for a single-phase circuit with a power factor of 0.50.

4. A manufacturer of conveyor systems in the United States is shipping a system to an overseas customer. The system requires a number of 5 HP motors. However, the customer intends to provide locally purchased motors, which will be watt-rated and which the customer has said will have efficiencies of 90% or better. What size motors, kW rated, will be required, assuming motors are available only in integral sizes?

5. Readings taken on a single phase 60 Hz line determined a RMS voltage of 277 V, RMS current of 10 A, and the current lagging by 25°. State the equation of instantaneous power for the circuit as a function of time starting from a point where $t = 0$.

6. Three equally sized, single-phase heaters are used to heat oil in a tank. The electrical source is 208/3/60 and each heater is stamped 6.4 A, 120 VAC. What is the power input from the electric heaters to the oil tank?

7. The nameplate on an induction motor states that the power factor is, "0.70." what is the lead/lag at full load of the phase currents with respect to phase voltage?

8. A 208/3/60 motor has a nameplate current of 5.3 FLA (full load amps) and a nameplate power factor of 0.75. If an oscilloscope were used to measure the lead/lag of line current at full load, what is the lead/lag of line current with respect to line voltage?

9. A three-phase, wye wound motor has nameplate data that states: 480/3/60, FLA: 12.3, PF: 0.78. What is the phase voltage, lead/lag of phase currents, line voltage, lead/lag of line current and wattage?

10. The following reading was taken on the line conductors going to a three-phase motor: Per oscilloscope: current lag with respect to line voltage: 20°. What is the power factor?

11. A new building is to be constructed, and it will require an electrical service for air conditioning, lighting, baseboard heating, and convenience outlets. A preliminary calculation indicated that the total maximum load will be somewhere in the vicinity of 50 kW. It was decided to use a 480/3/60 four-wire service, which would be configured within the building in a wye circuit, and the total power would be evenly distributed within the wye circuit. Power factor is estimated to be between 0.80 and 0.65. What is the range of currents in the conductors connected to the building?

12. Measurements on a three-phase 480 V delta connected load gave the following data. phase A–B: current of 4 A leading by 15°; phase B–C: current of 6 A lagging by 23°; phase C–A: current of 8 A lagging by 47°. What is the total power of the circuit?

13. A three-phase water pump motor at a power generating plant operates at a constant power output. A technician at the plant connected a power analyzer to the motor. He noted that currents averaged 15 A, and that they are balanced and lagging line voltage by 15°. Voltage was steady at 4160-3-60. The building has a permanently installed watt-hour meter that records the pump energy usage. If the technician returned the following day at the same time, he would expect the watt-hour meter to indicate an energy increase of how much?

14. Three heaters are used to keep an outdoor water tank from freezing during cold weather. The heaters were originally in a delta configuration with a 480/3/60 VAC service and each heater had an output of 4800 W. However, it was decided to reduce the heat output by connecting the heaters in a wye configuration. Determine the total heat output of the three heaters in the wye configuration.

15. The following readings were taken on the line conductors going to a three-phase motor. Current: 11 A. Voltage: 13,800 V. Current lead with respect to line voltage: 15°. Calculate the power, rounded to the nearest whole number.

16. A nominal 480/3/60 feeder serves two induction motors that have been operating at reduced load and that, consequently, have relatively low-power factors. An electrician used a power analyzer to measure currents and voltages in the feeder lines. He found that all three conductors had relatively identical currents at 32.6 A, which were lagging at 35° with respect to line voltages. The voltage on the day the measurements were taken was at 435 V. Determine the net power consumption at the time of the measurements.

17. A small machine shop has a four-wire 208/3/60 electrical service that is used for lighting, air conditioning, convenience outlets, and operation of a number of metal working machines on the shop floor. A single-phase wattmeter was used to separately measure power in the three phases, one at a time. An ammeter was used to measure the currents in the same three conductors. When taking the measurements, the current coils of the wattmeter were installed to measure currents in the conductors leading from the utility meter to the building's electrical panel board and the voltage leads measured potential from neutral to line voltage. So, the voltage measurements were all taken at 120 VAC. The recordings of the measurements were phase A–D, 8 kW, 95.23 A; phase B–D, 11 kW, 152.77; phase C–D, 7.9 kW, 101.28. Determine the power factors of the three phases: A–D, B–D, and C–D.

18. Using Equation 5.23, calculate the root mean square root of the function $v = 0.3t^2$ from $t = 0$ to $t = 60$ starting at $t = 0$ and sampled at $t = 0.0, 5.0, 10.00, \ldots, 60.0$.

# 6

# *Demand and Demand Response*

A charge for "demand" has long been a part of the bills sent to larger three-phase customers. For this reason, it behooves a customer to understand demand charges, why demand charges are in the bill, and what can be done to minimize those charges. A definition of the term "demand" may vary from one utility to another. Nevertheless, if the term appears on a customer's bill, it will have a value that is well defined by that particular utility.

The term "demand response" is of a relatively recent origin. Other terms that have appeared along with demand response are "smart grid" and "smart meters." An understanding of all of these terms is important to anyone who has an interest in three-phase electricity.

## 6.1 Demand Defined

In the English language, the word "demand" can be a noun or a verb. As a noun, the word is defined as a request, a petition, a plea, a request, or an ultimatum. As a verb, it is defined as to need, to command, or to insist. Used as an electrical term, the word "demand" is used to define the maximum use of electrical power within a billing period. One large electric utility explains demand as follows:

> Demand is the amount of electricity you require at a given time. For example, if you turn on ten 100-watt bulbs at the same time, your demand at that moment would be 1,000 watts, or 1 kilowatt. Because you pay for electricity demand as well as electricity use (kW-Hrs), you can save money by reducing demand. Even if you don't change the amount of electricity you use.

The measurement of demand is conducted by means of a demand meter that records electrical consumption within a predefined period of time. The demand charge part of the monthly bill is determined by the highest measured electrical demand recorded during the billing period. In some instances, demand is measured by energy consumption within a period of time. Other meters measure volt-amps within a period of time. From some time starting well before 1940, the electrical bills of the larger commercial and industrial users consisted of two parts. One part was for the energy, or watt-hours, used in a billing period. The second part of the bill was for the "demand" that was logged during the billing period.

The functioning of demand meters is discussed in more detail in Chapter 7.

### 6.1.1 Why Demand?

The general need for demand charges, by all utilities, is succinctly explained by the National Grid (which serves parts of Massachusetts, New York, and Rhode Island) in its position on demand charges for its New York commercial customers:

> What Is This Thing Called Demand. The price we pay for anything we buy contains the cost of the product plus profit, plus the cost of making the product available for sale, or overhead. In seeking to understand demand, we might equate it to this type of overhead expense. This is in contrast to charges National Grid customers pay for the electricity itself, or the "cost of product," largely made up of fuel costs incurred in the actual generation of energy. Both consumption and demand charges are part of every electricity consumer's service bill. Residential customers pay one rate of charges for electricity service, covering both consumption of electricity and demand. This simple, combined charge is possible because there is relatively little variation in electricity use from home to home. This is not the case among commercial and industrial energy users, whose electricity use—both consumption and demand—vary greatly. Some need large amounts of electricity once in a while—others, almost constantly. Complicating this is the fact that electricity cannot be stored. It must be generated and supplied to each customer as it is called for—instantly, day or night, in extremely variable quantities. Meeting these customers' needs requires keeping a vast array of expensive equipment—transformers, wires, substations and even generating stations—on constant standby. The amount and size of this equipment must be large enough to meet peak consumption periods, i.e., when the need for electricity is highest.

Other utilities may explain demand charges differently. Nevertheless, the reasons for demand charges by any utility the world over remain essentially the same and generally would echo the position stated by the National Grid.

### 6.1.2 Decreasing Demand

Demand charges typically are in the range of 30%–70% of a utility bill that is sent to customers of three-phase electricity. In many cases, demand charges can be substantially reduced by some modest changes in operations. Very often a modest change in the mode of starting-up equipment can result in surprisingly large monetary savings. Sometimes, merely staggering start-ups can reduce the demand charge. Shifting loads to another time of the day will generally assist. For those utilities that use kVA as a demand measurement, power factor correction can also have beneficial results.

---

## 6.2 Demand Response

### 6.2.1 Definitions

While the term "demand" has long been used by electric utilities, the term "demand response" is of relatively recent origins. The terms are entirely unrelated to one another. Following are two definitions that help explain the general use of the term "demand response."

*The U.S. Federal Energy Regulatory Commission (FERC)*

> Changes in electric usage by end-use customers from their normal consumption patterns in response to changes in the price of electricity over time, or to incentive payments designed to induce lower electricity use at times of high wholesale market prices or when system reliability is jeopardized.

*The U.S. Office of Electricity Delivery and Energy Reliability*

> Demand response provides an opportunity for consumers to play a significant role in the operation of the electric grid by reducing or shifting their electricity usage during peak periods in response to time-based rates or other forms of financial incentives. Demand response programs are being used by electric system planners and operators as resource options for balancing supply and demand. Such programs can lower the cost of electricity in wholesale markets, and in turn, lead to lower retail rates. Methods of engaging customers in demand response efforts include offering time-based rates such as time-of-use pricing, critical peak pricing, variable peak pricing, real time pricing, and critical peak rebates. It also includes direct load control programs which provide the ability for power companies to cycle air conditioners and water heaters on and off during periods of peak demand in exchange for a financial incentive and lower electric bills.

> The electric power industry considers demand response programs as an increasingly valuable resource option whose capabilities and potential impacts are expanded by grid modernization efforts. For example, sensors can perceive peak load problems and utilize automatic switching to divert or reduce power in strategic places, removing the chance of overload and the resulting power failure. Advanced metering infrastructure expands the range of time-based rate programs that can be

offered to consumers and smart customer systems such as in-home displays or home-area-networks can make it easier for consumers to change their behavior and reduce peak period consumption from information on their power consumption and costs. These programs also have the potential to help electricity providers save money through reductions in peak demand and the ability to defer construction of new power plants and power delivery systems—specifically, those reserved for use during peak times.

One of the goals of the Smart Grid R&D Program is to develop grid modernization technologies, tools, and techniques for demand response and help the power industry design, test, and demonstrate integrated, national electric/communication/information infrastructures with the ability to dynamically optimize grid operations and resources and incorporate demand response and consumer participation. To attain this goal, OE is supporting research, development and deployment of smart grid technologies, distribution system modeling and analysis, consumer behavior modeling, and analysis and high speed computational analysis capabilities for decision support tools.

It would be correct to say that the essential aim of the demand response programs is to allow electricity suppliers and grid operators to decrease stress to the electrical supply system by lowering peak demand on the electrical supply system. This lowering of peak demand is to be accomplished by either eliminating some consumption or by shifting some consumption to off-peak times. Participation in the programs is voluntary. Nevertheless, all customers can expect to benefit from the programs by way of a more dependable electrical supply. Some customers will be able to benefit from opportunities for cost reductions to their monthly electric bills. Utilities can expect a more efficient use of existing facilities as well as an improved financial return on capital investments.

There are at least three forms of demand response programs that have emerged with various utilities. While the precise definitions may vary somewhat, these are called emergency demand response, economic demand response, and ancillary services demand response. The mentioned programs differ in detail but the objective remains the same, namely, to use existing electrical gear to the maximum capability. Following are some explanations of the three mentioned types.

Emergency demand response is a program that is used to avoid interruption of service when demand approaches or exceeds the capability of the electric supply system. Participants sign-up for the program in advance and when called upon by the utility agree to voluntarily curtail electrical consumption. Monetary incentives are provided by the utility.

Economic demand response is a program that incentivizes customers to curtail or shift electric consumption to a time of day when the electric system is less taxed. The economic demand response programs are more recent than the conventional emergency demand response programs, and there is no universal description of what, exactly, these programs encompass. These programs, unlike the emergency demand response programs, allow the customer rather than the utility decide when to decrease electrical load.

Ancillary services demand response uses a variety of specialty services to help ensure secure and uninterrupted availability of the transmission grid. Through these programs, utilities contract with providers of ancillary services, as renewable power sources, to provide more or less power upon request by the utility.

## 6.2.2 Implementation Examples of Demand Response

While many utilities in the United States have embarked on demand response (DR) programs, the same can be said of many utilities located elsewhere. Demand response programs have become especially popular in both the EU countries and Great Britain. The common objective of these programs is to propagate policies that will foster the use of generating capabilities to the maximum possible extent. To this end, the utilities aim to somehow shift some electrical consumption from peak times to off-peak times—or by inducing customers to cut back on consumption during times of peak loads. Utilities intend to achieve demand response objectives by a variety of programs. Financial incentives are the main tool utilities used to achieve these objectives.

Following are specific examples of implementation of demand response programs that assist to explain how demand response functions in actuality.

### 6.2.2.1 NYISO (New York–Based Utility)

The utility has implemented demand response with four different programs, all of which were available at the same time:

> Emergency Demand Response Program (EDRP): This program is mostly for industrial and commercial customers. Participants must sign up in advance. Participants are paid by the utility for reducing energy consumption when asked to do so by NYISO. Reductions are voluntary for those in the EDRP program.
>
> ICAP Special Case Resources (SCR) program: Much as with the EDRP program, this program is also mostly for industrial and commercial organizations. Participation is voluntary. According to the terms of the agreement, participants in this program are required to reduce power when requested by the utility.
>
> Day Ahead Demand Response Program (DADRP): This program allows customers to bid proposed load reductions into the Day-Ahead energy market. Offers determined to be economically attractive are selected by the utility and paid at the market price. The DADRP program increases the amount of available electricity and at the same time moderates prices.

Demand Side Ancillary Services Program (DSASP): This program provides retail customers that can meet telemetry and other qualification requirements with an opportunity to bid their load curtailment capability to increase operating reserves. Participants are paid the market price at the time of use.

### 6.2.2.2 TXU Energy (Texas-Based Utility)

How the ("emergency response service") program works:

Based on previously accepted bids, participants who agreed to reduce load by 100 kW earned approximately $50 k per 4-month contract period. Payments are made for capacity whether a curtailment event is called or not.

TXU Energy will work with you to identify available load and determine a curtailment strategy to meet your business' requirements. Participants select the number of curtailment hours and negotiate a price for the load based on the time of year. If the bid is accepted, a 4-month contract is signed.

Events can be called at any time, but will only be called in case of emergency. Typically, events will only last 2–3 h and are limited to two events per 4-month contract period. Historically, there have only been two rolling blackouts in the past 20 years. Therefore, being called upon has been a rare occurrence.

Program administered by ERCOT with, potentially, no cost of entry to participate.

### 6.2.2.3 EnerNOC (Boston-Based Utility)

Introduction to Economic Demand Response in PJM

PJM's Economic Demand Response Program allows you to get paid for reducing electricity demand during high-priced hours. As a participant in the economic demand response market, you receive the same price as generators and also realize savings on your electricity bills. Unlike other forms of demand response where dispatches are initiated by the utility or grid operator, in economic markets, you're in control of if and when you reduce.

EnerNOC works with customers to develop response strategies tailored to their operating schedules and financial objectives.

Participants specify parameters such as reduction quantity, timing, price, and minimum response duration.

EnerNOC submits offers into the energy market for participants just like generators.

If offers clear, participants reduce demand and receive the hourly energy market clearing price.

How much advanced notice do I receive?

PJM offers two market options for participants in the Economic Demand Response Program:

Day-Ahead Market: Offers submitted by 12 pm the day before; clearing results posted approximately 4 pm the day before

Real-Time Market: Offers submitted up to 3 h in advance; participants dispatched with 2 h advanced notice

## 6.3 Smart Grid

### 6.3.1 Definition

The term "smart grid" is often used in association with the term "demand response." Nevertheless, there are distinct differences. Unlike the term "demand response," which can apply to a variety of systems as natural gas systems or transportation systems, the term "smart grid" is applicable to only electrical systems.

The Smart Grid Library succinctly defines the smart grid as "A bi-directional electric and communications network that improves the reliability, security, and efficiency of the electric system for small- to large-scale generation, transmission, distribution, and storage. It includes software and hardware applications for dynamic, integrated, and interoperable optimization of electric system operations, maintenance, and planning; distributed energy resources interconnection and integration; and feedback and controls at the consumer level."

The subject of "smart meters" is discussed in detail in Chapter 7.

### 6.3.2 Implementation Examples of Smart Grid

In the United States, serious efforts to incorporate smart grid technology began with the formation of the Federal Smart Grid Task Force under Title XIII of the Energy Independence and Security Act of 2007. Subsequently, the Smart Grid Investment Grant Program was started with authorization to offer substantial grants to participating utilities. In consequence, numerous improvements were made to grids within the United States. One highly visible evidence of the Government's efforts is the millions of smart meters that have been purchased and installed. In addition, utilities have installed specialized meters throughout their grid systems to monitor in real-time existing conditions of components of their transmission and distribution systems. Along with the development of the smart grids, smart meters have become an important part of that development. The smart meters provide the information that is needed by operators to make decisions pertinent to the operation of the grid.

### 6.3.2.1 Florida Power and Light

In 2013, Florida Power and Light (FPL) completed an $800 million project that installed 4.5 million smart meters and 10,000 smart devices to monitor the grid throughout its jurisdiction. The project was started in large part with a $200 million grant from the U.S. Federal Government. According to FPL, an additional benefit of the program has been the elimination of the need for 400 m readers. About 600 substations will be monitored more closely because of the program and various other benefits were forecasted for FPL's customers. In the long run, lower utility expenses will contribute to lower electric charges sent to their customers.

## Problems

1. Would it be correct to say that a demand meter is part of a program of demand response?
2. Do residential customers usually pay a demand charge?
3. If a customer of a utility is a participant in an emergency demand response program, is it the customer or the utility that decides when the customer is to decrease energy consumption?

# 7

## Instruments and Meters

Most devices used in the electrical industry to measure electrical properties are called "meters." For example, there are voltmeters, ammeters, as well as numerous other types of meters. Although most of these devices are truly instruments of some description, the term meter seems to be the more commonly applied term. By common usage, an electrical "meter" is a device that measures and sometimes records an electrical quantity as current, voltage, or resistance.

Electrical meters with a wide variety of capabilities are needed for installing, analyzing, and maintaining electrical systems. For example, megohm meters are needed to confirm that the insulation of a new system is adequate before the new installation may be safely energized. Phase sequence meters are used to confirm phase sequence and thereby help ensure proper connections to motors. After installation of an electrical system, ammeters may be required to measure system amperages in various conductors. The monitoring of voltage levels is needed to ascertain that potentials at an installation are within acceptable limits; for that purpose, either multimeters or permanently installed meters are required. Power quality meters can provide a higher level of monitoring of voltages. Wattmeters and power meters are special kinds of meters that allow a provider of electricity to charge for an electrical service.

Wattmeters and energy (watt-hour) meters are treated in the following paragraphs and some of the other common types of electrical instruments are reviewed later in this section.

### 7.1 Power Measurements

The measurement of circuit amperage or voltage can be accomplished with meters by an easily understood procedure. Power measurements can be a little more challenging. Whereas single-phase power and energy measurement can be conducted in a straightforward manner, the use of meters on three-phase systems requires greater knowledge and training. Three-phase meters have more connections than single-phase meters, and correct readings can be obtained only if the connections are made in the proper configuration.

### 7.1.1 Single-Phase Power Measurements

Power in a single-phase circuit can be measured with a single-phase watt-meter. By one means or another, a single-phase wattmeter measures current, voltage, and the lead or lag of the current with respect to the measured voltage. Internally, the wattmeter essentially performs a calculation in accordance with the algorithm

$$P = VI\cos\theta_P,$$

where
  V is the potential (voltage)
  I is the current (A)
  $\theta_P$ is the phase lead/lag angle between phase current and phase voltage (degrees or radians) for a lagging current, $\theta_P < 0$, and for a leading current, $\theta_P > 0$

In a single-phase circuit, $\theta_P > -90°$ and $\theta_P < +90°$.

One type of meter that has proved convenient for measuring the power of single-phase circuits is the handheld clamp-on type. A typical single-phase, handheld clamp-on power meter is shown in Figure 7.1. The feature of a clamp-on meter is that it does not require a cable to be disconnected, reconnected, disconnected, and finally, reconnected. Proper orientation of the

**FIGURE 7.1**
Clamp-on single-phase power meter. (Photo courtesy of Amprobe, Everett, WA.)

meter on the cable and correct connections of the potential leads are needed for valid power readings.

NOTE: The potential leads are not shown in Figure 7.1.

Of course, there are meters other than the clamp-on type that are available to read single-phase power. Bench meters can be used for the purpose. Also, permanently installed submeters are likewise available for the purpose. A permanently installed meter would have associated permanently installed toroidal current transformers on conductors.

### 7.1.2 Three-Phase Power Measurements

A number of parameters are required in the measurement of three-phase power. If a portable meter is being used for the purpose, proper positioning of the CTs and potential leads will be necessary.

Power is commonly measured with a wattmeter of some type. Wattmeters come in a variety of types and styles and are intended to meet a variety of needs. Among the classifications of portable types, there are the handheld wattmeters and the bench-mounted wattmeters. Some portable meters are capable of only measuring single-phase power, whereas others are capable of measuring both single-phase power and three-phase power. Portable meters are generally used for periodic measurements and are designed to be readily moved from one location to another. Portable meters might be used, say, for energy audits, to determine the various sources of energy consumption within a facility.

Electrical energy measurements are made with watt-hour meters that have a different function from that of a wattmeter. Permanently installed watt-hour meters are generally used for purposes other than those common to portable wattmeters. One well-known permanent watt-hour meter is the meter commonly installed outside a facility that measures the electrical energy delivered by a public utility. The purpose of a utility's watt-hour meter is to measure a customer's net energy consumption within a billing period so that a bill may be sent at a later date.

Aside from a utility's watt-hour meter, permanent power and watt-hour meters may also be installed within a facility by the owner for a variety of purposes. These types are called "submeters." Internal accounting could be one purpose for submeters. Another reason could be to allow a facility to bill its subcustomers. For example, marinas often have permanently installed submeters to bill the owners of boats that might dock at the slips at a marina.

Throughout the twentieth century, most watt-hour meters were built with a mixture of coils and mechanical parts combined in ingenious manners to measure energy usage over time. These meters were capable of a surprising accuracy that was typically better than 1%. To this day, a large number of installed wattmeters still are of this electromechanical design. More recently, wattmeters available from manufacturers are based on a digital design. Besides measuring energy consumption, many of these new designs offer a wide range of novel and useful capabilities. Well into the future, these new types of meters will become the preferred means whereby utilities can offer to their customers a menu of options. These new capabilities promise to benefit both utilities and their customers.

## 7.2 Measuring Power in Three-Phase, Three-Wire Circuits

A customer of a three-phase electrical service, three-wire or four-wire, will have a permanently installed meter that measures at least energy (watt-hour) consumption. The meter will be owned by the utility and provided for billing purposes. Depending on the area and the utility's policies, the meter may also measure other parameters for billing purposes. Larger users of electrical power are also often billed for reactive power factor and "demand."

Temporary measurements of a three-phase three-wire circuit may be conducted for any one of several possible reasons by any one of several methods. The preferred method is by means of a three-phase power meter, a power analyzer, or a power quality meter. Although more cumbersome, three-phase power may also be measured with a single-phase power meter or, depending on the application, several single-phase power meters. This option might be selected if, for example, a three-phase meter is not conveniently available. The selection of meters and the method chosen might depend on a variety conditions, including

1. Types and number of meters on hand
2. Balanced or unbalanced load
3. Rate of change of load
4. Frequency at which readings are to be made
5. Required accuracy

If only one single-phase wattmeter is on hand, there is only an occasional need for wattage measurements, and the loads change slowly; one single-phase wattmeter can usually provide adequate results. At the other end of the spectrum where there is a frequent need for measurements and the loads change rapidly, a three-phase meter, a power quality meter, or a power analyzer would be a better choice.

If one understands how single-phase meters can be used to measure power, it then becomes much easier to understand how a three-phase power meter functions. Power can also be determined by any means that permits measuring current, voltage, and current lead/lag.

### 7.2.1 Using One Single-Phase Wattmeter to Measure Power in a Three-Phase, Three-Wire Circuit

A possible procedure to measure the total circuit power of a balanced three-phase three-wire circuit would use the measurement of the power of only one phase. Since by definition all three phases have equal power consumption, the total power consumption is triple the power measurement of any one phase. This procedure would, of course, require access to the conductors that carry the phase currents as well as the conductors that deliver voltage potentials to the phases.

A procedure for measuring the phase power of a balanced delta circuit is shown in Figure 7.2. The phase potentials of a delta circuit are usually accessible, but the conductors carrying the phase currents may not be conveniently accessible. Usually, a different method would be preferred.

A procedure for measuring the power of one phase of a balanced wye circuit is shown in Figure 7.3. However, in many instances, the measurements depicted in Figure 7.3 may not be physically practical. Generally, the current of a wye circuit may be measured since the phase current is the same as the line current. However, access to the point where the phases tie together (point "d" in Figure 7.3) may not be readily accessible for a measurement. This would be the case, for example, if the circuit is a wye wound motor. Generally, connection of a potential lead to point "d" would not be required for a balanced circuit. This would be true of a wye wound motor since the currents of the three phases would be equal or nearly equal. Accordingly, the voltage can be calculated by the known relationship

$$V_P = \left[ 1/\left( \sqrt{3} \right) \right] V_L$$

Under some circumstances, a single portable (single-phase) wattmeter can be used to obtain adequate readings in an unbalanced three-phase three-wire circuit. As mentioned earlier, this arrangement would be practical only

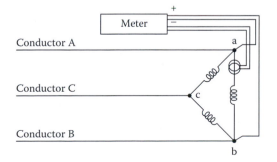

**FIGURE 7.2**
Measuring power of a delta circuit.

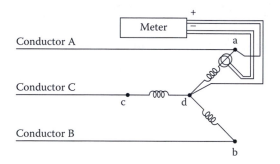

**FIGURE 7.3**
Measuring power of a wye circuit.

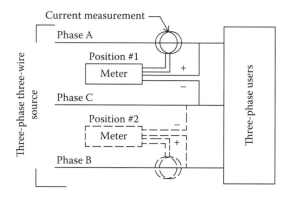

**FIGURE 7.4**
Use of a single wattmeter.

if the measured parameters remain constant throughout the period of time during which the measurements are made. The arrangement of Figure 7.4 shows a typical configuration that could be used to determine total circuit power with a single meter. To obtain a reading, the meter is first connected as shown in position #1. The current in conductor A is measured as well as voltage C–A. A reading is taken, and the meter is then moved to position #2. In position #2, the current in conductor C is measured as well as voltage C–B. The two readings are added to determine total circuit power. Alternate positions for measuring total power would involve the use of different conductors as explained later.

### 7.2.2 Using Two Single-Phase Wattmeters to Measure Power in a Three-Phase, Three-Wire Circuit

An effective and practical means of measuring power in unbalanced three-phase three-wire circuits involves the use of only two wattmeters in what is commonly called the "two-wattmeter method." Total power is determined

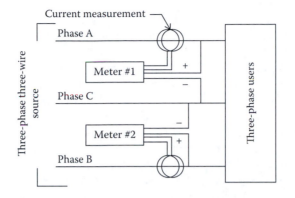

**FIGURE 7.5**
Use of two wattmeters.

at any selected instant by the sum of the readings indicated on the two watt-meters. The wattmeters determine power by measuring line parameters, that is, line potentials and currents. The two wattmeters are to be located in the three-phase lines that lead to the load as depicted in Figure 7.5. Only two currents are measured, and the voltages are measured as shown in the figure. In Figure 7.5, the two wattmeters are designated as meter #1 and meter #2. In one possible configuration, meter #1 measures the current in conductor A as well as voltage C–A. The positive voltage lead of meter #1 is connected to phase A, and the negative voltage lead is connected to phase C. Meter #2 measures current in conductor B as well as voltage C–B. The positive voltage lead is connected to phase B, and the negative lead is connected to phase C. An alternate 2 m configuration is shown in Figure 7.6, where the current coils are shown in phases A and C. In fact, the current measure-ments can be in any two of the three phases: A–B, A–C, or B–C. The posi-tive potential leads are in contact with the respective phase conductor and the negative leads joined together and in contact with any voltage. Proof of

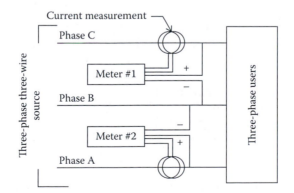

**FIGURE 7.6**
Alternate use of two wattmeters.

the two wattmeter method for three-phase three-wire circuits is presented in Section A.4 along with proof of the three wattmeter method, which is discussed later.

While an acceptable method of measuring total power of a three-phase circuit, the two wattmeter method has certain limitation. If the (line) current measured by meter #1 (of Figure 7.5 or 7.6) leads the reference potential of the meter by less than 90°, meter #1 will correctly indicate the value of measured power. (A line current that leads potential by 90° would correspond to a power factor in a balanced circuit of 0.50.) However, when line current leads the measured potential by more than 90°, problems may ensue depending on the type of meter used. If the wattmeter is of the electromechanical type, that is, one that uses internal coils to determine power, beyond 90°, there would be no indication of power level. In this case, a reading may be obtained by reversing potential leads and subtracting the reading of meter #1 from the reading of meter #2 since the reading of meter #1 would be a negative value. Some digital-type wattmeters would indicate a negative reading in which case there would be no need to reverse leads. With the two wattmeter method of measuring circuit power, it is possible to have some 20 or more possible incorrect connections. Yet, there is only one set of connections that will provide correct readings. For this reason, correct positioning of all CT connections and potential connections are needed to ensure valid readings.

### 7.2.3 Using Three Single-Phase Wattmeters to Measure Power in a Three-Phase, Three-Wire Circuit

It is possible, and at times desirable, to measure all three line currents of a three-phase three-wire circuit along with the three sets of line potentials. This might be the case, for example, if the three line currents are to be monitored in addition to total circuit power.

In the case of an unbalanced delta circuit, total power consumption can be determined by measuring the power of each phase. This could be done, typically, with the configuration represented in Figure 7.7. The total power consumption of the circuit would be the sum of the readings of the three wattmeters. As indicated in Figure 7.7, the current transformers are located in each of the three phases. However, in many instances, the arrangement of Figure 7.7 is not practical since it is often not convenient to read the currents of the phases of a delta circuit.

In the case of a three-phase three-wire wye circuit, three single-phase wattmeters could be used in some cases to measure total power in the configuration of Figure 7.8. The total circuit power is the sum of the phase powers. If the circuit is, say, configured in a wye configuration, as would be the case with a wye wound motor, it would generally not be possible to access the "d" point represented in Figure 7.8.

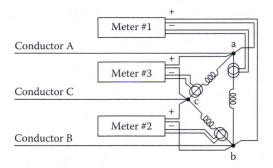

**FIGURE 7.7**
Measuring power of a delta circuit.

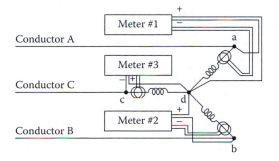

**FIGURE 7.8**
Measuring power of a wye circuit.

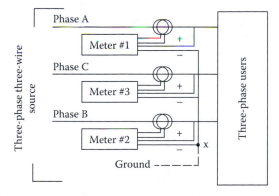

**FIGURE 7.9**
3Φ3W measurement.

An alternate method of measuring circuit power involves the use of three separate single-phase wattmeters to measure line parameters as represented in Figure 7.9. The users could be delta, wye, or any mix thereof. The positive potential leads of each of the meters is in contact with the respective phase in which the CT is located. The negative leads of the instruments are joined together. The negative leads can be connected to ground although a connection to ground is not necessary.

At first, it may seem strange to a reader that with the use of three wattmeters as depicted in Figure 7.9, all three negative leads of the instruments are joined together. This subject is discussed further in Section A.4, where it is explained that, in fact, this is the correct procedure. Regardless of the voltage (to ground) of the negative leads, the sum of the measurements of the wattmeters is the total circuit power. In addition, if the three-phase three-wire circuit is derived from a three-phase four-wire circuit, and the negative leads of the three wattmeters are connected to ground potential, the measurements of each wattmeter will indicate the respective power delivered by that phase, that is, phase A, phase B, and phase C.

The configuration represented in Figure 7.9 is an effective method of measuring the total power of a three-phase three-wire circuit. If only total circuit power is needed, the use of three wattmeters as depicted in Figure 7.9 offers no advantage over the two-wattmeter method treated in Section 7.2.2. In fact, the three-wattmeter method requires three current measurements instead of the two of the two-wattmeter method. On the other hand, the use of three CTs allows monitoring of the line currents as well. The use of a single three-phase wattmeter, rather than the use of three single-phase wattmeters, would generally be a more practical means to measure circuit power of a three-phase three-wire circuit.

Although the power of a balanced three-phase three-wire circuit can be measured with a single current measurement, unbalanced circuits require a minimum of two current measurements to establish a value of total circuit power. If each of the three phases is to be monitored for currents as well as power, then three current measurements are required. For this reason, wattmeters are available for use with one, two, or three CTs. A possible configuration using a single instrument is depicted in Figure 7.10.

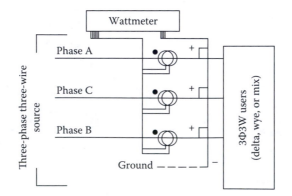

**FIGURE 7.10**
3Φ3W wattmeter.

**TABLE 7.1**

Power and Current Values Available: Three-Wire Circuit

| Type Load | Total Power | Line Currents | Number of CTs Required |
|-----------|:-----------:|:-------------:|:----------------------:|
| Balanced | ● | ● | 1 |
| Unbalanced | ● | | 2 |
| Unbalanced | ● | ● | 3 |

A summary of the possible configurations, the number of CTs required and the measurements of total power and line currents that can be provided is tabulated in Table 7.1. When making power measurements with a three-phase meter, valid readings can be obtained only if the CTs are positioned properly, and the potential leads are connected properly. Most CTs are marked so that they may be positioned in the correct direction. A common means of marking CTs is with a black dot or black square on the side of the CT that is to face the electrical source. Current transformers are discussed in detail in Section 2.4.2.

The algorithm used within a wattmeter to compute the various parameters is dependent on the number of CTs used to measure currents. The algorithms applicable to the various configurations for three-phase three-wire circuits are treated in Chapter 5.

### 7.2.4 Using a Three-Phase Wattmeter to Measure Power in a Three-Wire, Three-Phase Circuit

Earlier it was demonstrated how single-phase wattmeters may be used to measure the power of a three-wire three-phase circuit. If a person understands the principles involved in applying single-phase wattmeters to measure power, it then becomes much easier to understand the principles upon which a three-phase wattmeter operates. In fact, a three-phase wattmeter would almost always be the preferred method to measure three-phase power. The current transformers used with three-phase meters can be relatively expensive. For this reason, it is generally advisable to use no more CTs than what is required. As indicated in Table 7.1, a single CT will suffice if the load is balanced, and two will suffice for all other three-wire applications if there is no need to monitor currents in the three phases.

Digital, three-phase power meters function by making separate, single-phase measurements of currents and potentials. Depending on the application and the capability of the instrument, the single-phase measurements are then combined by algorithms contained within the envelope of the instrument to produce a variety of values. The available parameters may include phase currents, phase voltages, total power, phase powers, VAR, reactive power, and apparent power.

## 7.3  Measuring Power in a Three-Phase, Four-Wire Circuit

One single-phase wattmeter can be used to measure total power in a balanced three-phase four-wire circuit or in an unbalanced three-phase four-wire circuit if the loads are steady. The two-wattmeter method described in Section 7.2.2 for the three-phase three-wire balanced or unbalanced circuits cannot be used to measure power in a three-phase four-wire unbalanced circuit. For unbalanced three-phase four-wire circuits, a minimum of three single-phase meters are required. As with three-phase three-wire circuits, it is generally advisable to use a three-phase meter for measuring power in a three-phase four-wire circuit.

### 7.3.1  Using One Single-Phase Wattmeter to Measure Power in a Three-Phase, Four-Wire Circuit

In order to obtain an accurate measurement of total power using only one meter on a three-phase four-wire circuit, the measured parameters must remain relatively constant throughout the period during which the measurements are made. Otherwise, accurate readings would not be possible.

The total circuit power is the sum of the phase powers. It is noted that if the user is, say, configured in a wye configuration, as would be the case with a wye wound motor, it would generally not be possible to access the "d" point represented in Figure 7.3. So, in practice, there are many applications that could not use three wattmeters as represented in Figure 7.3. The more common usage of three wattmeters in a wye circuit would be in the configuration of Figure 7.4.

### 7.3.2  Using Three Single-Phase Wattmeters to Measure Power in a Three-Phase, Four-Wire Circuit

In Section 7.2.3, it was demonstrated that the power of a three-phase three-wire circuit may be determined by measuring the power of each of the phases, whether the circuit is a delta circuit or a wye circuit. Of course, the same remains true for a three-phase four-wire circuit. As with a three-phase three-wire circuit, the power of a three-phase four-wire circuit may be measured by measuring line parameters. Actually, that is generally the better way to do it.

Three single-phase wattmeters can be used to accurately measure the total power of a three-phase four-wire circuit. The CTs and potential measurements would typically be made as shown in Figure 7.11.

### 7.3.3  Using a Three-Phase Wattmeter to Measure Power in a Three-Phase, Four-Wire Circuit

If a three-phase four-wire circuit is balanced, then total power may be determined from a single current measurement. In reality, most three-phase four-wire circuits are unbalanced. For this reason, three current readings are

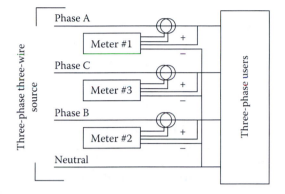

**FIGURE 7.11**
3Φ4W measurement.

normally required to measure total power. With a single three-phase meter to measure total power, the meter would be connected as represented in Figure 7.12.

A summary of the possible configurations for measuring power of a three-phase four-wire circuit is shown in Table 7.2. The number of necessary CTs is likewise shown in the table along with indication of possible current measurements.

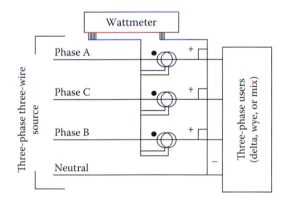

**FIGURE 7.12**
3Φ4W wattmeter.

**TABLE 7.2**

Power and Current Values Available: Four-Wire Circuit

| Type Load | Total Power | Line Currents | Number of CTs Required |
|-----------|:-----------:|:-------------:|:----------------------:|
| Balanced | ● | ● | 1 |
| Unbalanced | ● | ● | 3 |

## 7.4 Power Analyzers and Power Quality Meters

For many years, instruments described as "power analyzers" have been the traditional means of examining the characteristics of a three-phase circuit. Power analyzers come in a wide variety of shapes, sizes, prices, and with a wide variety of features. In general, power analyzers can provide numerous details characteristic of a circuit including parameters as currents, voltages, power factors, power, VARS, VA, frequency, phase sequence, lead/lag angles, and nonlinear current characteristics. Some power analyzers can display a circuit's phasor diagrams. Additional information measured by a power analyzer is generally available on a display on the instrument. The displays are especially helpful in finding circuit problems as harmonic currents and harmonic distortion of an electrical supply. Some power analyzers can transmit data serially over a cable to remote locations. Many models allow a printout of measured parameters. Although more costly than conventional wattmeters, in many instances, the greater cost of a power analyzer is justified by the greater amount of information that it provides. More or less by custom the term power analyzer is applicable to either portable handheld or bench-mounted instruments that are also portable but less so. Most power analyzers measure currents by means of clamp-on CTs that avoid the need to disconnect and then to reconnect conductors. In general, power analyzers are relatively expensive instruments.

In recent years, digital "power quality" meters have become available for the continuous monitoring of three-phase power sources. Many of the new generations of digital power quality meters or submeters are intended for permanent mounting in a panel. Unlike traditional, portable power analyzers, current measurements for the panel-mounted power quality meters are made with permanently installed toroidal CTs. The newer power quality meters can provide data comparable to that obtained with power analyzers. And the digital power quality meters are generally much less expensive than the traditional power analyzers. A power quality analyzer typical of the new generation of meters is shown in Figure 7.13. The instrument of Figure 7.13 can simultaneously display the traces of five voltages and two currents.

## 7.5 Watt-Hour Meters

Most watt-hour meters are used to measure electrical energy supplied by a utility to a user. Traditionally, watt-hour meters are mounted outside customers' buildings where a meter reader from the utility may visit the meters and take periodic readings. Meters of the past years were generally mounted in a conspicuous location that could easily be found by the meter reader and

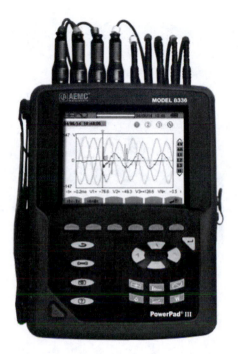

**FIGURE 7.13**
Power quality analyzer. AEMC Instruments Model 8336. (Photo courtesy of AEMC Instruments.)

where the theft of electricity might be more easily detected. Meters must be weatherproof and have a cover that has a utility seal that, if broken, might suggest tampering. A traditional, electromechanical watt-hour meter of the type used by utilities for many years is shown in Figure 7.14. To determine a customer's energy usage since the last reading, a meter reader would note the positions of the dials and manually record the energy usage. The utility would subsequently send a billing to the customer for energy usage since the last billing.

Watt-hour meters have been in use practically since the availability of electricity. For at least the first 100 years of their use, watt-hour meters have been of an electromechanical design of one sort or another. The most popular design has been based on the use of what is essentially a miniature motor that drives a disc. The rate of rotation of the disc is proportional to the power being used. Rotations of the disc are proportional to watt-hours of energy consumed, and the disc in turn drives a register that logs watt-hours (or, more exactly, kilowatt-hours) consumed. In the past, the registers of the type shown in Figure 7.14 had to be read and recorded by a meter reader who would travel from one customer to another. The process proved to be time consuming and prone to error. The meters of today are offering improved alternatives to the older types of meters.

**FIGURE 7.14**
Conventional watt-hour meter. (Photograph provided courtesy of Fry Electric Inc., Indiana-
polis, IN.)

Watt-hour meters are subject to governmental regulations that specify minimal accuracy properties of watt-hour meters. In the United States, each state determines its unique accuracy requirements. Many states require that new meters meet the accuracy requirements of ANSI Standard 12.1. Some states have very stringent accuracy requirements, and most states require an accuracy much better than 1%. Some states also require periodic in-service testing or sampling program to confirm that the type of meter used by a utility maintains sufficient accuracy after years of use.

Because many utilities have developing needs for innovative ways of billing customers, the traditional and mechanical watt-hour meters of the past have been found lacking. Smart meters, which are of a digital design, are gradually replacing the older conventional electromechanical designs in many parts of the world.

## 7.6 Demand Meters

As explained in Chapter 6, a demand charge is typically a large part of the bill sent by utilities to larger commercial and industrial customers with a three-phase service. Traditional demand meters of the type used in the past years have been of an electromechanical design and were designed to measure either a customer's peak usage of current or power (kW—not kWh). The choice of current or power of course would be determined by the respective utility as every utility has the right to formulate its method of billing.

**FIGURE 7.15**
Watt-hour meter with demand. (Photo courtesy of the Archman Corporation, Throop, PA.)

A typical electromechanical meter with both a kilowatt-hour reading and a demand reading is shown in Figure 7.15. The selected parameter, current or power, is measured throughout a predetermined period of time. A duration of 15 min has been the most popular time period, although some utilities use 30 min. A meter with a 15 min sampling period would actually take over 2800 readings within an average billing period.

Despite its title, a demand meter does not actually measure maximum power or current "demand." Rather, the value indicated by the pointer on the meter is in fact an integrated value. For example, if the energy usage during the first period of measurement (for a new billing period) was constant at, say, 20 kW, the red pointer would show 20 kW at the end of the period. Of course, subsequent measurements might move the red pointer higher. If the power level was at 40 kW for half the measurement period but 0 kW during the other half, the red pointer would still indicate 20 kW. Although the true peak usage was 40 kW within the measurement period, the integrated value during the time period would be recorded as only 20 kW.

Many of the traditional electromechanical demand meters are still in use, but many utilities are gradually replacing these older, electromechanical meters with the new, more capable smart meters that are digital based. Nevertheless, many utilities continue to charge customers for their maximum "demand" in the billing period.

## 7.7 Smart Meters

### 7.7.1 Definitions

One of the forerunners of the smart meters was the AMI metering system. The term advanced metering infrastructure (AMI) first appeared in the 1990s, and it was used in reference to a series of meters that contained what at the time were new capabilities. Mostly, the AMI meters would electronically record and store electric consumption by hour of the day, which data could be retrieved at a later date. Those early AMI meters did not have real-time, two-way communications capabilities. The term advanced metering infrastructure was still used (primarily by the U.S. Federal Government) in reference to what many began to call smart meters. Yet the later versions of the AMI meters began to incorporate features not found in the earlier AMI meters.

Another family of innovative meters that offered a number of features were the automatic reading meters (ARMs). The ARMs were available in a variety of models and with a range of features. Some allowed a meter reader to record data at a meter, but the instruments did not require a meter reader to first interpret the displayed numbers and then record the noted numbers. Rather, the data were downloaded to a handheld data recorder. Some advanced models did in fact offer radio transmission of data to a central gathering server.

The term smart meter describes a more recent family of meters that have two-way communications capability. Some utilities have purposely avoided use of the term smart meter. Public Service of Oklahoma, for example, uses the term advanced digital meter. Philadelphia Electric Company used the term smart meter for a time but eventually reverted to advanced metering infrastructure when referring to the meters. A typical three-phase smart meter is shown in Figure 7.16. Unlike the electromechanical registers used in the older meters, the smart meters have LED-type readouts.

Following is a typical definition provided by a utility.

**By the New York Public Service Commission (Reference 7.1):**

> Smart Metering—a concept embracing two distinct elements:
> Meters that use new technology to capture complex energy use information and communication systems that can capture and transmit energy use information as it happens, or almost as it happens.

The New York Public Services Commission further explains that the "Reasons for Using Smart Meters" are to

- Identify and implement operational strategies to control load factor and peak-load requirements and reduce energy waste
- Expand the capacity to manage operations in response to potential price volatility

**FIGURE 7.16**
Three-phase smart meter. (Photo courtesy of Itron, Liberty Lake, WA.)

- Understand and improve consumption patterns to secure better pricing from the retail electricity market
- Participate in demand response programs that pay end users to manage loads as needed to improve grid reliability during peak demand periods
- Measure and verify anticipated energy savings from energy-efficiency modifications, and
- Help monitor and address complex issues such as power quality

## 7.7.2 Operation

The term smart meter is commonly applied to not only electric meters but also to water meters and gas meters. The term smart meter as used here is used exclusively in reference to electric meters.

Smart meters are electronic digital instruments with a wider range of capabilities than the electromechanical designs of the past. Smart meters monitor electricity usage at a facility, and that data are stored internally in the meter. Periodically, a smart meter sends recorded data by radio signal to a receiver that is located nearby. A single receiver may gather data from dozens or hundreds of smart meters. Received data in turn are subsequently retransmitted to one of the utility's servers where it will be retrieved later for billing purposes.

The terms smart meter and smart grid are distinctly different concepts. Each refers to different purposes and different functions. Nevertheless, the two terms are often used in conjunction with one another. Smart grids are fitted with a variety of sensing devices that are intended to provide real-time information descriptive of the condition of an electrical distribution system. Smart meters have a different function. Primarily, the smart meters offer utilities information and a tool that can be used to shift some peak loads to off-peak times, thereby allowing better utilization of the utilities' generating and distributing system.

Smart meters often go by different names. Some utilities prefer to call the meters "AMI," "advanced metering infrastructure," "smart meter initiative," or merely "advanced meters." It appears that the variances in the names are a result of early, adverse press given to some "smart meters." That bad press in turn was largely traceable to two issues raised by consumers. The most frequently raised objection to smart meters was due to the fact that the meters periodically emit a radio signal for the purpose of transmitting recorded data. Some people feared the adverse health effects of being near transmitted radio signals. Another expressed concern related to the fact that the smart meters monitor customers' power usage and, therefore, customers' living habits. Some argue that for that reason the meters constitute an invasion of privacy.

Although smart meters present a means whereby a utility can induce customers to shift usage to off-peak times, the meters alone cannot accomplish this end. The reductions can take place only if a utility combines a program of incentives in combination with the measurement capabilities of the smart meters. The most common method used by utilities to reduce peak load is to offer reduced rates for electric consumption during off-peak times. This feature of the smart meters simulates to some extent the earlier TOU (time-of-use) programs offered by utilities. The TOU programs were used by utilities to provide reduced energy rates during off-peak periods.

Aside from the TOU initiatives, utilities have since devised a variety of programs with the intention of shifting energy usage away from peak times. It can be expected that some utilities' customers will take advantage of these programs, whereas others will find it too inconvenient or impractical to reschedule many if not most electrical loads.

In the event of an outage, an installed system of smart meters will immediately alert the respective utility to the condition. In addition, the smart meters will allow a utility to quickly determine the extent of a problem. As a result of the improved real-time intelligence, a utility will be better prepared to organize and more quickly dispatch line crews to restore electrical service.

### 7.7.3 Implementation Examples

Programs in at least 25 U.S. states called for deployment of some magnitude of smart meters. Some of these programs were extensive, whereas others could best be classified as trials. The following are typical examples of smart meter programs.

### 7.7.3.1 Ontario Power Authority, Canada

In conjunction with the installation of smart meters, the Ontario Power Authority introduced a policy of "time of use" for home owners and small business. Customers shifting electricity use to off-peak can bring a 46% rate reduction in energy costs for use during those periods. A reduction of 16% is offered for use during mid-peak times. The Authority also offers incentives to commercial and industrial customers for reducing use during peak periods. The installation of the smart meters facilitated implementation of the peak reduction program.

### 7.7.3.2 State of Pennsylvania

In 2008, the Pennsylvania State passed a bill that mandated installation of smart meters throughout the state by the year 2018. Armed with the new meters, the various utilities within the Commonwealth will be positioned to devise and offer a variety of programs all of which, it is anticipated, will benefit both the utilities and their millions of customers.

## 7.8 Submeters

Submeters are commonly used at marinas, condominiums, mobile home parks, RV parks, apartments, home owners associations, and the like. In a typical installation, the local utility would have a single meter at a facility that records net electrical consumption to the facility. A single party would be responsible for paying the utility for the electricity recorded by that single meter. In turn, submeters would measure the electrical usage of the various tenants or parties using electricity recorded by that single meter. The party that pays the single utility bill each month would be responsible for the submeters as well as collecting payments from the variety of electricity users. Some states have rules that determine which customers will be allowed to resell electricity.

Submeters might be used, for example, to track usage by internal departments or division within an industrial facility. A charge for electricity usage can then be levied against that user for accounting or internal billing purposes. Submetering can serve as an aid to conservation efforts. History has proved that if electrical consumption is charged to an identifiable individual or department, conservation efforts almost always follow.

Submeters are similar to conventional watt-hour meters in function; both designs of meters must accurately record energy use. Many submeters are not mounted outdoors and therefore do not have weatherproof cases typical of utility meters. Also, submeters often need not be tamper-proof to the degree required of utility meters as the meters can be located in a secure area. Submeters are available in a large number of models that offer a variety of

**FIGURE 7.17**
Submeter. (Photo courtesy of Simpson Electric, Lac Du Flambeau, WI.)

features and options. Designs are available for panel mounting, surface mounting or DIN rail mounting. Today, new submeters are invariably the digital type with remote CTs for measuring current. Accordingly, a large number of submeters can be grouped together to facilitate the convenient reading of the respective registers. A typical, modern, digital submeter is shown in Figure 7.17.

In some regions, the accuracies of submeters are regulated much as the accuracies of watt-hour meters. The State of California has rigid requirements for submeter accuracies. The State of Maryland publishes a list of approved submeters (Reference 7.2). Meters not on the State's list are disallowed.

## 7.9 Phase Sequence Meters

A phase sequence meter is a handheld instrument that allows a person to determine the phase sequence of a three-phase electrical source. A correct determination of the sequence is frequently important for a number of reasons. For example, the correct sequence is needed to properly connect a three-phase electrical source to a power distribution center or a motor control center. Both of these types of electrical gear are normally marked for the incoming electrical conductors to be connected in the sequence A to T1, B to T2, and C to T3. If this sequence is followed throughout to branch circuits,

**FIGURE 7.18**
Phase sequence meter. (Photo courtesy of Extech Instruments, Waltham, MA.)

motors connected to those branch circuits will rotate in the anticipated direction and in accordance with the direction of rotation shown on the motor. On the other hand, if the sequence of connection was A to T1, C to T2, and B to T3, the rotation of motors will be reversed and incorrect. Incorrect rotation of a motor can result in severe damage to property as well as injury to personnel.

It is true that color coding of cables will often indicate the sequence of phase rotation. Frequently in the United States the sequence "brown-orange (or violet)-yellow" is followed for three-phase circuit 480-3-60 conductors. Color coding is also often followed for lower voltages. However, color coding around the world vary. Above 480 VAC, black conductors are usually used and not color coded. Color-coded conductors, where found, may not accurately reflect the true phase sequence. When there is doubt, the phase sequence should be confirmed.

A typical phase sequence meter is shown in Figure 7.18. The photograph is of the Extech Instruments Model PRT200, which is a noncontact type of instrument. Phase sequence is determined with the meter merely by clamping the probes over the insulation of the conductors being analyzed.

## 7.10 Earth Resistance

As explained in Chapter 3, the grounding of electrical systems and gear is critically important for several reasons. Errors can occur in grounding systems. Sometimes a connection can be faulty from the day of installation

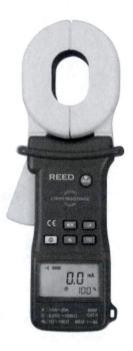

**FIGURE 7.19**
Ground meter. (Photo courtesy of Reed Instruments.)

and sometimes a connection that was initially adequate might become inadequate over time. Poor soil conditions are often the cause of an inadequate ground. The best means to confirm that a ground is viable is with a ground meter. After an initial installation, grounds should be tested periodically to confirm continuity. A suitable type of clamp-on ground meter is shown in Figure 7.19. To take a ground reading, the meter is merely clamped on the grounding connection and a reading taken. Both NEC and OSHA prescribe minimal values for ground resistance. If the resistance to ground is found to be unacceptably high, measures will generally be required to ensure a low resistance path is present. In the extreme, it may become necessary to excavate unsuitable soil and replace it with a more conductive soil.

## 7.11 Megohmmeter

After the installation of an electrical system, and before energization of the system, a megohmmeter is often used to confirm integrity of the insulation. Energization of an electrical system with inadequate or deteriorated

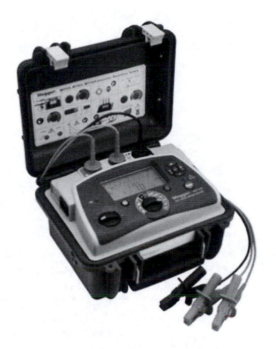

**FIGURE 7.20**
Megger instrument. (Photo courtesy of the Megger Corporation, Dallas, TX.)

insulation is inadvisable. Testing of a system's insulation is also suggested if electrical gear has been out of service for an extended period of time. This is particularly true of electrical equipment located in a high humidity environment. Testing of the insulation is especially important for high-voltage systems. A megohmmeter measures the impedance of insulation by the application of a voltage that might be in the range of 500–1000 V. An insulation impedance of 100 megohm would generally be considered as satisfactory. Lower readings would be questionable. Megohmmeters are often called "meggers." A typical megohmmeter is shown in Figure 7.20.

## 7.12 High-Potential Tester

A high-potential test is most commonly called the "hi pot test." It is also known as the "dielectric withstand test." The hi pot test is performed much for the same reason as a megohm test that is to verify the integrity of an insulation system. A hi pot test is performed by the application of a relatively high voltage and the resulting leakage current is noted. Some instruments are intended solely for the purpose of performing a hi pot test, although some can perform either a hi pot test or a megohm test. Under some circumstances, a hi pot test can result in the destruction of a system's insulation. So, care must be taken in the application of a hi pot test. The test should be conducted only by trained personnel using appropriate protective gear and in compliance with the manufacturer's guidelines.

## 7.13 Multimeters

A review of electrical instruments would not be complete without mention of the multimeter. A multimeter is an instrument that has the capability of measuring more than one parameter. Thus, the "multi" in multimeter. Most multimeters have the capability of measuring, at a minimum, dc voltage, (rms) ac voltage, and resistance. Many are also capable of measuring a number of other parameters as temperature, peak voltage, small dc currents, small ac current, ac frequency, capacitance, and reluctance. Some multimeters are of the handheld design, whereas others are intended for positioning on a bench. Years ago multimeters all had analog needles that pointed to the value of the measured parameter. Today, most multimeters, but not all, have digital readouts. For some types of measurements, as when the

**FIGURE 7.21**
Multimeter. (Photo courtesy of Triplett, Manchester, UK.)

measured variable is constantly changing value, the analog meters are still preferred. For power applications, the input impedance of the meter generally is immaterial. For electronic measurements, however, a very high-input impedance is needed. Otherwise, the current draw of the meter will cause a false reading. Probably the most common use of a multimeter by electricians is on troubleshooting missions. Practically every electrician will have a handheld multimeter at the ready. A typical handheld multimeter of the type an electrician might prefer for power circuits is shown in Figure 7.21.

## Problems

1. Would it be correct to say that when using a single-phase wattmeter, positioning of the potential leads is insignificant?

2. If the common unit for electrical power is the SI "watt" and the meter for measuring power is the "wattmeter," what is the industry unit for electrical energy and the type of meter used to measure electrical energy?

3. An electrician at a chemical plant was given two single-phase wattmeters and asked to measure the power of a 480-3-60 fan motor. At the location of the fan motor, the electrician was able to determine which conductors were the phase A, phase B, and phase C. He placed the CT of the first meter on the phase A conductor, the meter's positive (+) voltage lead to the phase A conductor, and the meter's negative (−) voltage lead on the phase C conductor. The CT of the second meter was placed on the phase B conductor. Where should the voltage leads of the second meter be connected?

4. State the algorithm that is used by a single-phase wattmeter to measure power. (Suggestion: Merely state the equation number of the applicable algorithm.)

5. What accuracy would be characteristic of the traditional electromechanical watt-hour meters? 0.1%, 1%, or 5%?

6. A designer of meters was given the assignment of designing a wattmeter to measure power in conductors going to three-phase loads. He was told the meter should be suitable for use with three-wire or four-wire circuits and loads that could be delta, wye, or a mix of any of these circuits. Identify the equation of the suitable algorithm. (Suggestion: Merely state the equation number of the applicable algorithm.)

7. How many CTs are required to measure power on a three-phase four-wire unbalanced circuit?

8. A demand meter installed at an industrial plant uses a 15 min sample period to determine the customer's demand charge. In one particular 15 min period, there was a 3 min period when the current was constant at 1200 A. For the next 5 min, the current was constant at 1800 A. During the remaining period of measurement, the current was 1300 A. If the 15 min period becomes the highest demand for the billing period, what will be the basis for the demand charge during that billing period?

9. A farmer wishes to monitor the energy usage of his 24 three-phase three-wire irrigation pumps. Because of the costs of CTs, he would prefer to purchase submeters that require the least number of CTs. Should he purchase meters suitable with one, two, or three CTs?

10. Would TOU be a feature available with a smart meter?

11. The manager of a large industrial plant decided that for accounting purposes, he was going to charge each department for its use of electrical energy. What type of meter would be best for the application?

12. A sawmill added a new, large three-phase-driven rotary saw along with a new circuit to serve the motor. What kind of meter would be used at startup to confirm that when power is applied to the motor, it will rotate in the desired direction?

13. Is a high-impedance multimeter necessary for troubleshooting three-phase power circuits? Explain.

14. A chemical plant had a 300 HP motor that it had not used for over 4 years, but there was interest in using the motor again in the near future. However, there was concern that the insulation might have absorbed moisture. What kind of meter should be used to confirm that the motor can be safely started?

15. A metal plating plant has a number of circuits that serve high-current thyristors that are used to generate dc voltage. There have been concerns that the thyristors might be causing distortion of electrical circuits that serve motors. What kind of meter would be used to inspect the quality of electrical circuits throughout the plant?

# 8

## Circuit Protection

### 8.1 General Requirements

The function of a circuit protective device is to protect a circuit from damage. Primarily, the devices act to avoid damage to insulation. Above all concerns, protective measures should be designed with the primary intent of preventing injury or death to personnel. Protective measures should also have as an objective the avoidance or minimization of damage to property.

A number of electrical devices are available for the purpose of protecting electrical circuits. The list of devices includes what are commonly called overcurrent protective devices (OCPD) that are used in a wide range of circuits. For large electrical gear, as transformers and generators, protective relays are used to detect a variety of potentially harmful conditions. (Protective relays are treated in Chapter 12.) The OCPDs are treated in detail in this section. According to the United States National Electric Code (NEC), there are two basic types of OCPDs, namely, circuit breakers and fuses.

A variety of codes and standards treat the subject of circuit OCPDs in great detail and provide appropriate guidelines. The United States National Fire Protection Association (NFPA) publishes a number of codes that deal with the safety of personnel and the protection of electrical assets. The most commonly used of the NFPA codes is NFPA Standard 70, the NEC, which prescribes the guidelines for electrical installations. Many of the chapters in the NEC treat electrical circuit protection. However, it is pertinent to note that not all installations within the United States are required to follow the guidelines of the NEC. Most large municipalities in the United States have passed regulations that require new installations follow the NEC but many small towns and urban areas have no such requirements. Power plants and industrial installations are exempted from following the requirements of the NEC. Many states within the United States impose codes and special procedures applicable to electrical installations, but many may not necessarily invoke the NEC. Many contracts for electrical installations require compliance with the NEC. Invoking the NEC has become a convenient method of setting forth the minimal requirements of an installation.

The NEC often makes reference to "the authority having jurisdiction." According to the NEC, if the authority having jurisdiction (AHJ) invokes the guidelines of the NEC, then the NEC is to be followed. If not, then there may be no requirement that the guidelines of the NEC are to be followed. Regardless, it is recommended that the guidelines of the NEC should generally be followed. In Canada, the applicable electrical code is the Canadian Electrical Code. Mexico uses the NEC as its primary guide and some South American countries likewise follow the NEC.

The requirements of the United States Occupational Safety and Health Administration (OSHA) impose a number of requirements on new installations and construction practices at worksites. NFPA 70E, standard for electrical safety in the workplace, is primarily concerned with practices intended to avoid injury to personnel involved with the installation or maintenance of electrical equipment. Assemblies that bear the Underwriters Laboratories (UL) mark follow the rules of UL, all of which are intended to certify that the assembly was fabricated in accordance with the guidelines of UL and, furthermore, in a manner to avoid electrical hazards. UL also publishes standards applicable to electrical components including circuit breakers and fuses that are generally followed by manufacturers of OCPDs.

The IEEE *Buff Book* (Reference 8.1), while not primarily concerned with the specifics of installation or the safety of personnel, is a well-recognized and authoritative textbook on the subject of circuit protection. The Buff Book consists of over 700 pages, and it treats the subject of circuit protection in great detail. The IEEE publication can be particularly helpful in resolving unique or special protection problems.

According to the IEEE *Buff Book*, the objectives of OCPDs are to

- Limit the extent and duration of service interruption whenever equipment failure, human error, or adverse natural events occur on any portion of the system
- Minimize damage to the system components involved in the failure

In order to meet the stated objectives, the *Buff Book* states, systems should include the following design features:

- Quick isolation of the affected portion of the system while maintaining normal operation elsewhere
- Reduction of the short circuit current to minimize damage to the system, its components, and the utilization equipment it supplies
- Provision of alternate circuits, automatic throwovers, and automatic reclosing devices

While the statements of the IEEE *Buff Book* are generally applicable to all electrical circuit protection devices, there are some unique considerations applicable to three-phase circuits. Regardless, it is true that an electrical circuit that is protected by the best available technology may not necessarily have the capability of reacting in a manner that would avoid injury or death in the event of shock to personnel.

Most circuits are to be isolated from a source of electricity in the event of a detected fault. However, this is not always the case. In some installations, there may be prevailing conditions that override the call for an immediate isolation of the circuit.

The two most common methods of protecting low voltage electrical circuits is by means of either circuit breakers or fuses. Each of these types of devices has unique characteristics and properties. In many instances, an electrical circuit may have both fuses and circuit breakers, one being upstream of the other or in a parallel circuit. Most low-voltage circuit breakers are furnished with an operating handle that also allows the circuit breaker to be used as a disconnect switch. The operating handle also permits convenient resetting after a fault condition has occurred. Large circuit breakers, which are not furnished with an operating handle, are operated by means of remote push-buttons that activate electromechanical devices within the circuit breaker that shift the CB's contacts from one position to the opposite position.

In some regards, fuses are not as convenient as CBs. To de-energize a circuit downstream of a fuse, a fuse must be removed from its holder. Removing and installing fuses in high-voltage circuits can present a dangerous condition to personnel who may not be adequately trained in the procedure. An overcurrent event opens a fuse and replacement of the fuse will be required to again energize the circuit. One blown fuse in a three-phase motor circuit can lead to undesirable single phasing of a motor. Briefly stated, circuit breakers and fuses each have unique features and drawbacks that usually cause one or the other to be a better fit to a specific application.

Protective devices of any type add expense to the cost of a circuit. A variety of protective devices available in the marketplace are capable of guarding against numerous types of postulated fault conditions. Some of these are described in the following paragraphs. A relatively inexpensive circuit that connects to relatively inexpensive gear may warrant only an inexpensive OCPD. On the other hand, a circuit that serves relatively expensive equipment, as a large and expensive motor or a large and expensive transformer, will warrant a greater expense and a larger number of protective devices. Personnel protection may become another cause for greater protection. Briefly stated, circuit protections should be commensurate to the application.

Several special terms are commonly used when referring to circuit overcurrent. Following are some of the commonly recognized terms:

- Overcurrent—current flow that is excessive for some reason. (An overcurrent condition could in consequence to a current above the design level, a short circuit or a ground fault.)
- Overload—current flow that is above predetermined levels. (The term is generally used in reference to a current flow that is somewhat

above desired levels—perhaps in the region of 1%–400% above an approved level.)

- Short circuit—an unintended connection allowing current flow: (1) from one phase to another, (2) from one phase to neutral, or (3) from one phase to ground (which would be the same as a ground fault). (Short circuit currents are drastically larger than overload currents—possibly in the range of 1,000—10,000 times larger.)
- Ground fault—a current flow from one phase to ground. (As with short circuit currents, ground fault currents can be much larger than an overload current.)

This chapter deals with the protection of three-phase electrical circuits but not necessarily motor protection circuits. The unique requirements of motor controls are treated in Section 9.2.

## 8.2 Ungrounded Systems

As explained in Chapter 3, most electrical systems are grounded, but there are some notable exceptions. In some applications, it is better to avoid isolation of a system in the event of a ground fault. There could be a number of reasons for having equipment ungrounded. One common reason would be to avoid interruption of operation of a system that would follow from a fault and the subsequent circuit isolation of an electrical system. Another reason could be to avoid propagation of a high-energy arc. In a grounded system, contact between live parts of the system and a ground will cause an overcurrent condition and an immediate isolation by action of an overcurrent protective device. In the case of an ungrounded system, contact between live part of the system and a ground has no immediate consequences and the electrical system continues to function. In an ungrounded system, an alarm, rather than a trip, is sometimes a better choice than circuit isolation.

## 8.3 Overcurrent

A prolonged overcurrent condition in an electrical circuit has the potential to cause significant damage to electrical equipment as well as other nearby equipment. The rise in temperature resulting from a relatively high current flow causes the greatest challenge in the design of electrical components of all types. Insulated cables are a typical example. An insulated copper wire of a certain diameter has a specific current carrying rating. At that specific rating, the temperature of the wire will rise to some extent because of current flow but no higher than a value that would prematurely deteriorate the insulation. So, it becomes important, for the purpose of ensuring an adequate life of an electrical product, that currents should not be allowed to rise above predetermined acceptable limits. The same general statement is true of motors, generators, transformers, and practically all electrical devices that have electrical insulation. A basic function then of an OCPD is to guard against high temperatures by preventing the high currents that cause excessively high temperatures. An overcurrent condition could be due to a current that is only slightly greater than an acceptable continuous current. Or, an overcurrent can be due to a fault condition that might be thousands of times greater than a circuit's continuous rating.

Aside from the potential damage that might result from high temperatures, there is also the prospect for mechanical damage due to the forces generated by short-circuit currents. The magnetic fields surrounding a short circuit current can be very strong, and in many, instances the current is capable of bending and twisting metallic components within its magnetic field. The internal components of transformers in particular are subject to mechanical stress and distortion that could result from a high level short circuit. Accordingly, a secondary function of an OCPD is to guard against forces that can cause mechanical damage.

## 8.4 Faults

A fault current is one that occurs inadvertently because of a short circuit and it is drastically greater than a circuit's continuous rating, perhaps by a factor of 1,000–10,000. The most common faults in electrical circuits are ground shorts. In a three-phase circuit, the most commonly considered faults are

- A ground fault from one phase to ground
- A ground fault of two phases to ground
- A short circuit from one phase to another phase
- A short circuit of all three phases, one to the other

These four possible configurations are represented in Figure 8.1. Any of these possible scenarios may result from one of several possible causes. The cause could be mechanical in nature with the result that a live conductor is brought in contact with either another live conductor or ground. An overvoltage

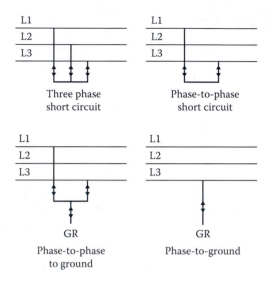

**FIGURE 8.1**
Possible short circuits.

condition may breakdown the circuit's insulation or insulation may deteriorate because of age. A fault might then follow. Of the mentioned types of faults, the phase to ground is by far the most commonly encountered form. A fault can also be classified as "bolted" or "arching." A bolted fault assumes that the resulting current is the same as would result from a conductor connected to either a ground or to another conductor by a bolted, low resistance connection. Bolted shorts can at times result in very high currents. An arcing short differs from a bolted short in that the current flow to either ground or another conductor is through air or another medium and the resistance to current flow is much higher than what would occur in a bolted short. In some regards, an arching short has the potential to cause considerably more damage than a bolted short. An arcing short might continue undetected and at a rate below a current rate needed to trip the circuit protective device. A continuing, arcing short can irreparably damage the insulation of all of the circuit that is common to the point of the fault. An arcing short can also start a fire or cause an explosion.

The computation of the maximum fault current is important so as to establish what the interrupting capability of the OCPDs should be. When an OCPD is selected, the interrupting capability should be one of the criteria. For example, a manufacturer may offer a 50 A breaker with a 10,000 A interrupting capability, 30,000 A capability, and 60,000 A capability. The higher the interrupting capability, the larger the physical size and the higher the associated cost. For this reason, it is generally desirable to use an OCPD that is adequate for the application but one that does not have a rating unnecessarily higher than what is needed. It is usually recommended that a short circuit study be conducted to establish the required interrupting capability of an OCPD. The subject of short circuit studies is treated in greater detail in Section 8.7.

## 8.5 Circuit Breakers

A circuit breaker is an electromechanical type of OCPD that is designed to protect an electrical circuit. In the event of a detected overcurrent condition, contacts within the circuit breaker are opened to deactivate the circuit protected by the circuit breaker. There are a large variety of circuit breakers intended for a multitude of applications and an array of configurations. Primarily, circuit breakers guard against both a circuit overload and a short circuit condition. For low-voltage applications, which are generally considered as below 1000 V, the commonly recognized types are the molded case circuit breakers (MCCB), the insulated case circuit breakers (ICCB), and the low-voltage power circuit breakers (LVPCB). Metal clad circuit breakers are used in higher-voltage applications.

Worldwide, the number of manufacturers of circuit breakers is small. No doubt this condition is largely due to the high costs of developing, testing, and manufacturing a wide range of products. All of the manufacturers of circuit breakers provide detailed information regarding their products and this information is needed by those persons considering devices for circuit protection. When viewing manufacturers' literature, it is pertinent to note that the recommendations, practices, and terms used by one manufacturer may vary considerably from those used by others in the industry. In this textbook, the terms and concepts used are largely common to all of the manufacturers of circuit breakers.

### 8.5.1 Molded Case Circuit Breakers

Molded case circuit breakers (MCCBs) are the most widely used type of circuit breaker for applications under 1000 V. This is due to the simple design of the devices and their relatively low cost. MCCBs have all of the operating parts contained within a case that is fabricated of a nonconducting material that is usually a thermoset plastic. Practically all MCCBs are fitted with an operating handle that serves as a disconnect switch. When the operating handle is in the "off" position the circuit is open and voltage is blocked from the downstream circuit. In the "on" position the circuit is closed and the CB is engaged to protect the downstream circuit. The energy required to open the contacts is provided when the operating handle is manually moved to the closed position. There are several types of common MCCBs. The most common types are the thermal-magnetic circuit breaker, the magnetic only

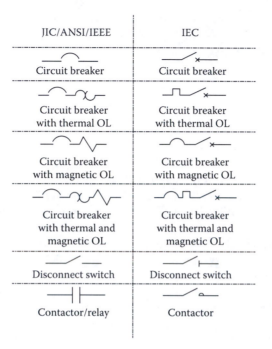

**FIGURE 8.2**
Types of MCCBs.

circuit breaker, and the electronic circuit breaker. (The electronic circuit breakers are also called the "solid state circuit breakers.") The thermal-magnetic MCCBs are used on all types of circuits although to a lesser extent on motor circuits. Magnetic-only MCCBs are intended primarily for motor circuits. The electronic CBs are less commonly used than the other types, mostly because of their higher costs, but are used to meet special circuit requirements not readily met with other types of circuit breakers.

The drawing symbols representative of the various types of CBs are shown in Figure 8.2. It is pertinent to note that (European) IEC symbols differ significantly from the (North American) JIC/ANSI/IEEE symbols.

### 8.5.1.1 Thermal-Magnetic Circuit Breakers

Thermal-magnetic circuit breakers are a very common type of MCCB, perhaps the most common. A MCCB is designed such that the monitored current passes through components contained entirely within the enclosure of the MCCB. Essentially, thermal-magnetic CBs house two types of sensing elements each of which has a distinctly different function.

The thermal components of a thermal-magnetic CB guard against an overload in the monitored circuit. As the monitored current increases, the temperature of the thermal element increases much in a manner that simulates

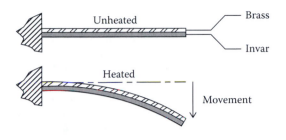

**FIGURE 8.3**
Bimetal element.

the temperature increases of the protected devices. The most common type of thermal-sensing element is the bimetallic element. As suggested by the name, a bimetallic element is constructed of two different types of metals, each having a different coefficient of expansion. When the monitored current passes through the bimetal element, each of the two metals expand but the metal with the higher coefficient of expansion will elongate more than the other metal. A typical bimetallic element is represented in Figure 8.3. Common metals used in bimetallic elements are invar and brass. Invar, which is an alloy of nickel and iron, has a relatively low coefficient of expansion at the temperature range within which it is used in MCCBs. Brass has a much higher rate of expansion with temperature. As current passes through the bimetallic element, it becomes heated due to the $I^2R$ effect and the brass grows with respect to the invar. Movement of the type shown in Figure 8.3 results. The bimetallic element is essentially a means of converting current flow to mechanical movement. At a point, the increasing current flow and the resulting movement of the bimetallic element trips the MCCBs contacts to the open position.

Unless a CB is marked "independent trip" or "no common trip," an overload in any of the monitored conductors will cause a trip of the MCCB and an opening of all contacts. Otherwise, only the affected circuit will be opened. Thermal magnetic CBs have an inherent time delay and an inverse tripping characteristic. In other words, the greater the overcurrent level the shorter the time to trip. Thermal magnetic CBs are available in designs of one pole, two poles, three poles, and, in some areas outside North America, four poles. Code requirements mandate that some ungrounded circuits must be isolated. Each pole must be monitored for an overload condition. A three-pole CB would have three separate thermal elements. A commonly recognized standard for CBs is the UL Standard 489 that calls for the calibration of the CBs at 40°C. Since the bimetallic element within a CB depends on ambient air for cooling, the trip setting of the CB will shift somewhat with shifts in ambient temperature. The higher the ambient temperature, the lower the value of the current required to cause a trip. According to the NEC code requirements, thermal magnetic CBs intended for service above 40°C must be rerated.

Under a short circuit condition, the "magnetic" components of a thermal-magnetic CB act to open the monitored circuit as quickly as possible. Each pole of the MCCB has two contacts: one that is stationary and one that is movable. When the MCCB is "closed" and the monitored current is within predetermined limits, the contacts are pushed tightly one against the other. The design of MCCB contacts is a specialty and every manufacturer strives to find rugged, durable yet inexpensive forms. Contact material must present a low resistance when the contacts are closed. Otherwise, a constant heating would result. Yet, when the contacts are opened, the heat generated by the arc must not melt or significantly deteriorate the surface of the contact material. A common material used for contacts is silver-cadmium oxide.

The common mechanical design of thermal magnetic CBs is such that the magnetic field of the monitored current acts to assist in the opening of the CBs contacts. The higher the current, the more forceful and greater the assisting forces. Since short circuit currents can be relatively high, the contribution from the magnetic field of the short current can be significant. Some thermal-magnetic MCCBs contain two sets of contacts electrically connected in series. One set of contacts is to open in consequence to thermal sensing and one to open when an instantaneous opening is required.

Some manufacturers provide what are termed a current-limiting circuit breaker. These types of CBs are for overcurrent protection on ac circuits subject to high fault currents. It would be fair to say these CBs are a specific form of the thermal magnetic CB.

### 8.5.1.2 Magnetic-Only Circuit Breaker

Magnetic-only MCCBs are used primarily in motor-control circuits. Unlike thermal-magnetic MCCBs, the magnetic only MCCBs have only magnetic elements and no thermal elements. Magnetic MCCBs are provided only to open the circuit in the event of a short circuit and are used mostly in motor protective circuits. Motor controllers alone guard against a motor overcurrent. Since motor controllers are designed specifically to guard against a motor overload, they are better suited to protect a motor than a thermal-magnetic circuit breaker. Magnetic CBs also include a manual switch that allows the CB to be used as a disconnecting means. Magnetic only CBs are set to trip at a current setting that by necessity must be above the locked rotor current of the motor. However, codes impose an upper limit on the settings of magnetic only CBs. Motor control circuits are treated in more detail in Chapter 9.

### 8.5.1.3 Electronic Circuit Breakers

Unlike thermal-magnetic circuit breakers, electronic circuit breakers do not contain thermal elements to measure current flow. Rather, current flow is measured by current transformers, contained within the enclosure of the

CB, and the determination to trip is made by electronic computing components. While generally more expensive than the thermal-magnetic circuit breakers, the electronic circuit breakers can fit a special need that cannot readily be met with other types of circuit breakers. Electronic circuit breakers are fitted with a variety of adjustments that allow shaping of the time–current characteristics to meet unique circuit requirements. Typical variables that can be controlled by adjustments on electronic circuit breakers are as follows:

- Continuous current: The allowable level of the breaker's continuous current rating that will not trip the CB. Typically adjustable in the range of 20%–100% of the breaker's continuous rating.
- Long time delay: Controls the variation in tripping due to a current level above the allowable continuous current, namely, a current in the "overload" region.
- Instantaneous pickup: Determines the level of current at which the breaker will open without a delay of any magnitude. The level of instantaneous pickup is typically in the region of 2–40 times the breaker's continuous rating. The instantaneous setting overrides all other settings.
- Short time pickup: Determines the level of current below which a trip is not required. Typically adjustable in the range of 1.5–10 times the breaker's continuous current rating. The setting causes a delay in tripping that permits coordination with downstream circuit breakers.
- Short time delay: Time delay in the short time pickup range, typically in the region of 0.05–0.2 s. The time delay facilitates coordination.
- Ground fault pickup: Determines the level of ground fault current that will result in a trip. (Since electronic circuit breakers measure three phase currents, an algorithm within the breaker can readily determine that a ground fault has occurred and then call for a trip.)

A significant feature of the electronic circuit breakers is that the devices facilitate coordination with other breakers by purposely delaying a trip, thereby allowing the downstream CB to open. In a way, an intentional delay seems contradictory to the purpose of an OCPD. There are arguments pro and con for this practice. It is difficult to generalize since circuits vary greatly in design and purpose. Obviously, a delay in trip allows a greater amount of energy to pass through the CB before the circuit is opened. This is certainly true but in many instances is immaterial. If both the upstream CB and the downstream CB are capable of interrupting a short circuit current, then why not allow the downstream CB to complete the task while holding back action by the upstream CB? The results would enhance, or ensure, the prospects of coordination. Of course a delay in response by the upstream CB would allow a greater amount of energy to pass through a CB to the downstream circuit. Should there be a short in that circuit, energy let-through would not have been restricted to the lowest possible level. This would be a consideration if an arc flash is a concern. On the other hand, if the two CBs are located in the same panel, which is very often the situation, experience indicates that the chances of a short in that circuit are very low. Briefly stated these are some of the considerations in the use of electronic circuit breakers.

Some electronic circuit breakers are capable of additional features not commonly available in other types of circuit breakers. Special features available in some models of electronic circuit breakers are as follows:

- Shunt trip: A shunt trip permits tripping of the circuit breaker from a remote circuit as determined by logic external to the CB. For example, a remote ground fault detector could be used to trip an electronic circuit breaker in the event of detected fault.

- Undervoltage trip: An undervoltage trip can be provided with the CB to trip the CB in the event of a detected low-voltage condition. (Some electronic circuit breakers can be provided with a built-in undervoltage trip that can be a relatively inexpensive way to provide an undervoltage tripping feature.)

- Remote on–off–reset: This accessory allows remote activation of the CB to "on," "off," or "reset." (This feature is especially desirable if a person might be in proximity to the OCPD and there is a potential for an arc flash.)

- Auxiliary contacts: Auxiliary contacts to remotely indicate the CB's position as "closed" or "open."

### 8.5.1.4 Sizing MCCBs

Several considerations enter into the sizing of a molded case circuit breaker for a specific application:

- The CB must remain closed when the monitored current is below the CB's nominal overload rating. From a safety consideration, no problem is created in the monitored circuit if the circuit breaker inadvertently opens while the monitored current is below its nominal rating. Nevertheless, inadvertent openings are undesirable because of the inconveniences and, sometimes, safety issues that might result from the deactivation of a circuit.

- A CB must isolate the circuit if the monitored current rises to or exceeds the overload value—although after a time delay that is a function of the degree of overcurrent.

- The sizing of a CB should be such that it has an adequate interrupting capability, that is, the capability to open the circuit against a fault current.
- Coordination is possibly an issue.

### 8.5.1.4.1 Sizing for Normal Load and Overload

A specific example might better explain some of the principles that are involved in the sizing of a CB for a normal load and for an overload.

> **Example 8.1**
>
> Consider a case in which a circuit breaker is to be selected to protect a three-phase heater with a continuous current rating of 40 A. For the purposes of this example, the voltage is immaterial. Consider the use of #8 AWG conductors to be extended from the MCCB to the heater. Procedures outlined in the applicable code will determine the allowable amperage of the conductors. Assume that the authority having jurisdiction (AHJ) requires that the guidelines of the NEC are to be followed. Under some conditions the NEC allows #8 AWG copper conductors rated 75°C to continuously carry 50 A. However, derating of conductors is possibly required for a number of reasons. Assume that a derating factor of 0.94 is applicable because of an expected high ambient temperature. Applying the derating factor, the current carrying capability of the #8 AWG conductors (called the conductor's "ampacity" by the NEC) is reduced to 47.0 A. The code allows selection of the next higher standard size breaker, which in this case is 50 A. If manufacturer's catalogs are reviewed, it will be found that a number of three-pole 50 A thermal-magnetic circuit breakers are available. However, the NEC has a rule concerning the loading of a MCCB. That rule states that unless the MCCB is suitable for continuous amperage at the rated value (and so clearly marked) the MCCB is not to be used with a continuous current above 80% of its rating. Using a 50 A MCCB with a nominal rating of 50 A, the usage would be at $(40/50) \times 100\%$ or 80%. So the installation would satisfy that rule of the NEC. One of the variations is the interrupting capability of the CB. So, having determined that a nominal 50 A breaker will suffice, a MCCB with an adequate interrupting must be selected. The procedure for determining the required interrupting capability of a circuit breaker is discussed in detail in Sections 8.5.1.4.2 and 8.7.
>
> **NOTE:** The sizing of conductors and OCPDs is a subject that is treated in wiring codes as the NEC. The NEC is a very detailed and long document. Its criteria are revised every two years to reflect more recent knowledge and the changing needs for safe electrical installations. Often, local and state codes impose additional requirements. These codes provide critically important guidelines that are intended to ensure a safe electrical installation. The brief computations shown here illustrate the importance of codes and that familiarity with the code is needed to properly select conductors and OCPDs.

### 8.5.1.4.2 Sizing for Interrupting Capability

Protective devices are described in part by their capability to interrupt fault currents. The highest current that can be interrupted is called the device's "interrupting rating" or its "ampere interrupting rating" (AIR). The CB must be capable of opening its contacts without permanent damage in the event that a fault current should occur at the level of the CB's AIR. As would be expected, interrupting ratings of a CB are drastically higher than the rated overload current. Listed interrupting currents for MCCBs are typically in the range of 5,000–200,000 A. How to determine the required interrupting capability?

It is generally accepted that for three-phase applications, the only way to properly determine required interrupting currents of a CB is by means of a short-circuit analysis. The IEEE provides several documents that outline procedures for conducting a short-circuit analysis (References 3.3 and 8.1). Some manufacturers of CBs also provide brief descriptions of methods for determining interrupting requirements. There are also a number of PC programs available that can serve as a helpful guide in determining interrupting requirements. The PC programs also offer a convenient means to document selections. Some textbooks (Reference 8.1) suggest that the utility supplying the electrical service, upon request, may provide the interrupting capacity at the meter. Such values would constitute an upper limit as electrical gear, mostly transformers and conductors, within a facility would provide additional impedance. It will generally be found that small users of a three-phase service may have an interrupting requirement of, say, only 5000 A. However, most three-phase users will require OCPDs with much higher interrupting capabilities.

It is gravely important that a circuit protective device should be adequately sized to interrupt the highest current that may result from a short circuit. Should a fault current exceed the rating of a protective device that device may violently explode thereby generating a potentially hazardous condition. It is particularly important to note that the cause of a MCCB's explosion could be the result of an individual moving the MCCB's operating handle from the off position to the on position. In a scenario of this type, an individual would be in close proximity to the exploding device and in an especially dangerous position.

### 8.5.1.4.3 Series Ratings of MCCB

In general, it is desirable to have coordination between OCPDs located in series. If coordination is not necessary in many instances the associated interrupting rating of a CB can be selected on the basis of the series rating of two OCPDs. A series rating allows use of an OCPD with an interrupting rating less than the available short circuit current. The reduced rating is applicable to the OCPD downstream of another OCPD. The merits are that the downstream CB would generally be less expensive and physically smaller.

Series ratings can be determined only through reference to a manufacturer's listing of the available series ratings.

### 8.5.1.4.4 Coordination

The coordination of two OCPDs actually has nothing to do with circuit protection. The lack of coordination may result in an inconvenience and possibly a dangerous condition. Nevertheless, it is assumed that the two OCPDs of concern have been properly selected to protect their respective downstream circuits. Yet, coordination is generally preferred and is in fact mandated for some applications. So, it is fair to state that coordination may be an issue in the selection of OCPDs. For this reason, the subject of OCPDs coordination is considered here, under the subject of "Circuit Protection."

### 8.5.1.5 Coordination of Circuit Overcurrent Protective Devices

### 8.5.1.5.1 Coordination Defined

According to the NEC, the definition of "selective coordination" is "Localization of an overcurrent condition to restrict an outage to the circuit or equipment affected, accomplished by the choice of overcurrent protective devices and their ratings or settings." Most personnel who deal with protective devices merely use only the term "coordination" rather than the full "selective coordination." Some individuals in the industry have coined the phrase "total selective coordination" to describe a configuration that is coordinated throughout a selected range of possible fault currents. It can be argued that this phrase is a misnomer since the NEC definition implies that "selective coordination" requires coordination throughout the entire range of possible fault currents. It is true that many possible configurations can be coordinated throughout a specific range of possible fault currents but to satisfy the NEC definition of "selective coordination," coordination must be throughout the entire range of possible currents.

A well-designed electrical circuit will include suitable coordination of the circuit protective devices. If the coordination is correct, a fault condition will result in the isolation of the affected circuit but no other circuits downstream of the upstream OCPD. The principles of "selective coordination" can best be explained with a simple example.

#### Example 8.2

Consider the three-phase circuit arrangement represented in Figure 8.4. In Figure 8.4, there are four overcurrent protective devices (OCPDs) shown. For the present, assume the four OCPDs could be either circuit breakers or fuses. A feeder delivers power to main overcurrent protective device OCPD-M, which protects Circuit M. Circuit M delivers electrical power to three downstream OCPDs: OCPD1, OCPD2, and OCPD3.

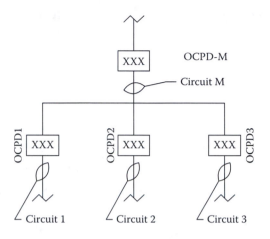

**FIGURE 8.4**
Circuit with OCPDs.

OCPD1 protects Circuit 1, OCPD2 protects Circuit 2, and OCPD3 protects Circuit 3. Should a fault occur, say, in Circuit 1, OCPD1 will trip (open) and separate Circuit 1 from Circuit M. Without proper coordination, OCPD-M may also trip thereby deenergizing not only Circuits 1, but Circuits 2 and 3 as well. In consequence to the trip of OCPD-M, the devices served by Circuits 2 and 3 might have been unnecessarily made unavailable. If there had been coordination between OCPD-M and the three sub-OCPDs, only OCPD1 would trip in the event of a detected fault on Circuit 1 whereas OCPD-M, OCPD 2, and OCPD3 would hold closed and Circuits 2 and 3 would remain energized.

If the OCPDs of Figure 8.4 are not coordinated, a short on any circuit will trip the respective OCPD and all three circuits will be isolated. Another form of a lack of coordination would be an overcurrent condition involving fuses when only one fuse opens and allowing the other two phases to remain energized. A condition of this description presents another set of unique problems as discussed in the following paragraphs.

The traditional method of determining CB coordination involves a procedure that examines the time–current characteristics of proposed protective devices. If the circuit is to have selective coordination, as required by the NEC for some applications, all possible short currents are assumed for downstream protective devices. Then it is determined if those currents and the associated time delays would result in a trip of an upstream protective device. If the analysis determines that a short of any magnitude would result in both protective devices opening, then there is a lack of coordination. If coordination is truly required, different OCPDs should be considered for the application.

Coordination is primarily concerned with the loss of availability and not damage to property. However, in some installations the lack of coordination can have serious and adverse consequences. Starting in 1993, the requirements for protective circuit coordination began to appear in codes. The first requirements were for elevator circuits. In subsequent years, additional requirements were imposed for other applications as essential electrical systems for health care facilities. Otherwise, coordination is not mandated for circuits. Code requirements for coordination have been the source of controversy, liabilities, and in many instances, the reasons for exceptions by the respective authority having jurisdiction (AHJ).

### 8.5.1.5.2 Circuit Breaker Time–Current Characteristics

The manufacturers of circuit protective devices provide the time–current characteristics (TCC) for their products. (Circuit breaker time–current characteristics are also commonly called "trip curves.") Time–current characteristics are applicable to both circuit breakers and fuses. An understanding of these TCCs will be of assistance in determining coordination of prospective OCPDs. The time–current characteristics for a circuit protective device presents the time of opening at an assumed specific current. A typical time–current characteristic for a molded case circuit breaker is shown in Figure 8.5. The curve shows a thermal-magnetic CB will hold closed indefinitely below 1.0 multiple of current. If the current rises slightly above 1.0 multiple, the trip point may be slowly approached and after a delay a trip might result. A current significantly above 1.0 multiple will bring about a trip after a relatively short delay. The higher the current above 1.0 multiple, the shorter the time delay to the trip. The time–current characteristics of an OCPD provides the opening times for the device throughout its useful range. Manufacturers determine the values of TCCs by tests conducted in accordance with standards by UL or IEC. Since the two agencies have different testing methods, manufacturers generally identify which standard is applicable to their published data.

The ordinate of an OCPD's characteristics show the time to break on the ordinate. Plots of TCCs are generally made in log–log scales. The abscissa typically shows numbers that are multiples of the nominal breaker size. Since manufacturers usually offer MCCBs for a wide range of service, it is common to find only one time–current characteristic plot for, say, a family of MCCBs. For example, for a 50 A MCCB, the number "1" on the ordinate is the equivalent to $1 \times 50$ A, or 50 A and $10 \times 50$ is the equivalent to 500 A. The characteristics of Figure 8.5 show both a thermal sensing region and an instantaneous region. The thermal region in the representation is from approximately "1.0" to "10.0." For several reasons, MCCB time–current curves are generally shown with a range, or band, of values. The left, or lower, side represents the minimum trip time and the right, or upper, side of the band represents the maximum trip time. The band of published time–current characteristics illustrates that MCCBs are not precise devices. The principles involved in the application of MCCBs can best be illustrated by a specific example.

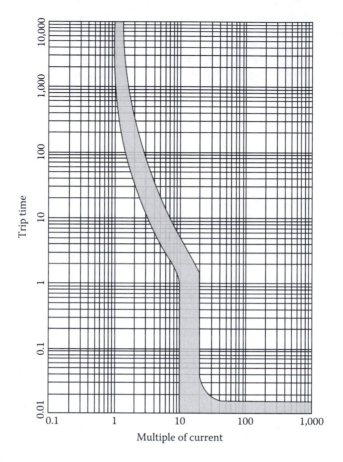

**FIGURE 8.5**
MCCB TCC.

### Example 8.3

Figure 8.5 is modified to show the specific values of a TCC for a nominal 50 A MCCB. The specific values are shown in Figure 8.6. The currents are shown on the abscissa and the times to trip are shown on the ordinate. The TCC of Figure 8.6 shows that the 50 A MCCB will hold indefinitely for currents below a value somewhere between 50 and 55 A. According to the TCC, values above 55 A will definitely cause a trip but after a time delay. The TCC of Figure 8.6 is duplicated in Figure 8.7 where it is shown that a current of 250 A will cause a trip of between 4 and 20 s.

### COORDINATION PRACTICES

As mentioned earlier, the traditional method of establishing CB coordination is by comparing the associated time–current characteristic. Again, the principles can best be explained by means of specific examples that examine OCPD coordination.

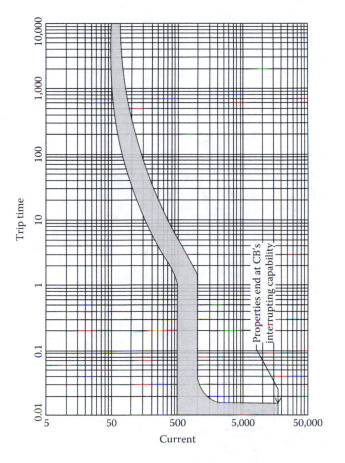

**FIGURE 8.6**
TCC for 50 A MCCB.

### Example 8.4

### USING TCCS

To consider a typical coordination study, assume the circuit in Figure 8.4 is to use all thermal-magnetic MCCBs for the OCPDs as represented in Figure 8.8. The specific OCPDs become CB-M, CB1, CB2, and CB3. Assume the conditions of Example 8.1 for the downstream CBs, namely CB1, CB2, and CB3. As stated in Example 8.1, the normal loads of Circuits 1, 2, and 3 are 40 A and CB1, CB2, and CB3 are all nominal 50 A circuit breakers. The normal current through CB-M is 3 × 40 or 120 A. The conductors connecting CB-M to CB1, CB2, and CB3 must be capable of at least 120 A since CB-M must be sized to handle at least 120 A. According to the NEC, the AWG #1, 75°C conductors are capable of 130 A so the AWG#1 conductors merit consideration. Allowing for an elevated ambient temperature, as in Example 8.2, a derating factor of 0.94 is applicable. The ampacity of the conductors connecting CB-M to CB1, CB2, and CB3

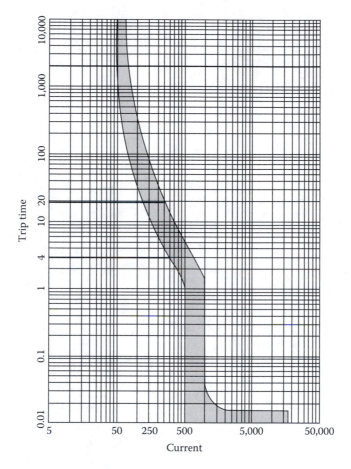

**FIGURE 8.7**
Trip times at 250 A.

then becomes 0.94 × 130 or 122.2 A. The next higher standard rating of MCCBs is 125 A. However, in applying the 80% rule (per Article 210.20 A), the allowable current must not exceed 0.80 × 125 or 100 A. Thus, a 125 A MCCB will not be adequate. Consider next a 150 A MCCB. In this case, 0.80 × 150 is 120 A and this value meets the NEC code's criteria. This computation suggests CB-M should be a 150 A breaker. The use of a 150 A breaker will require conductors extending between CB-M and CB1, CB2, and CB3 to have an ampacity greater than the AWG #1 conductors that were originally considered. The 75°C AWG #2/0 conductors have a rating of 175 A before application of a derating factor and after a 0.94 derating have an ampacity of 164.5 A. So, the AWG #2/0 conductors will suffice. Manufacturers' catalogs indicate that there are 175 A thermal-magnetic MCCBs available with terminals capable of accepting the AWG #2/0 conductors. Therefore, the AWG #2/0 conductors will suffice. If the configuration in Figure 8.8 is to be coordinated, CB-M must not trip in the event of a trip at CB1, CB2, or CB3.

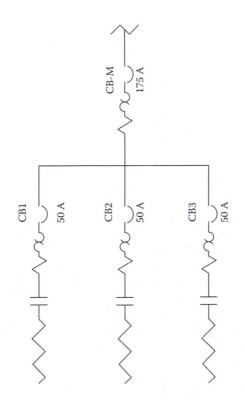

**FIGURE 8.8**
Example 8.4.

The next step requires an evaluation of the coordination of the two types of MCCBs under consideration. The traditional method of evaluating coordination is to examine the TCCs for the two CBs under consideration. The TCC for the prospective 50 A CB is shown in Figure 8.6 and the TCC for the 175 A CB is shown in Figure 8.9. The TCCs for both the 50 A CB and the 175 A CB are shown together in Figure 8.10. The representation in Figure 8.10 indicates that for currents up to 2000 A the two CBs are coordinated. In other words, a current of 2000 A or less will allow the 50 A breaker to open while the 175 A breaker will hold closed. However, at currents above 2000 there is ambiguity. A current of, say, 10,000 A will cause one or possibly both MCCBs to open within 0.017 s, but it cannot be predicted which will open first. Therefore it cannot be said that the two MCCBs under consideration are selectively coordinated. If coordination is required, then an alternative configuration will be required. (The "boot" characteristic of CB TCCs is often the source of a problem when efforts are made to confirm coordination.)

**NOTE:** In this example, the TCCs for both the 50 A MCCB and the 175 A MCCB are shown in a single drawing, Figure 8.10. This was done to allow a convenient comparison of the two TCCs. In practice, it is usually not convenient to show the two TCCs on the same drawing. Rather, the

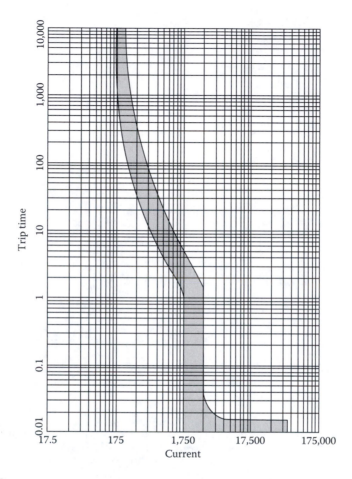

**FIGURE 8.9**
175 A CB TCC.

common method of comparison involves having one of the TCCs on tissue paper so that it can be positioned over the other TCC. In this manner the two TCCs may readily be compared to determine coordination.

**Example 8.5**

**USING COORDINATION TABLES**

In Example 8.4, it was found that by using the manufacturer's TCCs it could not be confirmed that coordination would exist with the considered thermal-magnetic MCCBs under consideration. However, a study of the TCCs may not be the final answer. There are alternatives. Manufacturer's TCCs are generated from test results conducted in accordance with the

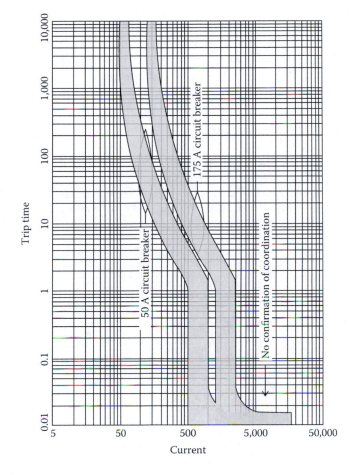

**FIGURE 8.10**
50 A and 175 A TCCs.

applicable guidelines of either UL (which would be UL Standard 489) or
the IEC. These guidelines call for testing of a single CB. While the TCCs
may not suggest that any two CBs may be coordinated, such may not
necessarily be the condition. There are a number of reasons to suspect
that a CB tested with another CB of a higher rating may very well display
coordination. In fact, this is often the case. For this reason, most manufacturers today offer coordination tables that show which specific CBs,
when used in series, will be found to be coordinated. These coordination
tables have proved to be a great convenience to persons responsible for
demonstrating coordination of OCPDs. In summary, it would be fair to
say that a TCC study may not demonstrate coordination, whereas a coordination table may actually show coordination. This was found to be the
case for the 50 A and the 175 A breakers of Example 8.4.

**Example 8.6**

## USING AN UPSTREAM ELECTRONIC CIRCUIT BREAKER

Some of the methods of obtaining circuit breaker coordination are discussed before. Another, often mentioned, method involves the use of an electronic circuit breaker upstream of a thermal-magnetic circuit breaker. By use of an upstream electronic circuit breaker, coordination can be demonstrated using the respective TCCs. In the way of illustration, consider the circuit of Figure 8.8 and assume the upstream CB is to be an electronic circuit breaker with a rating of 175 A, essentially the CB of Figure 8.11. The downstream CB will be the same 50 A CB used in Example 8.4. The resulting TCCs are shown in Figure 8.12. As is apparent in Figure 8.12, the TCC for the 175 A electronic circuit breaker is to the right and above the TCC of the 50 A thermal-magnetic CB thereby demonstrating coordination of the two breakers.

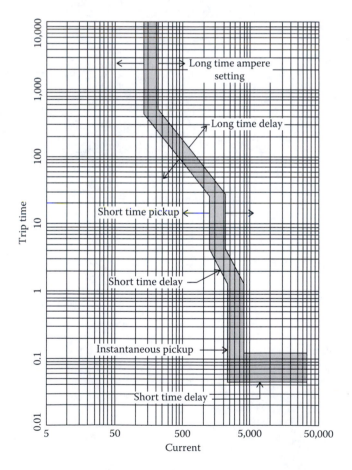

**FIGURE 8.11**
175 A electronic TCC.

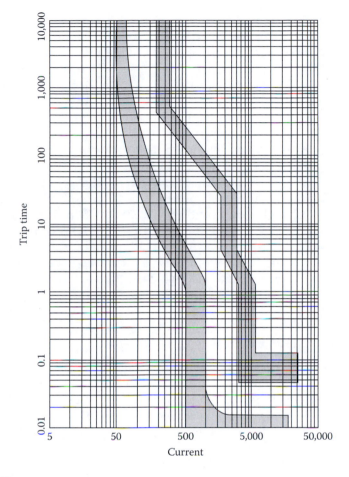

**FIGURE 8.12**
Coordinated CBs.

## 8.5.2 Insulated Case Circuit Breakers and Low-Voltage Power Circuit Breakers

Insulated case circuit breakers (ICCB) are similar in many regards to molded case circuit breakers and are manufactured to meet the same standards common to molded case circuit breakers. Insulated case circuit breakers are physically larger than MCCBs and are generally capable of interrupting higher energy levels. Some models of ICCBs are suitable for potentials up to 1000 VAC, and some are capable of continuous currents of 6400 A. As suggested by the name, insulated case circuit breakers are contained within an insulated case. Much as electronic molded case circuit breakers, ICCBs are furnished with a number of adjustments that allow shaping the time–current characteristics to meet the needs of a specific installation. ICCBs are also available in designs with drawout capability. That is to say, the ICCBs can be readily isolated from the circuits for testing, maintenance, or for the readjustment of settings. The isolation is accomplished merely by moving the ICCB "in" or "out" and no conductors need to be disconnected or reconnected. With the drawout feature an ICCB can be withdrawn a few inches from its "in" (or "closed") position to a "test" (or "open") position within its cradle (mounting mechanism), thereby separating the CB from both the incoming cables and the outgoing cables. ICCBs are most commonly used in motor control centers and power distribution centers.

Low-voltage power circuit breakers (LVPCB) are similar to ICCBs but generally offer a few more features. LVPCBs are generally available for the same range of voltages and currents as insulated case circuit breakers although some models offer higher interrupting currents. LVPCBs are manufactured and tested to standards that are different from those used for MCCBs and ICCBs, which appears to be the primary explanation for the different titles. Much as the electronic CBs, the time–current characteristics of LVPCBs can be programmed to meet special TCC needs and, as ICCBs, are available in a drawout design.

### 8.5.3 Metal-Clad Circuit Breakers

Metal-clad circuit breakers are intended primarily for interrupting currents at energy levels that start where the ICCBs and LVPCBs end. At the upper end, metal-clad circuit breakers are capable of interrupting the current generated by a nuclear power plant, which could be in the region of 1300 MW. By definition, a metal-clad circuit breaker is housed entirely within a metal enclosure. Metal-clad circuit breakers are fitted with the drawout feature that facilitates periodic testing and maintenance. Metal-clad circuit breakers are fitted with a variety of mechanisms designed to extinguish the arc that occurs when the breaker's contacts are opened. In some designs, compressed air is used to assist in blowing the arc away from the path between the breaker's contacts. Other designs use vacuum contacts and yet others use special gases to quench the arc. Because of their large physical size, metal-clad circuit breakers cannot react with the speed of the smaller insulated case or low-voltage power circuit breakers. Medium-sized metal clad circuit breakers cannot open in less than approximately three to five cycles. Larger breakers require more time yet to interrupt a circuit. Coordination is not an issue with metal-clad circuit breakers. Because of the forces required, metal-clad circuit breakers are indexed to the ready position by motors within the breaker's cubicle. Most breakers are fitted with "antipumping" logic that prevents the breaker from repeatedly opening and closing. A metal-clad circuit breaker can be "racked in" so that it is operative or it can be "racked-out" for testing or maintenance.

## 8.6 Fuses

Fuses have been used to protect electrical circuits almost since the advent of electrical circuits in the nineteenth century. Today, both fuses and circuit breakers are acceptable types of overcurrent protective devices. A fuse differs from a CB in the way it protects a circuit. A fuse contains a metallic link that provides a conducting path through its internal parts and it allows current flow through the metal at current levels below the rating of the fuse. Increases in current flow warm the metal, and when the current flow rises to a relatively high level, the metal melts to open the circuit. In past years, zinc had been used as the common metal in fuses. If fuses with zinc were cycled numerous times at normal currents, the zinc would fatigue over time and inadvertently open. Today, fuses use metals that melt at a higher temperature and fatiguing of the metal is no longer a concern. A fuse is a sacrificial device that requires replacement once it has opened to interrupt current flow. Whereas CBs provide a convenient and compact means of disconnecting a circuit, a circuit protected with fuses lacks the same convenience. A blown fuse requires the procurement of a replacement fuse. To disconnect a fuse-protected circuit, either a disconnect switch is required upstream of the fuse or the fuses must be removed to disconnect the circuit. While the replacement of a fuse can entail delay and inconvenience, the use of fuses in lieu of circuit breakers can often translate to considerably less initial cost. The common symbols for a fuse, as used in electrical diagrams, are shown in Figure 8.13.

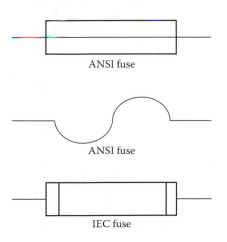

**FIGURE 8.13**
Fuse symbols.

Both circuit breakers and fuses are available for a wide range of voltages and currents. Much as a circuit breaker the behavior of a fuse can be described by a time–current characteristic. The time–current characteristics for fuses have an overall different appearance than those of circuit breakers. The time–current characteristics for circuit breakers have many curves and bends as shown in Figures 8.5 through 8.12. The TCCs for fuses more approximate a straight line. Typical TCCs for a 50 A fuse and a 175 A fuse are shown in Figure 8.14. Note that the fuse TCCs do not have the "boot" shape in the low opening times typical of the thermal-magnetic circuit breaker TCCs. In some ways, this characteristic simplifies the task of establishing essential coordination. Note that the times to open at the bottom of the TCC are less ambiguous than what was characteristic of the 50 and 175 A MCCB characteristic curves considered in Figures 8.6 and 8.7.

The TCCs in Figure 8.14 indicate that coordination of the two fuses can be demonstrated on paper. It will be apparent that the coordination of two

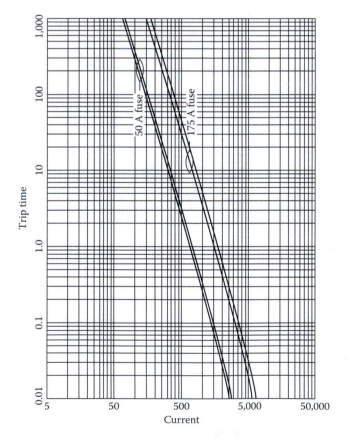

**FIGURE 8.14**
TCCs for fuses.

fuses can be readily confirmed since the two curves essentially are side by side and do not overlap at any point. It will be apparent that with a fuse downstream and a CB on the line side, selective coordination will be present. However, with a thermal-magnetic CB on both the line side and the load side there would be an overlap and coordination could not be confirmed solely from an analysis of the TCCs. Most fuse manufacturers provide coordination tables that identify the combination of fuses that will coordinate one with the other. When selecting only fuses, the coordination tables make the selection process very convenient. However, when a mix of fuses and circuit breakers are involved, use of the respective TCC's will generally be required to confirm coordination.

Fuses are available with a variety of time–current characteristics. The two most common types are the quick-acting and the time delay. The time delay fuses are particularly suited to motor circuits. A quick-acting fuse used on a motor circuit, unless of a relatively high current rating, would open due to a motor's inrush current when the motor is started.

## 8.7 Short Circuit Study

It is often accepted that the process of selecting an OCPD should include a short circuit study, the purpose of which is to determine the maximum fault current the OCPD might be required to interrupt. A fault on the load side of an OCPD could result in a relatively large current and the OCPD must have the capability to safely interrupt that current. The maximum fault current calculated by the short circuit study then becomes the OCPD's required interrupting capability and necessarily one of the criteria used to select an OCPD.

There are several commonly recognized methodologies for determining a fault current. Some of the more common procedures are the ohmic method, the point-to-point method, and the per-unit method. The computations involved in a comprehensive short circuit study conducted manually can be relatively intricate, challenging, and time consuming. The services of an experienced professional electrical engineer might be required along with an associated expense. Computer programs specifically tailored to short circuit studies can be very helpful and provide a standardized means of documenting calculations. There are also various published aids available to assist in the effort and many manufacturers of OCPDs also offer technical papers that can be of assistance. Aside from a full blown short circuit study that considers all of the possible considerations, in many instances there are alternative methods that suffice. None of the mentioned, and highly detailed, methods of calculating short circuit currents is reviewed here although a few pertinent considerations are treated.

The normal procedure that is followed in conducting a short circuit study begins with generation of a one-line diagram of the circuit under consideration. A one-line diagram places on paper the important components of a short circuit study and it provides a visual basis for the effort to be undertaken. The one-line diagram is to include the source of the electrical power, which in most cases is a utility's transformer. An on-site generator might be another possible source. Utility engineers might have a different set of considerations. Generally, the transformer is the most important element of a short circuit study as it is most often the primary source of the fault current at a facility. The size of the transformer also limits the short circuit current. It should be added that motors or in-house capacitors in the circuit can add to a fault current. The one-line diagram that is made to show all of the possible contributors of current is to also include the assumed point of the short. Motors or capacitors, if in the circuit, should be clearly shown.

A typical one-line diagram of the type necessary for a short circuit study is shown in Figure 8.15. The representation of Figure 8.15 includes the

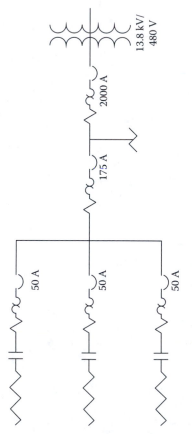

**FIGURE 8.15**
One-line diagram for short circuit study.

**TABLE 8.1**

Typical Short Circuit Currents

| Type of Installation | Typical Short Circuit Current |
|---|---|
| Small residential—100–200 A service | 10,000–15,000 A |
| Small commercial building—400–800 A service | 20,000–30,000 A |
| Larger commercial and manufacturing building system—2000–3000 A service | 50,000–60,000 A |
| Commercial building directly connected to utility grid | 200,000 A or greater |

same configuration that was used in Example 8.4 except that the representation is expanded to show a transformer and an assumed load. While the transformer supplying electrical power to the circuit determines the largest part of a short current, the impedances of devices between the transformer and the assumed point of a short circuit will decrease the value of the current made available at the transformer. The longer the distance from the transformer to the point of the short circuit the greater the intermediate impedance. Taking into account the impedance will result in at least some reduction in the required interrupting capability of the OCPD. It is pertinent to consider the prospective range of interrupting currents that might be determined by a short circuit study. Table 8.1 tabulates a summary of typical short circuit currents at various types of installations (Reference 8.2).

In most residential or small commercial installations, most of which have a single-phase service, the persons selecting OCPDs typically select an interrupting rating based on what has normally proved to be satisfactory in the past. Transformers in residential areas tend to be of approximately the same size. Because of the likeness of installations, rarely would a formal short circuit study be involved in the selection of OCPDs for a residential installation. Unlike residential applications, transformers serving three-phase facilities vary much more in size from one installation to the next. In larger commercial or industrial installations, it is likely that the transformer supplying electrical power might be capable of a relatively high interrupting current and the impedances along the path to some of the assumed shorts might not be all that high. The consequences could be very high interrupting currents. So, a short circuit study of some degree is appropriate for practically all three-phase installations. Without an accurate determination of the necessary interrupting capability of the OCPDs, there would be the danger of the OCPDs possibly not being correctly sized.

As mentioned in Section 8.4, there are several types of short circuits that might occur in a three-phase circuit. While the phase-to-ground short circuit is by far the more common type of short circuit, the type that draws the highest fault current is the three-phase short circuit. Of all of the possible types of faults, the bolted three-phase symmetrical fault is considered to result in the highest short circuit currents. So, an assumed symmetrical short can be assumed to be a worst-case condition. (It can safely be assumed that a detailed short circuit study will indicate a reduced maximum fault current.) For that reason, the three-phase short circuit condition is most often used as the basis for calculating symmetrical short circuit current. Adjustments will then be needed to account for asymmetrical currents. How then to proceed to determine interrupting current? A good place to start is with the transformer as the transformer's parameters are pertinent. Aside from the types of short circuits mentioned, a short circuit can be either symmetrical or asymmetrical. If the fault is symmetrical, it is assumed that all three phases are affected equally and all three phases have equal currents. If only some phases are affected the fault is considered to be asymmetrical. The asymmetrical condition would draw the highest current and therefore merits evaluation.

Parameters pertinent to the transformer supplying power are critical to any short circuit study. Much of this data are available on the transformer's nameplate. (However, according to UL requirements, transformers under 15 kVA are not required to be furnished with a nameplate.) The data on a transformer generally include the intended primary voltage, the secondary voltage, the full load secondary current, the "impedance percentage" (or "impedance voltage"), and very likely additional data. The impedance percentage value is of importance in determining the value of current that the transformer can deliver to a short circuit. The impedance percentage is sometimes shown on the nameplate as merely "%", "Z%," or "% p.u." The impedance percentage is defined as the percent of rated input voltage required at the primary terminals, with the secondary shorted, to produce full load current at the secondary. Having the impedance percentage, the maximum current available at the transformer's secondary for assumed symmetrical conditions can be readily computed. Common percentage impedances for three-phase transformers are shown in Table 8.2.

A sample calculation of the maximum transformer's secondary fault current based on a transformer's percentage impedance is shown in Example 8.7. A word of caution: Approaching a transformer to obtain

**TABLE 8.2**

Common Percentage Impedances

| Transformer Size (kVA) | Percentage Impedance (%) |
| --- | --- |
| <200 | 3 |
| 200–500 | 3–4.5 |
| >500 | 5–9.5 |

nameplate data can be dangerous and should not be attempted by untrained persons. In most instances a call to the utility is all that is needed to obtain needed data.

**Example 8.7**

A sample calculation is performed to determine maximum symmetrical current that can be derived from a transformer based on a transformer's impedance percentage. In the way of illustration consider a transformer with a 480/3/60 secondary, a 3 MVA rating, and an impedance percentage of 5%. Let $MVA_f$ be the fault MVA, MVA the transformer's rated (nonfault) mega volt-amperes, $I_f$ the fault current, and $V_s$ the transformer's secondary voltage.

$$MVA = 3,000,000 = \left(V_s I \sqrt{3}\right)$$

The secondary rated current is

$$I = 3,000,000 \div (V_s)\left(\sqrt{3}\right) = 3,000,000 \div (480)\left(\sqrt{3}\right) = 3,608.4 \text{ A}$$

If the primary is at rated voltage, the secondary current would become

$$I_f = (100 \div 5)(3,608.4) \text{ A} = 72,168 \text{ A}$$

So, the maximum (symmetrical) current available from the transformer is 72,168 A.

The computation of Example 8.7 assumes full voltage at the primary terminals throughout a short circuit condition that would almost always never be the case. In reality, a short on the secondary of a transformer will normally result in significant voltage sag on the primary. All other factors being equal, a decrease in the primary voltage will decrease the available current at the secondary. In this regard the current so determined would be conservative, that is, higher than what could in reality be expected with full rated voltage at the primary. While a transformer is the usual source of a short circuit current, a generator might also be the source.

When conducting a short circuit study of a circuit that includes a motor, it is important to consider the contribution of the motor acting as a generator. The current added by a motor acting as a generator adds to the short circuit current that is provided by the transformer. Generally single-phase motors and motors below 50 HP are excluded from consideration. The contribution of current from a motor, if used, is typically computed as FLA X 4. The same is true of capacitors that might be common to a circuit. The energy stored in capacitors can add to the severity of a short circuit and that energy must be considered in determining short circuit current.

It was stated before that the assumption is that the short circuit is symmetrical. Once the maximum possible current available to a short has been determined, a person is in position to decide the next move.

Example 8.7 demonstrates a method of calculating the maximum possible current available from a transformer. Will that value then become the maximum interrupting current of the OCPDs? Not necessarily so. It may be necessary to use OCPDs with an interrupting capability higher than what is determined as the largest current available from the transformer as well as any motors or capacitors in the circuit. The explanation harks back to the method used to test and rate OCPDs. The limitations of OCPDs are related to the asymmetrical currents that occur at the time of a short circuit. A few words of explanation!

### 8.7.1 Asymmetrical ac Currents

As stated in the previous section, the purpose of a short circuit study is to determine the value of the current that might occur in the event of a short circuit. The intention of the study is usually for the purpose of sizing a circuit's OCPDs. A short circuit in an ac circuit normally results in currents that are asymmetrical and that condition introduces a set of unique problems that are not necessarily addressed in a short circuit study. Nevertheless, asymmetrical currents must be considered in the selection of OCPDs.

When typical ac equations are employed, symmetrical voltages and currents are assumed. For example, the following equations are valid only if the RMS ac voltages and currents are symmetrical:

$$V = IR \,(\text{single phase})$$

$$P = I^2 R \,(\text{single phase}) \qquad\qquad (1.6)\ (\text{three phase})$$

$$P = \left(\sqrt{3}\right) V_L I_L \,(\text{PF})$$

By definition, a symmetrical ac current is symmetrical about a zero level and the direction of the current reverses twice every cycle. An asymmetrical current is one that is not symmetrical. Examples of both symmetrical and asymmetrical currents are shown in Figure 8.16.

When a short circuit occurs in an ac circuit, the current that results is normally asymmetrical to some extent. This is the case because of the circuit's reactance elements. The immediate, resulting asymmetrical current is larger than what would result if the circuit's elements were purely resistive. A typical asymmetrical current that might result from an ac short circuit is represented in Figure 8.17. The asymmetrical current in Figure 8.17 is a

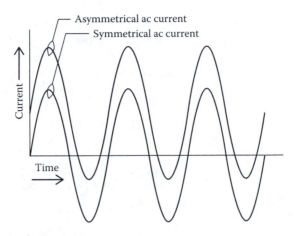

**FIGURE 8.16**
Typical symmetrical and asymmetrical current plots.

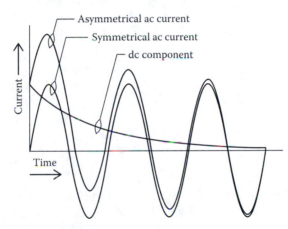

**FIGURE 8.17**
Asymmetrical current showing symmetrical and dc components.

combination of an ac component and a dc component. The dc component decays in time and its value with time can be described by the equation,

$$I_{dc} = I\left[1 - e^{(-t/T)}\right],$$

where
  $I_{dc}$ is the dc voltage
  ($t$) is the time (s)
  T is the time constant (s)

  The time constant, T, determines the decay rate. When $(-t/T) = 1$, the value of $I_{dc}$ will have decreased by 63.2% from its initial value. In Figure 8.17, there are three full, symmetrical ac cycles represented. For a 60 Hz circuit, each cycle would be (1/60) s, 0.0166 s, or 16.66 ms in duration. The 63.2% decrease occurs at 0.01388 s. So, the time constant of the dc component in Figure 8.17 is 13.88 ms. The greater the lag of current with respect to voltage in the circuit, the lower the power factor and the longer the time constant of the dc component.

Power factor in a linear circuit defines the lead or lag of current with respect to the applied voltage. If a reactance diagram is made of a circuit, the power factor is cos (X/R), where X is the magnitude of the reactive element and R is the resistive component. (A problem with the use of power factor is that it does not define lead or lag. For example, a power factor of 0.5 indicates a lead or lag of 30° but from the value of power factor alone it is not possible to determine if the current is leading or lagging. That information must be obtained separately.) In short circuit studies, the terms "X" and "R" are commonly used and the expression X/R often appears. Obviously if the power factor is known the ratio X/R can be calculated. The ratio X/R is the tangent of the angle that describes the power factor: If the angle describing the lead or lag of current with respect to voltage is $\theta$, then,

$$\cos\theta = PF, \quad \text{or}$$

$$\theta = \cos^{-1} PF$$

$$\tan\theta = X/R, \quad \text{or}$$
$$\theta = \tan^{-1} X/R$$

In short circuit studies, the ratio X/R, rather than power factor, is commonly used with reference to transformers. The ratio is used to determine the capability of an OCPD to interrupt current. The higher the X/R ratio the higher a circuit's reactance and the greater the difficulty an OCPD has in interrupting a current. In this regard, a circuit's X/R ratio is related to a short circuit study. The rating of any OCPD is limited to a specific X/R ratio. Circuit breakers bear an interrupting rating that is based on test results that were conducted in accordance with the guidelines of a standard. Molded case circuit breakers are tested in accordance with either a North American or a European standard. Table 8.3 lists a summary of the minimal X/R ratios that are mandated by the UL standard for tests on MCCBs (Reference 8.4). The IEEE provides the minimal requirements for low-voltage power circuit breakers (Reference 8.5). The applicable X/R ratios for power circuit breakers per IEEE are shown in Table 8.4.

A review of Tables 8.3 and 8.4 suggests the obvious question: What if a circuit has an X/R ratio higher than that used in the tests that determined the OCPD rating? That would be a more severe condition than what the device has been proved to be capable of handling. For X/R ratios above an OCPDs

**TABLE 8.3**

Properties per UL for MCCB

| Breaker's Interrupting Rating (A) | Test Circuit Power Factor | Test Circuit Equivalent X/R Ratio |
|---|---|---|
| 10,000 or less | 0.45–0.50 | 1.73 |
| 10,001–20,000 | 0.25–0.30 | 3.18 |
| Over 20,000 | 0.15–0.20 | 4.90 |

**TABLE 8.4**

Properties per IEEE for LVPCB

| Type Breaker | Test Circuit Power Factor | Equivalent X/R |
|---|---|---|
| Unfused breaker | 0.15 | 6.60 |
| Fused breaker | 0.20 | 4.90 |

rating, a derating must be applied. Adjustments can be made by means of a multiplication factor (MF or FM). The multiplication factor is either applied to the symmetrical current that has been calculated (by a short circuit study) and an OCPD selected accordingly. Or the symmetrical rating of a prospective OCPD is derated by dividing its rating by the multiplication factor. How to determine the multiplication factor? The referenced standards provide multiplication factors for X/R ratios that are higher than those used for testing. The multiplication factors are shown in Tables 8.5 and 8.6.

Obviously the value of the X/R ratio is needed in order to determine a multiplication factor from Table 8.5 or Table 8.6. If the transformer in question is supplied by a utility, very likely that utility can provide the X/R data. Otherwise the information very well might be available from the transformer manufacturer. In the absence of the X/R ratio, the IEEE standard allows the use of a multiplication factor of 2.7 for the value of peak current. In comparison to the values of multiplication factor listed in Tables 8.5 and 8.6, a value of 2.7 seems

**TABLE 8.5**

Multiplication Factors ($F_M$) for MCCRs (UL-489)

| Power Factor | X/R | ≤10,000 A Interrupting Rating | 10,001–20,000 A Interrupting Rating | >20,000 A Interrupting Rating |
|---|---|---|---|---|
| 0.50 | 1.73 | 1.000 | 1.000 | 1.000 |
| 0.40 | 2.29 | 1.078 | 1.000 | 1.000 |
| 0.30 | 3.18 | 1.180 | 1.000 | 1.000 |
| 0.25 | 3.87 | 1.242 | 1.052 | 1.000 |
| 0.20 | 4.90 | 1.313 | 1.112 | 1.000 |
| 0.15 | 6.59 | 1.394 | 1.181 | 1.062 |
| 0.10 | 9.95 | 1.487 | 1.260 | 1.133 |
| 0.05 | 19.97 | 1.595 | 1.351 | 1.215 |

**TABLE 8.6**

Multiplication Factors ($F_M$) for LVPCB (Reference 8.5)

| Power Factor | X/R | Unfused Breaker | Fused Breaker |
|---|---|---|---|
| 0.20 | 4.90 | 1.000 | 1.00 |
| 0.15 | 6.60 | 1.000 | 1.07 |
| 0.12 | 8.27 | 1.04 | 1.11 |
| 0.10 | 9.95 | 1.07 | 1.15 |
| 0.085 | 11.72 | 1.09 | 1.18 |
| 0.07 | 14.25 | 1.11 | 1.21 |
| 0.05 | 20.00 | 1.05 | 1.26 |

**TABLE 8.7**

Typical X/R Ratios for Self-Cooled Power Transformers (Reference 8.3)

| Range of MVA | Range of X/R |
|---|---|
| 0.05–0.50 | 1–4 |
| 0.5–2 | 4–8 |
| 2–10 | 8–16 |
| 10–50 | 16–30 |
| 50–200 | 30–45 |

very high. It may in fact be applicable if the transformer is relatively large. Typical X/R ratios for transformers are shown in Table 8.7.

Consider an example in which a calculated (symmetrical) interrupting current is corrected to allow for the asymmetrical aspect.

**Example 8.8**

In Example 8.7, a transformer with a 3 MVA rating was considered and it was determined that the transformer was capable of a symmetrical short circuit current of $I_f = 72{,}168$ A. Assume that an unfused LVPCB is under consideration for the application. Assume further that it was found that the transformer's X/R ratio is 10.0. Assume also that for the moment a short circuit study is not to be conducted. Rather, a simple calculation is to be done to consider only the short circuit current available from the transformer without regard to the impedance of circuit elements. According to Table 8.6, a multiplication factor of 1.18 is applicable. Correcting the symmetrical current, the required interrupting capability of the MCCR must be at least

$$I_{int} = 72{,}168 \times 1.18 = 85{,}159 \text{ A}$$

If the LVPCB is intended for use in, say, the circuit of Figure 8.15, then all of the LVPCBs must have an interrupting rating of at least 85,159 A. A detailed short circuit study would most likely increase the current value when the contributions from motors and capacitors are added. Circuit impedances and the findings would most likely lower the values to some extent. (A notable caveat: If circuit breaker coordination is not a requirement then the series ratings of the involved circuit breakers can be considered. This combination would lower the required rating of each circuit breaker.)

### 8.7.1.1 Summary of Sizing for Interrupting Capability

In this section, consideration is given to the effects of asymmetrical currents on the interrupting ratings of MCCBs and LVPCBs. Briefly stated, short circuit currents are almost always asymmetrical and for this reason computed symmetrical currents must be corrected to allow for the higher energy due to the asymmetrical aspect of current flow. For applications requiring ICCBs and metal-clad CBs, it is recommended that reference be made to the respective manufacturer's literature to correctly select a CB for a specific application.

This discussion is centered on the North American practices for the selection and rating of OCPDs. The IEC methods are somewhat different. To satisfy IEC requirements a different set of computations is required. Wiring codes as the NEC have specific requirements applicable to the selection of OCPDs for circuit protection. It is recommended that in those applications where the code is applicable that the requirements of the codes should be followed.

## 8.8 Damage Curves

In electrical devices, an overcurrent condition will increase to some extent the temperature of the metal conductors contained within an electrical device as a consequence to the $I^2R$ effect. The device's insulation will in turn be heated by the metallic conductors and if the resulting temperature rise is excessive, the insulation can be either seriously deteriorated or, in the extreme, rendered useless. In consequence to failed insulation, a short circuit may result either immediately or after a delay. For this reason the possible effects of an overcurrent condition warrant a careful evaluation. This generality is true of cables, transformers, generators, capacitors, and other electrical devices that might be at risk to an overcurrent condition. These electrical devises, much as the aforementioned OCPDs, have an associated unique time–current curve that is known as the device's damage curve. Currents exceeding the respective damage curve are considered as harmful to the device to some degree and should be prevented by a suitable OCPD. An example will demonstrate some of the principles that are involved.

**Example 8.9**

In the way of a specific example, assume that three #6 AWG, THHN (90°C) copper conductors are to be installed in a conduit as a part of a three-phase circuit. The conduit is air cooled. The three conductors will be the only conductors in the conduit. According to NEC Article 310.13, the uncorrected ampacity of the conductors is 75 A. The normal temperature is expected to be no greater than 30°C, and no derating is required.

Reference is made to IEEE Standard 242-2001, which states that an overcurrent greater than 10 s is considered as intermediate to long term and an overcurrent lasting less than 10 s is considered a short circuit current. Each of these two time periods is treated separately.

## 8.8.1 Currents Lasting 10 s or Longer

The values of computed overcurrent vs. time are dependent on the assumed value of maximum insulation temperature during a short. The IEEE document maintains that calculations are based on what is said to be a conservative value of 150°C. Following the guidelines of IEEE 242-2001, the values of overcurrent were computed for a #6 AWG (4.12 mm diameter) copper conductor (which has a nominal ampacity of 75 A per NEC) and the values are listed in Table 8.8.

### 8.8.1.1 Currents Lasting 10 s or Less

To determine the maximum allowable short circuit currents of cables, IEEE 242-2001 refers to Standard P-32-382-1999 of the Insulated Cable Engineers Association (ICEA). According to Section 9 of IEEE 242-2001, the applicable equation for determining maximum short circuit time is

$$t = 0.0297 \, \log_{10}[(T_f + 234)/(T_o + 234)] \, (A/I)^2,$$

where
  A is the conductor area (cmils)
  $t$ is the time of short circuit (s)
  I is the short circuit current (A)
  $T_o$ is the allowable operating temperature of the insulation (°C)
  $T_f$ is the maximum allowable insulation short circuit temperature (°C)

**TABLE 8.8**

Allowable Overcurrent Values per IEEE 242-2001 for #6 AWG Conductor

| Time (s) | Overload Current |
|---|---|
| 10 | 618 |
| 100 | 209 |
| 1,000 | 99 |
| 10,000 | 84 |
| 18,000 | 83 |

**TABLE 8.9**

Time vs. Temperature per IEEE 242-2001

| Max Current (A) | Time (s) (250°C) | Time (s) (150°C) |
| --- | --- | --- |
| 10,000 | .0356 | .0157 |
| 5,000 | .1425 | .0603 |
| 3,000 | .396 | .167 |
| 2,000 | .891 | .377 |
| 1,500 | 1.584 | .670 |
| 1,000 | 3.56 | 1.508 |
| 750 | 6.33 | 6.035 |

The values of overcurrent are dependent on the assumed value of maximum acceptable insulation temperature. General Cable uses a maximum insulation temperature of 250°C for 90°C conductors. This temperature is higher than the 150°C temperature used in IEEE Standard 242-2001. The computed values using the two temperatures are listed in Table 8.9.

Calculated values of current vs. time for an assumed #6 AWG conductor are listed in Table 8.9 and plotted in Figure 8.18 along with the TCC for a prospective 40 A thermal-magnetic MCCB OCPD. The plots indicate that, except for the "boot" area of the MCCB, either the 150°C assumed insulation temperature or the 250°C assumed insulation temperature would be acceptable. In the boot area, acceptability is dependent on the interrupting capability of the MCCB. With the 150°C assumed temperature, there is no problem up to an interrupting current of 10,000 A. With the 250°C assumed temperature, there is no problem up to an interrupting current of 15,000 A. The damage curves tell much about the prospective combination of the selected MCCB and the #6 AWG conductor. What if the damage curves suggest a problem? Perhaps the "boot" area of the MCCB suggests a problem with interrupting current. Or, perhaps the assumed temperatures are considered too high for an application that, say, is critically important. There are several alternatives that could be considered. Use of a fuse instead of a MCCB would avoid concern related to the "boot" area of the MCCB. Another alternative would be to use a larger conductor as the next larger size, namely a #4 AWG (5.18 mm diameter) conductor.

A #4 AWG conductor is considered in Figure 8.19. Use of the larger conductor shifts the short circuit withstanding capability to 15,000 A for an assumed maximum of 150°C and higher for the temperature of 250°C. The greater amount of copper in the larger conductor acts as a heat sink thereby reducing the insulation temperature for a given short circuit current.

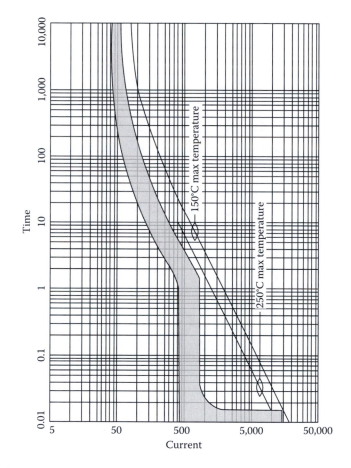

**FIGURE 8.18**
#6 AWG conductor and 40 A thermal-magnetic MCCB.

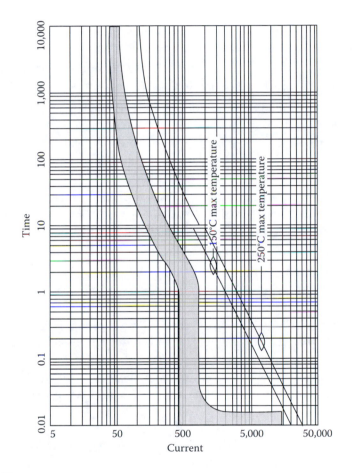

**FIGURE 8.19**
#4 AWG conductor and 40 A thermal-magnetic MCCB.

## Problems

1. Name three types of electrical short circuits.

2. If a live conductor comes in contact with ungrounded equipment would an overcurrent protective device generally cause an immediate isolation of the circuit?

3. Name two possible adverse effects on electrical components that might result from a prolonged overcurrent condition.

4. What is the difference between a bolted fault and an arcing fault?

5. What are the types of circuit breakers used for low voltage applications?

6. Sketch the common representation of a JIC circuit breaker with thermal-magnetic overloads and an IEC circuit breaker with thermal-magnetic overloads.

7. What type of metal that has a low coefficient of expansion is often used in thermal overloads?

8. Name a significant feature of electronic circuit breakers that often will justify the higher cost of the devices.

9. Name three criteria in the sizing of a MCCB.

10. What is the AIR of a MCCB?

11. If coordination is not necessary, can two OCPDs, each having a lower interrupting rating when used singly, be used in series to obtain a higher interrupting capability?

12. If circuit breaker CB1 is upstream of circuit breaker CB2 and the two are coordinated, will CB1 hold closed under all possible trips of CB2?

13. Assume that a 100 A rated, molded case circuit breaker is under consideration for an application and that the manufacturer's TCC is per Figure 8.4. What would be the expected range of trip times if the current is 200 A?

14. Are electronic circuit breakers capable of shunt trips? Cite a possible specific role of an electronic circuit breaker used for shunt trips.

15. What would be the merits of an ICCB with a drawout feature?

16. Would it be expected that a metal-clad circuit breaker on a 1000 MW circuit could interrupt the current in less than 1 ms?

17. Name both a feature and a disadvantage of using a fuse.

18. What are the two common types of fuses?

19. What would be a typical short circuit current for a small commercial business?

20. When determining maximum short circuit current, what data on a transformer's nameplate is needed?

21. According to testing by UL, what is the X/R ratio for a MCCB with a short circuit rating of 15,000 A?

22. Assume a MCCB is under consideration for use in a circuit that is supplied by a 480/3/60 transformer and the maximum interrupting current must be determined. Rather than conducting a short circuit study, it was decided to merely use the maximum possible asymmetrical current available from the transformer. The transformer is rated 700 kVA, the impedance percentage is 7%, and, according to the utility, the transformer's X/R ratio is 7. What interrupting rating should be used in selecting the MCCB?

23. If the 250°C damage curve is assumed as applicable to the #4 AWG conductor in Figure 8.19, what would be the maximum interrupting current that would be compatible?

# 9

## Motors and Motor Protection

### 9.1 Motor Designs

Electric motors of all types have brought about improvements to the quality of life of millions of people around the world. Motors drive pumps that deliver potable water to homes and businesses. Sewage plants depend on a variety of motors. Motors drive air conditioners for dwellings, hospitals, and high-rise office buildings. Without motors, whole populations would find life very different as well as drastically less pleasant. Motors account for half of all generated electricity worldwide (Reference 9.1). In North America alone, there are over one billion motors (Reference 9.1).

There are numerous types of motors that are manufactured for a wide range of voltages and applications. Many of the fractional horsepower motors are for use with a single-phase electrical source although some fractional HP motors also operate on polyphase power. Single-phase motors are especially common in North America and are used primarily to power household appliances as clothes washers, refrigerators, and household air conditioners.

Any motor, single-phase or three-phase, consists of a stator and a rotor. The rotor is the rotating component of the motor that delivers the developed power. The stator is integral to the motor case and its function is to provide a rotating magnetic field that, in turn, induces torque and power in the rotor.

In a single-phase motor, there is no natural, rotating magnetic field. For this reason, a single-phase motor must use some means to generate a rotating field. Several methods are commonly used. One common arrangement uses a capacitor in either a "capacitor start" or "capacitor run" configuration. A capacitor start motor uses the capacitor only to start the motor. Once the motor is up to speed, a centrifugal switch disconnects the capacitor from the circuit and the motor continues to run without the capacitor. A capacitor run motor retains the capacitor in the circuit during starts as well as during normal operation.

Split-phase, single-phase motors use a second set of coils only for starting and are disconnected after the motor is up to speed. Some single-phase motors also use what is called a shaded pole to create a rotating field. Except for residential air conditioning applications, few motors above one horsepower are the single-phase type. A three-phase electrical service is much more suited to the larger, integral horsepower motors.

Unlike a single-phase motor, a three-phase motor makes use of the natural, rotating field characteristic of three-phase electrical power. For this reason, a three-phase motor needs no special adaptations for starting or operation. The three conductors of a three-phase service, say conductors A–B–C, will provide a rotating field that naturally spins in one direction. Reversing any two of the three-phase conductors, say to A–C–B, will reverse the direction of rotation of the field. This feature provides a convenient means of reversing the rotation of a three-phase motor that, say, is found to be rotating in the incorrect direction.

There are several types of commonly used three-phase motors that are discussed in the following paragraphs.

### 9.1.1 Synchronous Motors

Synchronous motors are used for a small number of unique applications but are not nearly as prevalent as asynchronous motors. A synchronous motor operates at only synchronous speed. These motors are used only with loads that have light starting resistance. Unlike asynchronous motors, synchronous motors have a leading power factor and for this reason are sometimes used at a facility to assist in correcting a lagging power factor. Synchronous speeds vary depending on the line frequency and the number of poles as shown in Table 9.1.

### 9.1.2 Asynchronous Motors

Motors defined as "asynchronous" operate at a speed somewhat below synchronous speed. The difference between operating speed and synchronous speed is called the "slip." The slip of an asynchronous motor will increase with an increase in load. There are several classifications of asynchronous motors although the three-phase induction motor is by far the most commonly found form of three-phase motor worldwide. There are two common

**TABLE 9.1**

Synchronous Motor Speeds (RPM)

| Line Frequency (Hz) | 2 Pole | 4 Pole | 6 Pole | 8 Pole | 10 Pole | 12 Pole |
|---|---|---|---|---|---|---|
| 60 | 3600 | 1800 | 1200 | 900 | 720 | 600 |
| 50 | 3000 | 1500 | 1000 | 750 | 600 | 428.5 |

forms of the induction motor, namely, the squirrel cage motor and the wound rotor. Of the two, the squirrel cage design is the most prevalent.

### 9.1.2.1 Wound Rotor Induction Motors

The wound rotor induction motor uses slip rings to conduct current to the rotor. In past years, wound rotor motors were popular for variable speed applications. However, the variable frequency drive (VFD) motors have largely replaced the wound motors for new variable speed applications. Slip rings are also subject to wear, necessitating replacement, and are generally considered a maintenance nuisance.

### 9.1.2.2 Squirrel Cage Induction Motors

The squirrel cage induction motor is a simple and rugged motor design. It is free of the slip rings found in wound rotor motors. The design has become the most common form of three-phase motor for a variety of valid reasons. Of all of the types of motors, squirrel cage motors account for 90% of installed capacity (Reference 9.1). The squirrel cage induction motor has a rotor consisting of laminated steel sheets with copper or aluminum conductors imbedded in its periphery. The conductors at the periphery run longitudinally along the rotor. The thin steel sheets used for the rotor minimize eddy currents that would otherwise cause undesirable heating of the member as well as a decrease in efficiency. Squirrel cage induction motors are common in sizes from fractional horsepower to 10,000 HP. A three-phase, squirrel cage induction motor in a cast iron casing intended for severe duty is shown in Figure 9.1.

**FIGURE 9.1**
Ruggedized three-phase motor. (Photo courtesy of Baldor, Fort Smith, AZ.)

### 9.1.2.3 Variable Frequency Drive Motors

In years gone by, a variable frequency drive commonly consisted of a fixed speed electric motor that by various mechanical devices drove an output shaft that operated at a variable speed. These mechanical configurations required high maintenance and were not all that precise in speed regulation. More recently, variable speed drives (VFDs) have largely replaced the older mechanical configurations. By means of semiconductor components, VFDs convert normal ac or dc power to a variable frequency ac output that drives an ac motor. The new generation of VFDs has good speed regulation and requires minimal maintenance. VFD motor sizes are becoming increasingly larger, and the applications are likewise increasing in number. VFD motors are available for single-phase and three-phase applications and motor sizes from fractional sizes up to 600 HP and infrequently beyond. The VFD motors are usually asynchronous but synchronous motors are also used.

### 9.1.3 Motor Nameplates

Most motors bear a nameplate that describes the respective motor's critical characteristics. Worldwide, the two most commonly recognized nameplates follow the criteria of standards published by either the National Electrical Manufacturer's Association (NEMA) or the International Electromechanical Association (IEC).

### 9.1.3.1 NEMA Motor Nameplates

In the United States, the requirements of the National Electric Code (NEC) are necessarily followed where an authority having jurisdiction invokes the code. In those applications where the requirements of the NEC are invoked, motors are required to be marked with a minimum of specific information (Reference 9.2). The requirements of the NEC are reflected in the NEMA Standard MG-1 series, which dictate that the markings shall be provided on a motor. The most common means of marking a motor is with a metallic nameplate that is permanently affixed to the motor casing. A typical NEMA motor nameplate is shown in Figure 9.2.

A motor nameplate states the manufacturer's data that is applicable to the motor to which the nameplate is affixed. Motors vary greatly in characteristics and capabilities. Some motors are intended for intermittent and relatively light duty, whereas other motors must perform under severe, demanding, and long-lasting conditions. Of course, the cost of manufacturing a motor depends on the needed capabilities. The more severe the conditions, the higher the price. The nameplate on a motor becomes the manufacturer's statement that a motor is suited for an application. The data on a motor

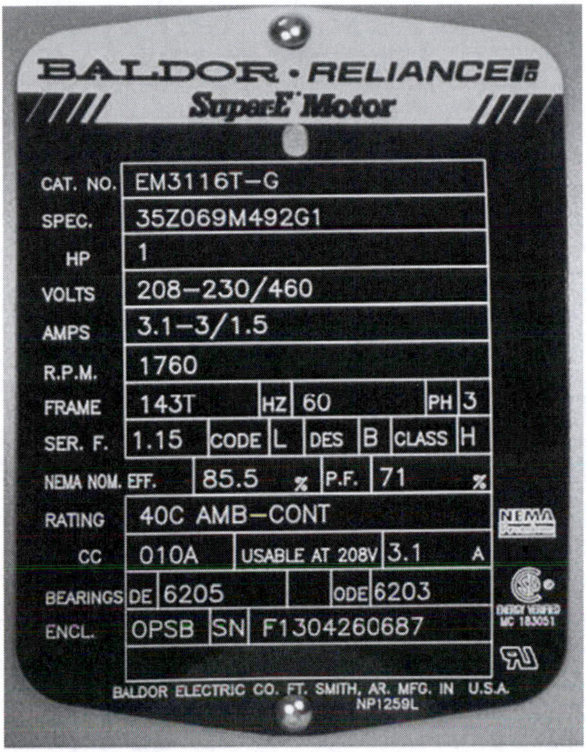

**FIGURE 9.2**
NEMA motor nameplate. (Photo courtesy of Baldor, Fort Smith, AR.)

nameplate is also needed to correctly size conductors, conduit, and motor protective devices. Nameplate data are also helpful whenever a motor is to be replaced.

The NEMA Standard meets the minimal marking requirements of the NEC and imposes some additional requirements. As applicable to three-phase ac motors, both the NEC and NEMA standards require that the following minimal information is to be provided on a motor nameplate:

1. Motor manufacturer's name
2. Rated volts and full load amps (FLA)
3. Rated frequency and number of phases
4. Rated full load speed
5. Rated temperature rise or insulation system class
6. Time rating
7. Rated horsepower if 1/8 HP or more (not required on polyphase wound-rotor motors)

8. Code letter or locked rotor amps (LRA) if rated 1/2  HP or more

9. Design letter for design (B, C, D, or E)

10. Thermally protected motor are to be marked "Thermally Protected" or simply "TP"

The NEMA standard also requires

11. Service factor

12. Frame size

13. Efficiency

Aside from the information required by the NEC or NEMA, motor nameplates often contain supplementary information as

14. Manufacturer's model number

15. Maximum ambient temperature

16. Enclosure type

17. Manufacturer's serial number

18. Bearing data

19. Connection diagram

20. Power factor

While most of the stated requirements are obvious, some cannot be understood without reference to the applicable NEMA standard. Following are explanations applicable to the less obvious requirements of the standard:

1. Rated volts and full load amps: The rated voltage provided on the motor nameplate is the nominal voltage at which the motor is intended to operate. Motors often are purchased for a voltage that is slightly lower than the service voltage. For example where a 480 VAC service is provided to a facility, motors are often purchased with a 460 VAC rating. The 10 VAC difference between the nominal service voltage and the motor ratings is to account for a voltage drop from the meter to the motor leads. Motors should be suitable for starting and, generally, continuous operation at a voltage that is within +10% to –10% of the stated voltage rating. The full load amps is the motor current when the motor is operating at the stated full load rating and is shown as "FLA."

2. Rated frequency and number of phases: Most motors are marked 50 (Hz) or 60 (Hz) and sometimes 50/60. Variable speed drives (VFDs) operate within a range of speeds and the frequencies corresponding to the speeds will be given. A numeral "1" designates single phase and a "3" designates three phase.

3. Rated full load speed: Synchronous speed for synchronous motors and a speed somewhat below synchronous speed for asynchronous motors. The speed of asynchronous motors will vary with load. The stated speed on the nameplate is the speed at full load. Typically, three-phase asynchronous induction motors operate at 96%–99% of synchronous speed at full load. At reduced loads, the speed will rise toward synchronous speed.

4. Rated temperature rise or insulation class: The recognized insulation NEMA insulation classes are A, B, F, and H. The temperature limitations for a 20,000 h motor life are A: 105°C; B: 130°C; F: 155°C; and H: 180°C.

5. Time rating: Motors intended for continuous operation are marked "CONT." Motors intended only for intermittent operation are marked with the respective time period on the nameplate. Stated intermittent time periods are usually from 5 to 60 min.

6. Rated horsepower: The NEMA standard uses the imperial system of units, which is also commonly called the inch–pound system. In the imperial system, 1 HP = 550 in.-lb/s. Also, 1 HP = 745.7 W = 0.7457 kW.

7. Code letter or locked rotor amps: There are a total of 19 recognized code letters of the alphabet from "A" to "V" and which identify the range of kVA per HP for a motor. (The letters "I," "O," and "Q" are not used.) From A to U, the values of kVA per HP vary from 0 to 22.39. Code letter V is for values of 22.4 and greater. The locked rotor amps (LRA) is the current that would result with the rotor locked in-place. Of course the LRA current is also the inrush current that would be present at the instant voltage is applied to a motor.

8. Design letter: Design letters are Design Letters A, B, C, and D. Letter A motors have average to high starting torque and a high corresponding inrush current. Letter B motors have average starting torque and inrush less than Letter A motors. Letter C motors have high starting torque, low inrush, but high full load slip. Letter D motors, much as the Letter C motors, have high starting torque, low inrush, and high slip.

9. Thermally protected: A thermally protected motor has a temperature sensing element imbedded internally, which, on a high-temperature condition, will act to interrupt voltage to the motor thereby protecting it from an excessive temperature. Some thermal protective motors reset automatically, whereas others must be manually reset. A motor that is not thermally protected must depend on other protective means to prevent an overtemperature and potentially damaging condition. (Motor protection is discussed in Section 9.2.)

10. Service factor: The service factor is the number 1.0 or a higher number that defines the actual, short-time power capability of a motor. For example, a motor with a rating of 100 HP and a service factor of 1.15 has the capability of safely producing 1.15 × 100 or 115 HP for limited periods of time. However, long-term operation above the motor rating is inadvisable since it will decrease motor life to some degree. Conversely, a motor operated only at rated horsepower will have an extended service life. A service factor of 1.15 is very common in a wide range of horsepower ratings. Motors with low power ratings often have service factors greater than 1.15.

11. Frame size: Frame size according to the NEMA standard actually describes the mounting dimensions. The frame size has nothing to do with the actual motor casing diameter, length, height, or sturdiness. The dimensions corresponding to a particular "size" (i.e., dimensions) are contained in the NEMA standard. According to the standard, dimensions are given in inches.

12. Efficiency: Motor efficiency is the ratio of input power to output power expressed as a percentage. Large three-phase premium motors, say in the region of 300 HP, typically have efficiencies in the region of 96%. Smaller three-phase motors, say around 1 HP, are usually less efficient and typically have efficiencies closer to 85%.

13. Enclosure type: The NEMA standard defines a number of motor enclosure types. Since motors are used in a wide variety of applications, it can be expected that there is a wide variety of enclosures. Some of the more common motor enclosures are as follows:

    a. ODP—open drip proof: The design allows air to circulate through the motor for cooling, but the internals are protected from drips that are up to 15° off of vertical. Typically used for indoor applications in relatively clean, dry locations. Generally the least expensive type of motor enclosure.

    b. TEFC—totally enclosed fan cooled: No airflow through the motor. An external fan blows air over the exterior of the motor to provide cooling. This type of motor is not watertight. Outside air and moisture can enter the motor, but generally not in adequate quantity to impair performance. Typically used for outdoor and dirty locations.

    c. TENV—totally enclosed nonventilated: Similar to a TEFC but has no cooling fan. It is dependent on natural convection and radiation for cooling.

    d. TEVF—totally enclosed with vented flange.

    e. TEAO—totally enclosed air over: A special motor used for fans. It has no integral fan, but uses the airflow from a fan driven by the motor for cooling.

f. TEBC—totally enclosed blower cooled.

g. TEWD—totally enclosed wash down: Designed to withstand high pressure wash downs or other high humidity or wet environments. Available on TEAO, TEFC, and TENV enclosures. Common in plants where food is prepared.

h. WP—weather protected.

i. EXPL—explosion-proof enclosures: The motor is designed to withstand an internal explosion of specified gases or vapors and not allow the internal flame or explosion to escape from the casing. Typically available on TEFC or TENV enclosures.

j. HAZ—hazardous location: For use in various hazardous locations as defined by the NEC.

### 9.1.3.2 IEC Motor Nameplates

Outside the United States, motor nameplates mostly follow the criteria of the IEC. While Germany, France, and Great Britain have unique standards, their standards closely parallel the IEC motor standard.

The objective of IEC motor standards is identical to that of the NEMA standards, which is to provide a standard type of guideline for data applicable to a motor. However, there are significant difference between the standards of the two organizations. The IEC requirements for nameplate data are published in the IEC 60034 series of standards. (There are a number of applicable standards within the 60034 series that deal with nameplate data. For example, Standard 60034-1 treats rating and performance. Standard 60034-11 treats thermal protection. Standard 60034-12 treats starting performance.)

Outside the United States, the SI system of units is commonly used and the IEC reflects that practice. While the NEMA standard was the primary guide for North American motor manufacturers for many years, increasing globalization has influenced many of these manufacturers to also recognize, and in many instances follow, the IEC guidelines. Consequently, many motors manufactured today in North America are shipped with IEC nameplates.

Both the NEMA standard and the IEC standard use "code" designations to identify features of a motor. The significance of the various codes can be understood only through reference to the respective standard. Unfortunately, the code designations are less than similar. With the IEC nameplates, more frequent reference is needed to the standard. Following are some of the notable differences of the IEC motor standard as applicable to nameplate data:

*Power rating*: The IEC unit of shaft output power is the SI unit of watts or kilowatts. Output in kW can be converted to horsepower by the relationship, 1 kW = 1.3410 HP.

*Insulation class*: Much as NEMA, IEC use a system of codes to designate the insulation class. The IEC classes for code A, B, F, and H are similar to those of the NEMA classes but slightly different. In addition, IEC has a Class "E." According to the IEC code, the temperature rises for an assumed ambient of 45°C and a 20,000 h motor life are A: 60°C; E: 75°C; B: 80°C; F: 100°C; and H: 125°C.

*Duty cycle*: The NEMA standard speaks of "time" capability whereas the IEC describes duty cycles as a way of describing the same property. The IEC has eight duty cycles that are identified as S1 through S8. The S1 duty cycle of the IEC standard is the continuous capability that is identical to the "continuous" rating in the NEMA standard.

*Cooling*: Unlike the NEMA standard that speaks of motor "enclosures," the IEC standard refers to the means of "cooling." Again, IEC uses an extensive system of codes to describe the enclosure and cooling means. Some of the more common IEC codes are

IC01: A motor design that allows air to flow in and out of the motor and the air is forced by a fan

IC40: Equivalent to the NEMA TENV motor

IC41: Equivalent to the NEMA TEFC motor

*Service factor*: The IEC standard does not use service factors. If, say, a power slightly greater than 100 kW is periodically needed for an application, the next larger manufacturer's size must be selected.

*Frame size*: Much as the NEMA standard, the IEC frame sizes describe mounting dimensions. The IEC standard provides dimensions in millimeters, and there are no IEC frame sizes that are identical to NEMA sizes. So mounting an IEC motor where a NEMA motor had been used, or vice versa, can be a problem.

*Enclosure*: The IEC standard uses a two-digit number to describe the ability of the enclosure to resist the intrusion of objects and water. The first digit goes from "0" to "5" and it describes the resistance to the entry of objects. The second digit goes from "0" to "8" and it describes the resistance to water intrusion. In both cases, the "0" designates the least protection and the highest number the greatest protection.

*Design type*: Much as the NEMA standard, the IEC standard describes the torque vs. speed characteristics by a letter code. The IEC codes are entirely different from those of the NEMA standard.

*Power factor*: The power factor is not required on the IEC nameplates but it may be shown. Sometimes the entry on IEC nameplates is stated as "cos" since power factor is the cosine of the current lag/lead with respect to phase voltage.

## 9.2 Motor Controls

The term "motor controls" encompasses a number of devices. While requirements and practices vary around the world, common devices used in a motor control circuit typically include

A disconnect switch

Branch circuit overload protection

Motor circuit conductors

Motor controller

Motor overload protection

The NEC prescribes minimal control circuit requirements for motors in Section 430. To persons who do not deal with motor control circuits on a regular basis, some of the terms used in the industry may seem confusing. Accordingly, some of the commonly recognized terms warrant explanation and are described in the following paragraphs.

### 9.2.1 Motor Controller Terms

There are several terms commonly used relative to the starting, operation, and protection of motors:

*Disconnect switch*: Most motor circuits have a disconnect switch upstream of the motor circuit. The disconnect switch allows personnel to completely and safely isolate a motor circuit from all electrical sources. Most disconnect switches are fitted with a means that accommodate a padlock that can lock the switch in the open position. When a circuit breaker is provided, the circuit breaker may be used as the disconnect means.

*Motor branch circuit overcurrent protection*: According to codes, a motor circuit is required to be protected by an overcurrent protective device as a circuit breaker or a fuse.

*Motor circuit conductors*: Motor circuit conductors must satisfy several criteria. First, conductors must be adequately sized to carry the full load amperage of the motor. Conductors must also be of a size that prevents an excessive voltage drop to the motor terminals. Applicable codes set forth specific requirements.

*Motor controller*: A motor controller is a device that controls the operation of a motor. A motor controller might be a manual controller or an automatic means that permits the starting and stopping of a motor. In its most simple form, a motor controller could be merely a manual switch that delivers electric power to a small motor.

*Manual controller*: A manual controller is a device that includes a set of contacts that are closed and opened by manual activation. Typically, manual controllers are activated by the depression of a button or the rotation of a knob. As used with a motor, a manual controller permits the manual starting and stopping of a motor.

*Motor overload protection*: To prevent premature failure of a motor, it is important that the temperature of the insulation not be exceeded. High temperatures in general will shorten motor life, and excessive temperatures may lead to a failure in the near term. The means used to avoid high insulation temperatures vary. Many small motors have inherent protection; switches within the motor will interrupt current under a detected high-temperature condition. Some large motors have temperature detectors imbedded in the motor's insulation. Under a high-temperature condition, the embedded temperature detector activates an external circuit that in turn interrupts current to the motor. The more common arrangement depends on current measuring elements external to a motor. Specific designs vary but the objective is to interrupt motor current when that current rises to a level that is considered analogous to an excessive temperature condition within the motor insulation. In many common configurations, overload detecting elements open a contactor that interrupts the current to the motor. Overload protection devices are commonly called "overloads."

*Motor starter*: To some extent, the word "starter" as used in "motor starter" is a misnomer since the function of a motor starter is not only to "start" a motor but to continually protect it from an overcurrent after the motor has been started. Motor starters also go by the name, "magnetic motor starters." A conventional motor starter consists of a contactor as well as an overload relay that is provided to detect motor current and which is to act to interrupt power to that motor upon a detected overcurrent condition. Motor starters can be manual or automatic. A manual motor starter is energized by a person in close physical proximity to the device. An automatic motor starter is closed by the activation of the control circuit. Reset of automatic motor starters is by either local manual or automatic. Motor starters are classified as either NEMA rated or IEC rated.

*Contactor*: Contactors are used in motor control circuits where power must be repeatedly applied to a motor and subsequently interrupted. A contactor consists of a set of contacts and an electromechanical coil that, in response to a remote activation, closes and opens a set of contacts. Contactors are an important element of most motor control circuits. The contacts of a contactor must be sized to cycle a motor on and off numerous times. However, those contacts may not necessarily be adequate to interrupt a line short circuit. For this reason, an overcurrent protective device as a circuit breaker or a fuse would be positioned upstream of a contactor, and it must have the capability of isolating the motor circuit under a detected short circuit condition. Contactors for motor applications are usually rated according to horsepower (per NEMA) or kW (by IEC). Unless specifically stated otherwise, most contactors are of the nonreversing design. Reversing contactors are for use with ac motors that must be periodically reversed in the direction of rotation. Contactors are discussed in greater detail in Chapter 12.

*Combination starters*: The term "combination starters" defines an assembly of electrical devices mounted in an enclosure. The assembly includes a motor starter, motor overload protection and protection against either a short circuit or a ground fault. Some combination starters also include a disconnect switch.

*Motor Control Center*: A facility that has a large number of motors will often have a motor control center to house motor controls of a common voltage in a central, protected location. A motor control center (MCC) consists of a steel cabinet that may house the controls for a dozen or more motors. High amperage feeder conductors bring electrical power to the MCC and conductors lead from the respective cubicles in the motor control center directly to the connection boxes on the controlled motors. Motor control centers are generally custom fabricated to meet a specific order. A typical custom fabricated motor control center is shown in Figure 9.3. The MCC cabinets of Figure 9.3 are for 480/3/60 service and are at a 3 MW cogen and LNG facility. Note the high resistance grounds, by Post Glover, mounted above the center cabinet.

A typical three-wire diagram of a manual motor control circuit for a three-phase motor is shown in Figure 9.4. To start the motor in Figure 9.4, the "start" pushbutton is depressed. This action energizes relay M, which locks-in power

**FIGURE 9.3**
Enclosures with motor control centers. (Photo courtesy of Cordyne, Inc., Houston, TX.)

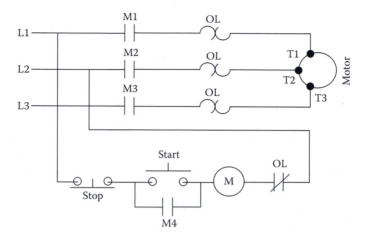

**FIGURE 9.4**
Manual motor control circuit.

to relay M through contact M4. The motor can be stopped by depressing the "stop" push button, which momentarily interrupts voltage to relay M unlatching contact M4. Of course, the motor would likewise be stopped if there is a loss of electrical power, which would deenergize relay M, unsealing the circuit through contact M4. Should electrical power subsequently be restored, the motor would not automatically be restarted since relay M had been deenergized and contact M4 had opened. This is the most common and preferred scenario since it is undesirable to simultaneously start all of the motors at a facility whenever power is lost and then restored. The current surge associated with the start of numerous motors could be very large, unnecessary, and highly undesirable. An overload condition would also stop the motor and prevent restarting until the overload sensors have reset.

### 9.2.2 Types of Motor Starters

Starting a motor causes a stress of some degree to a motor for the reason that the starting action is associated with a high inrush current. Since the starting inrush current is significantly greater than the full load current, there can be local heating that can over time lead to the gradual deterioration of winding insulation. In fact, it is often said that the life of large motors is determined not by the duration of operation but, rather, by the number of starts. Smaller motors are less affected by the number of starts.

There are several accepted methods of starting motors. In general, the method selected depends on several factors including the nature of the load and the motor characteristics. Common methods of starting motors are (1) across the line, (2) star-delta, (3) reduced voltage starters, (4) part winding, (5) variable frequency drives, (6) soft starters, and (7) reversing starters. Following is a brief description of each type of common starting means:

*Across-the-line starter*: An across-the-line (also called direct on line) motor starter, when closed, directly connects a motor with the electrical source. Across-the-line starters are used most often with smaller motors and motors that, because of the nature of the load, will quickly come up to operating speed. An across-the-line starter is the least expensive of the motor starters but also causes the greatest inrush. Nevertheless, across-the-line starters are the most common type of motor starter used for smaller motor sizes.

*Star-delta starter*: A common type of reduced voltage starter is the star-delta starter. Star-delta starters are used in a two-step action that reduces inrush currents. The first step activates the motor winding in a star configuration. In the star configuration, the inrush is considerably less than what it would be in a delta configuration. After the motor has been started and is partly up to speed, the electrical source is connected in a delta configuration. The changeover to the delta activation is usually determined by a timer. Once at operating speed, the motor continues to operate in the delta configuration. Star-delta motor starters are commonly used on large motors and motors with a high inertial load (as centrifugal gas compressors).

*Reduced voltage starter*: A reduced voltage starter typically uses either an auto-transformer or series inductors that reduce line voltage to the terminals of the motor. After a predetermined period of time, line voltage is connected to the motor terminals.

*Part winding starter*: Most induction motors are manufactured with at least two parallel windings. A part-time winding starter activates only one set of the windings to start a motor. After a period of time, all windings are activated to bring the load up to full speed. As with the star-delta starter, the inrush is drastically reduced with this method compared to what it would be using an across-the-line starter. Much as star-delta starters, part-time starters are common on large motors with high inertial loads.

*Variable frequency drive*: VFDs provide an alternate means of avoiding high inrush currents. When started at a low frequency, the associated inrush current is relatively low. As frequency and speed increase, motor current also increases but below a rate that would be encountered with an across-the-line starter.

*Soft starter*: Soft starters are being used more frequently for motors that are started against a heavy load. Soft starters use thyristors to provide current to a motor in bursts while the motor is slowly brought up to speed. In consequence, the net current averaged over a period of time is reduced and considerably less than what would occur with an across-the-line starter. Soft starters are limited to lower-powered motors.

*Reversing starter*: A reversing starter is used for motors that are capable of reversing direction while spinning in an opposite direction. A starter that is not capable of reversing action is identified as a nonreversing starter.

### 9.2.3 Ancillary Motor Protection

A large motor represents a large financial commitment by its owner. For this reason, the larger the motor, the more incentive there is to invest in a level of protection not justified for smaller motors. In critically important applications, small motors may also warrant a higher degree of protection. Motors may be protected by a variety of ancillary devices including protective relays or merely interlocks in the control circuit.

#### 9.2.3.1 Motor Protective Relays

According to ANSI/IIEEE Standard C37.2, some of the protective relays used in the electrical industry have a device number for uniformity, simplicity, and prompt identification. The ANSI device numbers are commonly used in schematic wiring diagrams to identify the types of protective relays that are used. Some of these device numbers are shown in the following paragraphs together with a brief description of the pertinence to motor protection. Various motor protection means that are often selected for large or critically important motors are as follows:

*Overcurrent*: Whereas smaller motors may be protected against an overcurrent condition by motor controllers, large motors may have the function performed with a protective relay that interlocks with the motor's control circuit. ANSI Device Number 50, Instantaneous Overcurrent Relay, or Device Number 51, ac Inverse Time Overcurrent Relay.

*Undervoltage*: According to the NEMA standard (Reference 9.3), motors should be capable of operation and starting at voltages within 10% of nameplate voltage. Starting a motor below 90% of rated voltage is especially undesirable since the condition may expose the motor windings to locked rotor current for a prolonged period of time while the motor fails to rotate. If undervoltage protection is provided, a motor is prevented from either starting or operating when detected line voltage drops below the undervoltage setting. ANSI/IEEE Device Number 27, Undervoltage Relay.

*Overvoltage*: While not considered as potentially damaging as an undervoltage condition, an overvoltage condition nevertheless can lead to excessive heating of a motor's windings. While an overvoltage protective device might be integral with undervoltage protection, the two are also available separately. ANSI/IEEE Device Number 59, Overvoltage Relay.

*Ground fault detector*: A ground fault detector acts to isolate a motor from its electrical power source in the event of a detected ground fault. ANSI/IEEE 57, Short-circuiting or Grounding Relay.

*Adverse phasing*: Problems can develop in the quality of the three-phase power delivered to a motor. It is possible that one phase can either completely disappear or be drastically unequal in magnitude to the other two phases.

The phase sequence might also be reversed. Depending on the motor and the application, serious consequences can follow. ANSI/IEEE Device Number 46, Reverse-phase or Phase Balance Current Relay, or Device Number 47, Phase-Sequence or Phase-Balance Voltage Relay.

*Bearing problems*: Bearing problems are a common cause of motor failures. The most common cause of bearing failures is due to friction in the bearing and subsequent overheating. If detected early, very often the bearing as well as the motor may be saved. ANSI/IEEE Device Number 38, Bearing Protective Device.

*Start inhibit*: Starting a motor with windings that are already at an elevated temperature can contribute to degradation of the insulation. This is particularly true of large motors as the windings in large motors dissipate heat more slowly than the windings in smaller motors. For this reason, many motors have interlocks in their control circuits that inhibit restarts until after a period of time. The inhibit is typically accomplished by means of a time delay relay that begins timing from either the last start or from the time the motor was stopped. ANSI/IEEE Device Number 2, Time Delay Starting Relay.

NOTE: The protection relays listed here are some of the specific types protective relays used for the protection of motors. Protection relays intended for general, electrical applications are discussed in Chapter 12.

### 9.2.3.2 Interlocks

Protective relays can detect specific problems related to an electrical circuit and provide electrical signals that can be interlocked with the control circuit. In addition, external interlocks are frequently entered into a motor's control circuit. The specific interlocks of course depend on the application. For example, a large motor with forced lubrication may have an interlock in its control circuit that confirms adequate lubrication and prevents motor operation if the lubrication motor is not running. A motor that drives a compressor might have an interlock that stops the motor in the event of a detected high vibration at the compressor.

---

## Problems

1. How can the direction of rotation of a three-phase motor be reversed?
2. What is the synchronous speed of a 50 Hz, four-pole motor?
3. Which of the following types of motor has slip rings: wound rotor or squirrel cage?

4. Which NEMA and IEC standards prescribe what is to be included in a motor nameplate?

5. Would a squirrel cage, 60 Hz, two-pole motor operate at exactly 3600 RPM?

6. A 200 HP motor with a service factor of 1.2 can safely be operated intermittently up to what horsepower?

7. Which NEMA frame sizes match the IEC frame sizes?

8. What type of NEMA motor frame would be suitable for a plant where sausage is made and which is hosed down every evening after the day shift?

9. Which NEMA insulation class has the highest temperature rating?

10. What is the primary function of the star-delta starters, the reduced voltage starters, and the part-time starters?

11. What is the IEEE device number of an undervoltage protective relay intended to detect a low voltage condition?

# 10

## Power Factor Correction

Lagging power factors cause difficulties for both the utilities that must transmit and deliver electric power as well as the utilities' customers. Electrical equipment that has a low power factor necessarily has associated higher currents for a given energy usage. In the way of a specific example, consider the computed values of current for a hypothetical load of 1000 kW with various power factors as tabulated in Table 10.1.

As indicated in Table 10.1, the lower the power factor, the higher the current for a given amount of transmitted energy. Higher currents in turn require larger transformers, larger cables, generators, and auxiliary gear. To avoid transmitting power at a low power factor, utilities often install capacitor banks at distribution centers. To encourage customers to correct low power factors, utilities often impose a penalty on customers with a low power factor.

## 10.1 Low Power Factor Penalties

A charge imposed by a utility on a customer for a power factor below a preferred value seems entirely warranted. Low power factors necessitate greater capital expenses by the utility. Some utilities forgive small users for a low power factor penalty. Some utilities charge only large users, while other utilities forego the charge entirely regardless of size. In many areas, the charges can represent a large part of a bill. For those who are charged, the definition of a "low power factor" will vary from one utility to another. Some utilities charge for a power factor below 0.95. Other utilities start at lower power factors and some as low as 0.85. The longer electricity is supplied to a customer at a low power factor, and the lower the power factor, the greater the penalty that appears in the monthly billing. It is difficult to generalize on potential savings that may be obtained with the use of capacitor banks because penalties vary greatly. Nevertheless, for a medium-sized commercial building or a medium-sized industrial facility, an annual penalty for low power factor in the range of $10,000–$100,000 is not uncommon. With a proper installation of capacitor banks, the penalty can be entirely eliminated.

While a charge for a low power factor may in many instances be a reason for raising power factor, in other instances, there are different reasons.

**TABLE 10.1**

Current vs. Power Factor for a 1000 kW, 4160 VAC Load

| Power Factor | Current (A) | Increase in Current (%) |
|---|---|---|
| 1.00 | 138 | 0 |
| 0.95 | 146 | 6 |
| 0.90 | 254 | 11 |
| 0.85 | 163 | 18 |
| 0.80 | 173 | 25 |
| 0.75 | 185 | 34 |
| 0.70 | 198 | 34 |
| 0.65 | 213 | 54 |
| 0.60 | 231 | 67 |

Transformers are rated by VA and not kW. So, if a transformer is approaching full rated VA capacity and the power factor is less than unity, raising the power factor will allow an increase in kW. In most cases, the continued use of an existing transformer is a much less costly alternative to replacing an existing transformer.

## 10.2 Low Power Factor Causes

Low power factors can be caused by a number of electrical energy-consuming devices including fluorescent lighting, transformers, arc furnaces, welding equipment, and induction motors. Mostly the common causes of low power factors are induction motors, which always have a lagging power factor. If uncorrected, most commercial buildings, schools, and the like will have power factors in the range of 0.80–0.90. Industrial facilities mostly have uncorrected power factors below 0.80.

## 10.3 Capacitors for Power Factor Correction

In those installations with a lagging power factor, corrections can be made by the use of capacitors wired in parallel with the load. When wired parallel to a device with a lagging current, the result can be an improvement in net power factor, namely, one that approaches unity. The result in many instances could be a reduced monthly electric bill. Very often the return of investment (ROI) for the cost of an installation is significant, and the payback period can be as short as a few months.

Capacitors used to correct a lagging power factor are generally installed either at a single user, as a motor, or in banks that could be installed on a bus or at a service entrance. Banks of capacitors could be either of a fixed value or automatically switched in steps. Capacitor assemblies for three-phase applications would of course have a minimum of three capacitors. In banks, there could be a large number of capacitors to allow switching between various levels.

While it is true that capacitor banks can be used to correct a lagging power factor, it goes without saying that the installation must be done in a correct manner to be of benefit. There are associated pitfalls that are to be avoided. Done poorly, an installation can cause more problems than it solves. Generally, it is advisable to use the services of a consultant or firm with broad experience in designing capacitor bank systems for power factor correction.

## 10.4 Capacitors Defined

What, exactly, is a capacitor? A capacitor is an electric component that can store an electrical charge. Capacitors are used in both dc and ac circuits. All capacitors contain external conductors, or terminals, that connect to conductors within the envelope of the capacitor. Inside the capacitor envelope, thin layers of conducting metal or foil are connected to the terminals and are separated by layers of a nonconducting dielectric material. The dielectric material acts as an insulator between the two opposing layers of conducting material. The smaller the separation between the conductors, the greater the capacitance value. Likewise, the greater the surface area of the conductors, the higher the capacitance value. In capacitors of the size used for power factor correction, the dielectric material is commonly thin sheets of polyester or polypropylene. The layers of conducting foil and dielectric material are usually rolled into the shape of a tube. The assemblies are often enclosed within a metallic cylinder or metallic, rectangular-shaped box. Some designs contain fuses within the enclosure. For safety reasons, capacitor assemblies intended for use on three-phase systems generally contain resistors that drain the electrical charges from capacitors within several minutes after power is removed.

The value of capacitance is defined by the letter "C." The value of C is the ratio of the electric charge, Q, to the potential, V. The units of Q are coulombs and the units of capacitance are farads (F).

## 10.5 Capacitor Locations

Where to locate capacitors to correct a lagging power factor? Since motors are the most common source of a lagging power factor, it would be natural to consider the installation of capacitors near each motor and to wire the capacitors in parallel with the motor. An installation of that description would certainly be a prospective solution and one that merits consideration. Other possible locations are near the service entrance or connected to buses that have low power factors.

### 10.5.1 Capacitors near a Motor

Some of the prospective connections to a circuit serving a motor are illustrated in Figure 10.1. Each of the possible locations is identified as location A, B, C, and D. Each prospective location has merits.

*Location A*: This location has a particular merit in an installation that has long conductor leads to the motor. Installation of the capacitor bank near the motor would reduce currents in the branch lines to the motor and cause less voltage drop to the motor leads. Overloads must be sized to account for the net lower line currents. Comments #1 and #2.

*Location B*: The overloads can be sized in accordance with normal procedure as the capacitor currents do not pass through the overloads. Comment #2.

*Location C*: This location averts the possibility of an excessively high voltage that could be caused by stopping and then starting the motor shortly thereafter. This is a better position for motors that are jogged frequently, multispeed motors, and reversing applications. Comment #3.

*Location D*: Location D allows sizing of the circuit breaker, motor starter, and overloads according to conventional procedure. The location can also provide power factor correction for other motors that would be a more cost-effective approach to low power factor correction. In position D, a dedicated disconnect switch and dedicated overload protection are required for the capacitors. Comment #3.

*Comment #1*: Position A would not function well if the motor loading is variable. At part load, the capacitor bank would cause the current to lead excessively.

*Comment #2*: In positions A and B, there is the danger if the motor starter is opened and then closed shortly thereafter. The motor could act as an asynchronous generator when the contactor is closed and possibly cause

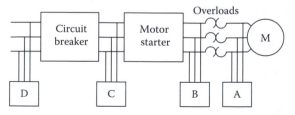

**FIGURE 10.1**
Prospective capacitor bank locations.

an excessively high voltage in the circuit after the contactor. The problem can be averted if interlocks are provided to prevent a restart of the motor within a predetermined period of time after the motor has been stopped. Some problems might occur if the capacitors are switched on and off with the motor. This is a relatively simple installation that functions best with fixed capacitors. Capacitors should not be installed in positions A or B with soft-start starters.

*Comment #3*: Positions C and D might require switching the capacitor bank off when the motor is not operating. Otherwise overcorrection might result, and this would permit a multistep capacitor bank to be used with a motor that has a variable loading. In positions C and D, there is no question of the capacitor bank interfering with the function of the motor starter to protect the motor.

### 10.5.2 Capacitor Bank at Service Entrance or Bus

Rather than have individual capacitor assemblies dedicated to a motor, it is often better to have a capacitor bank further upstream. For large installations with a number of buses, capacitor banks might be better located common to a bus that has a low power factor. For smaller installations, a capacitor bank near the service entrance might be the better choice. For capacitor banks intended for installation common to buses or at the service entrance, the automatic type of capacitor bank is a better choice. The automatic configurations consist of a number of capacitors that can be switched in or out in steps depending on the power factor. Controllers measure the power factor and determine when to engage more or less capacitors.

## 10.6 Harmonic Mitigation

More and more users of three-phase electrical power are using electrical devices that cause harmonic distortion of the electrical source. Devices that are known to cause harmonic distortion are

1. Fluorescent and mercury lighting
2. Welders
3. Electric arc furnaces
4. SCR's used to generate a dc supply
5. Variable speed dc drives
6. Variable frequency ac drives

Typical results from harmonic distortion are shown in Figure 10.2 along with representation of a normal sine wave electrical supply. Harmonic distortion is highly undesirable for several reasons. First, it can cause overheating of

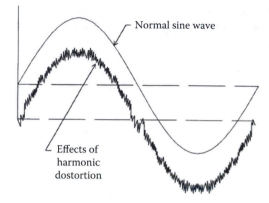

**FIGURE 10.2**
Harmonic distortion.

motors, transformers, and conductors. Harmonic currents can cause nuisance trips, blown fuses, and damaged capacitors. Sharp spikes of harmonic distortion can interfere with communications, and some instances will cause automatic controls to act spuriously. (Harmonic currents are also discussed in Section 5.4.)

The addition of capacitors to correct a low power factor at a facility can enhance the potential for interaction between the capacitors and the service transformer. In a facility that already has some harmonic distortion, the addition of the capacitors can worsen the distortion unless precautions are taken. The result can be harmonic currents of significant magnitude flowing back and forth between the transformer and the capacitors. The phenomenon is known as "harmonic resonance." Without compensation, a typical facility with power factor correction may experience harmonics in the vicinity of the 5th to the 13th harmonic. If capacitors are to be added to a facility to correct power factor, it is recommended that steps be taken to mitigate the potential for harmonics.

The most common procedure followed to minimize the effects of harmonics is with the use of inductors wired in series with the capacitors. A typical configuration of capacitors and inductors is shown in Figure 10.3. The inductors are sized in recognition of the values of the capacitors.

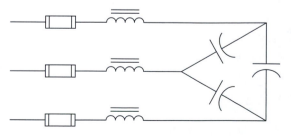

**FIGURE 10.3**
Capacitors and inductors.

Properly done, the net effect is that the combination of inductors and capacitors is detuned away from resonant conditions and the potential for harmonic currents would be minimized. Fuses are shown in Figure 10.3 as these would be required according to code to protect both the capacitors and the conductors. Resistors are generally provided to drain electric charges from the capacitors, but these are not shown in Figure 10.3.

It is pertinent to note that in some applications, capacitors are added solely for the purpose of correcting a system with harmonic distortion and not necessarily for the purpose of power factor correction. In some applications, the two objectives can be met simultaneously with the installation of capacitors.

## 10.7 Special Considerations

Power factor correction for any of the following motors presents some unique problems: autotransformer, series resistance, part winding, wye-delta, or reactor. For applications of these kinds, it is advisable to consult with a specialist who has specific experience in the field.

Switching of capacitors on and off with contacts that are of an electromechanical design can generate voltage spikes that can be a source of difficulties. Designs that use thyristors to provide "zero switching" have proved to be especially beneficial in eliminating this source of voltage spikes.

## 10.8 Sizing Capacitors

A sample calculation to determine the required sizing of a capacitor bank was performed in Example 4.5. In the assumed example, it was calculated that the installation of a capacitor bank with current of 86.61 A will return the service line current to unity power factor (from a lagging power factor of 0.50) and reduce the service current from 100 to 50.00 A. It is mentioned in Example 4.5 that the power consumption would be unaltered. So, in the example, there would be a resulting reduction in the utility's charge for electrical service.

Some manufacturers of power factor correction equipment size a capacitor bank according to the kVARs required to correct a lagging power factor. This method is a convenient approach to determining the required size of a capacitor bank. Using this approach, the time-consuming calculations of the type performed in Example 4.5 would not be required. To determine capacitor kVAR for the assumed values of Example 4.5, a diagram is first needed to represent a circuit's imaginary power (kVA), the real power (kW), and the reactive power (kVAR). Relative to Example 4.5, the respective values are computed as follows:

Imaginary power (kVA): For a three-phase circuit, imaginary power is

$$kVA = VI\sqrt{3} \div 1000$$

$$= \left[(480)(100)(1.732)\right] \div 1000$$

$$= 83.14 \text{ VA}$$

Real power (kW): For a three-phase circuit, real power is defined as

$$kW = VI\sqrt{3}\,(PF) \div 1000$$

$$= \left[(480)(100)(1.732)(0.50)\right] \div 1000$$

$$= 41.57 \text{ kW}$$

The reactive power (kVAR): For any circuit, single-phase or three-phase

$$(kVA)^2 = (kW)^2 + (kVAR)^2$$

$$= (83.139)^2 = (41.569)^2 + (kVAR)^2$$

$$(kVAR) = 71.999 \text{ VA}$$

$$= 72.0 \text{ VA (required size of the capacitor bank)}$$

These values are shown drawn to scale in the diagram of Figure 10.4.

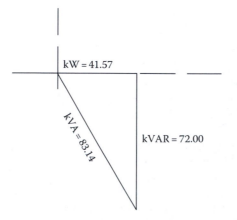

**FIGURE 10.4**
Diagram of kW, kVA, and kVAR.

## Problems

1. A motor at a plant has nameplate data of 200 HP, 460-3-60 VAC, FLA of 240 A, and PF of 0.80. Three AWG 250 conductors, which have an ac resistance of 0.054 $\Omega$/1000 ft, extend to the motor. Each of the three conductors is 300 ft in length. What is the voltage drop at full motor load in each of the conductors?

2. With the motor of Problem #1 running at full load, what is the net energy loss in the three conductors connected to the motor?

3. What is the kVAR value required of capacitors, per phase, to correct the power factor of the motor in Problem #1? Assume the capacitors are to be installed adjacent to the motor and the PF is to be corrected to unity at FLA.

4. Assume a facility has a 480/3/60 service supply but has been experiencing a lagging power factor of 0.75. An automatic capacitor bank is being considered for installation near the service entrance to correct the low power factor to unity. Peak power was measured at 850 kW. What would be the capacitor bank current at peak load?

5. A13.8 kV transformer is rated 1,300,000 VA. However, it was noted that at some peak loads, the VA was 1,234,567 with PF of 0.81. This condition comes within 65,433 VA of the transformer's rating. If a capacitor bank is installed to restore PF to unity what would the VA margin become?

6. Name two types of materials commonly used as the dielectric material in capacitors used for power factor correction.

7. What is the difference between fixed capacitors and an automatic bank of capacitors?

8. A 20 HP motor has nameplate data of 230/3/60 VAC, FLA 54 A, and PF of 0.83. What value of capacitors, in farads, is required per phase to restore PF to unity?

# 11

## *Tariffs*

As is the case with many English words, the word "tariff" has more than one definition. A common definition is the tax imposed on imported goods. The second definition is the rate structure charged by a party for services or goods. This second definition is used by many utilities in reference to their published charges for electrical service. Every utility will have a rate structure that it charges for its electrical service, but not every utility will necessarily identify that rate structure as a "tariff." Some utilities use the term "rates" when referring to their schedule of charges.

Tariffs for electricity vary greatly around the world and often within a country. A large number of factors determine the cost of electricity in any particular region. The two major expenses that determine the cost of a customer's electricity are the cost of energy and the cost of capital.

Obviously, the cost of fuel is an expense to a utility. If a utility's fuel costs rise, the utility must raise its fees to help pay for the greater expense. Worldwide, the costs of both fossil and nuclear fuels have been steadily rising, and likewise, the costs of electricity have been trending upward. Some sources of electrical energy as hydro, solar, and wind are free. There is no energy expense associated with these sources, but there is a capital expense that must be reimbursed. The cost of capital is the other large expense that every utility must address.

## 11.1 Residential Tariffs

Residential users of electricity have limited options available that may be taken to reduce the monthly billing for electricity. Time of use (TOU) has become one option that has been offered to some residential customers. By reducing electrical usage during peak periods, a home owner with a TOU option can save on the electrical bill. Some utilities also offer cost-reducing programs whereby a home owner can agree in advance to allow the periodic interruption, by the utility, of the operation of certain appliances during peak periods.

A small percentage of home owners have taken the step of installing solar panels or wind turbines to offset the monthly electrical bill.

## 11.2 Commercial and Industrial Tariffs

Commercial and industrial customers who have a three-phase electrical service are charged for their service according to a tariff that is very different from the residential schedule. Unlike residential customers, the customers of three-phase electricity are typically classified into one of the several possible tariff categories. And within a particular classification of a tariff schedule, there will be a number of possible options available to users for reducing the monthly billing. Some utilities have a large variety of tariffs. Some California utilities have dozens of tariff schedules. Typical categories of tariffs offered by these utilities are

- Residential
- Residential TOU
- Maximum demand of 1000 kW
- Commercial A/C programs
- Demand bidding programs

A three-phase customer would be well-advised to understand the possible options available. Often a few minor, and sometimes inexpensive, changes can be made to significantly reduce the cost of purchased electricity.

In most areas, the cost of a three-phase service in any tariff category will have several components to it. The most common categories are

- Energy (kWh)—peak
- Energy (kWh)—off peak
- Demand (kW, current, or VA)
- Power factor penalty

### 11.2.1 Energy Charges

Customers of a three-phase service very often can reduce the charge for energy (kW-h) by one or more methods. With the more frequent appearance of the smart meters, the number of options is generally increasing. Shifting energy consumption to off-peak times is very often one of these options. However, many three-phase customers find it impractical to shift electrical usage as other considerations dominate. Sometimes, savings can be found in other charges appearing on the monthly electrical bill.

### 11.2.2 Demand Charges

Many larger customers of a three-phase service are billed for electrical "demand." The term "demand" may appear on the monthly billing although the charge may appear under a different title. "Billing capacity" is one such term sometimes used in lieu of "demand." Typically demand charges are a measure of the power (kW), current, or VA consumption integrated over a predetermined time period of time. The rates for demand vary greatly from one utility to another and are typically in the range of $0–$6/kW. For many years, demand was measured by an electromechanical demand meter that determined the demand charge for a billing period. The demand meter had only one method of measuring demand regardless of the time of day or time of year. However, with the advent of the smart meters, some utilities are adjusting demand charges up or down depending on time of day and time of year.

Following are some specific examples of demand charges.

### 11.2.2.1 NorthWestern Energy, Montana, and North Dakota

The utility uses a 15 min period for measuring demand. Customers are cautioned that operating a motor for a short period during a billing period will result in a demand charge that is a high percentage of the billing. It is stated in the way of example that if a customer runs an irrigation pump for only 5 h during a billing period, the demand charge would be 98% of a $301 billing. On the other hand, if the pump were operated for the entire billing period, the demand charge would be only 24% of a $1243 billing.

### 11.2.2.2 Roanoke Electric Cooperative

The utility cites two cases in which the kWh consumption is identical but the demand charge is different.

*Case #1:* 20 kW load for 50 h results in 1000 kWh of energy usage and a 20 kW demand charge. The billing for the period would be $297.70.

*Case #2:* 2 kW for 500 h also results in 1000 kWh of energy usage but a demand charge of only 2 kW. The billing for the period would be $144.70.

### 11.2.2.3 Georgia Power and Light

Georgia Power and Light use an intricate method to calculate the billing of those customers with demand meters. Their policy is briefly explained as follows:

> Georgia Power is a summer peaking utility because most of its customers have their highest usage during the months of June through September. The Company's greatest operating expenses occur during these months. In order to assist with the cash flow of both Georgia Power and its customers, the Georgia Power and Light tariff uses a billing demand, also called a "carryover" or "ratchet" demand. The billing demand is a calculated number, based on the peak demands set in the current month, plus the previous eleven months. Each month, the twelve month "sliding window" moves forward in time, adding the current month and dropping the same calendar month from the previous year.

The policy of billing is further explained as follows:

> Georgia Power's billing system calculates the customer's billing demand each billing month. Summer months are the billing months of June through September, while winter months are the billing months of October through May. The determination of summer or winter month depends on the date the customer's meter is read. For example, a meter read on June 15 indicates the summer billing month of June. Using the twelve month "sliding window," different criteria are used to determine billing demand, based on whether it is a summer or a winter month. Seasonal and carryover benefits of the Georgia Power and Light tariffs allow customers to increase demand in winter months and off peak times to certain levels without impacting their billing demand. In order to determine the appropriate Georgia Power and Light tariff, the billing system considers 60% of a winter demand or 95% of a summer demand, whichever is higher, in the twelve month window. The 100% of the current month criteria is not used in this determination.

A customer's bill is calculated through reference to a number of tables that are dependent on a number of factors including net energy consumption in the billing period.

These examples of billing practices related to demand illustrate how a customer of electricity could be severely taxed for a very limited operation of electric equipment. If that operation occurs within a billing period when electrical energy consumption within the billing period might be on only a few occasions that demand charge could be especially hurtful. Owners of irrigation pumps and air conditioning chillers are two categories of electric customers who are frequently surprised with a high demand charge on a billing that shows up near the beginning or end of a season of general heavy use.

Sometimes a slight modification of electrical usage practices can result in significant savings in the demand charge. Demand charges are discussed in detail in Chapter 6, and the operation of demand meters is explained in Chapter 7.

### 11.2.2.4 Power Factor Penalty

Most customers of three-phase electrical power have electrical gear that is inclined to cause a lagging power factor. Motors, which are predominantly of the induction type, are the major cause of a lagging power factor. Transformers and some lighting ballasts can also be a cause. Most, but not all, utilities charge a penalty for a low power factor. Billing practices used by utilities who charge a penalty vary greatly. So, generalizations on the subject are not easily made.

Following are some examples of penalties charged by utilities for a low power factor. (Note that billing practices followed by utilities often change from year to year.)

### 11.2.2.4.1 *Clark Public Utilities, Vancouver, WA*

Adjustment of Demand for Power Factor: Demands will be adjusted to correct for average power factors lower than 95%. Such adjustments will be made by increasing the measured demand 1% for each 1% or major fraction thereof by which the average power factor is less than 95% lagging.

Metered real energy measured in kilowatt-hours (kWh) and metered reactive energy measured in kilovolt ampere-hours (kVARh) are used to calculate the average monthly power factor as follows: Power factor = kWh/(square root (kWh squared + kVARh squared))

### 11.2.2.4.2 *Southwestern Electric Power Company*

Should the average lagging power factor during the month be determined to be below 90%, the customer's kilowatts of billing demand will be adjusted by multiplying the kilowatts of billing demand by 90% and dividing by the average laggings power factor.

### 11.2.2.4.3 *Florida Power and Light*

This large Florida utility does not charge a power factor penalty.

To reduce or eliminate power factor penalties, capacitor banks may be installed by a user. The costs of installation and maintenance are the owner's responsibility. Capacitor banks and their use are treated in detail in Chapter 10.

## 11.2.3 Miscellaneous Charges

A bill for three-phase electrical power may contain a listing of a number of miscellaneous charges, most of which are fixed by the utility and nonnegotiable. In addition to the common charges for energy, demand, and low power factor the following are separate listings sometimes found on electrical bills:

Meter charge—this is a charge for having an active electrical service to a facility.

Minimum billing—a minimum charge for an active electrical service.

Transmission charge—a charge for the transmission of electrical power.

Nuclear decommissioning—a fund set aside for the eventual decommissioning of a utility's nuclear plant.

Customer charge—a charge to help pay for overhead expenses as mail, maintenance, line service, and reading meters.

Environmental charge—a fee to pay for equipment specifically mandated by regulators to protect the environment.

Fuel charge—a separate listing of fuel costs which often vary from 1 month to the next.

Taxes—remuneration for taxes a utility must pay.

## 11.3 Summary of Common Electric Tariffs

A user of three-phase electrical power would generally be well-advised to have an experienced professional review the monthly electric bills to consider changes that might be made to reduce charges.

## Problems

1. A small sized commercial customer uses electric duct heaters to heat the facility during winter months. The owner estimated that in one especially cold month the portion of the monthly electric bill attributed to the heaters was 7200 kWh. The utility charges an energy rate of $0.065/kWh. How much of the bill is due to the heaters?

2. A warehouse has an older type meter that measures both energy use and demand. Assume the ventilation fan is the only electric usage in the building and the 20 kW fan is operated only 8 h a day on 22 days of the month. Assume a $0.10/kWh energy charge and a $4/kW demand charge. What would a typical billing be?

3. Consider the aforementioned billing practice of Southwestern Electric Power Company with regard to penalty for lagging power factor. Assume a customer has a billing for a month of 17,600 kWh and the energy rate is $0.062/kWh. The average power factor during the period was 0.86. What is the adjustment to the billing for the lagging power factor?

4. Reference is made to the tariff schedule of Clark Public Utilities, Vancouver, WA. Assume that a customer in a different utility area has a similar tariff schedule and the monthly bill that states the following: (1) energy consumption of 10,000 kWh, (2) energy charge of $0.065/kWh, (3) measured demand of 150 kW, (4) demand charge of $3.80/kW, and (5) kVARh of 4,000. What part of the monthly billing is for energy, demand, and power factor penalty?

# 12

## Relays and Contactors

### 12.1 Relays and Contactors: General

Relays of various types are an integral and necessary part of the control and protection of larger electrical gear as transformers, motors, and generators. The more expensive the electrical gear, the greater the number of relays that will generally be used. The basic function of a relay is to isolate one voltage level from another level. For example, a relay operating with a coil voltage of 120 VAC might be used to call for the closure of a 4160 VAC circuit breaker that operates a large motor.

There are numerous types and classifications of relays, some of which are intended for very special and unique applications. The most common type of relay is the control relay. Other types of commonly used relays include the time delay relays, protection relays, solid-state relays (SSRs), and reed relays. Generally, control relays are intended for controlling relatively low current applications, more or less in the range of 300–2000 VA. Mercury-wetted relays were popular years ago and may still be found in existing control circuits. However, today few organizations allow the use of mercury in any form due to its high toxicity. As a result, mercury-wetted relays are seldom used today.

By common usage, the term "contactor" is applied to a special type of device that is functionally similar to the relay. Yet, a contactor is not normally called a "relay." Rather, the term contactor is used to describe a type of device that is used very much in the same manner as a relay.

Relays of all types are generally integrated into a logic circuit together with other relays and controls to perform specific control and sequencing functions. Typically, the logic required of a control system is described by logic diagrams and ladder diagrams. The proper selection of specific devices to perform the functions described by a diagram can be a challenging task. If selections are not made with due recognition of the capabilities and limitations of electrical devices, the result can be the premature failure of components. Those premature failures can result in the loss of availability of expensive electrical gear and, in some instances, the loss of protection.

The most commonly used relays and contactors are reviewed next.

## 12.2 Electromechanical Control Relays

By definition, electromechanical control relays are constructed of both electrical and mechanical components. One of the basic electrical parts of the design is a coil that consists of numerous windings and that, when energized, produces a magnetic field in a ferromagnetic core. The resulting force moves components that in turn shift a set of electrical contacts. When the magnetic field is collapsed, the contacts revert to the position corresponding to the de-energized state of the magnetic field. The two most common types of relay-actuating mechanisms are the hinged armature and the plunger. An armature-type actuator is shown in Figure 12.1. When the coil is energized, the magnetic forces generated by the coil act to close an air gap beneath the armature. The resulting motion of the armature shifts an adjacent set of contacts. When the coil is de-energized, collapsing the magnetic field, a spring retracts the armature and it reopens the air gap, reversing the position of the contacts.

A plunger-type mechanism is depicted in Figure 12.2. When the electrical coil is energized, the magnetic forces draw the plunger into the center of the coil against the forces of a spring. Movement of the plunger shifts a set of contacts. When the coil is de-energized, the spring forces the plunger out of the center of the core of the coil and thereby reverting the contacts to the position corresponding to the de-energized position.

Relay coils are available in both dc and ac voltages. In all designs of actuators energized by an ac voltage, the magnetic field must be biased with a metallic ring called a "shading ring." Without the shading ring, the magnetic field would reverse direction numerous times every second and would not be functional. With the shading ring, there is no reversal of the magnetic field.

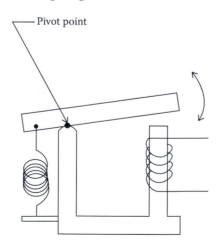

Pivot point

**FIGURE 12.1**
Armature-type actuator.

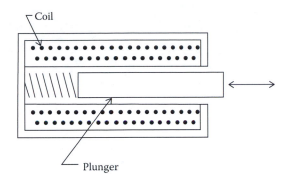

**FIGURE 12.2**
Plunger-type actuator.

If the activating voltage is dc, a shading coil is not needed as the magnetic field would be consistently in one direction. Because of the time required to shift either an armature or a plunger from one position to another, electromechanical relays are relatively slow-acting devices. Closing and opening times are in the range of 5–15 ms.

Reed relays are considered a type of electromechanical device since they contain components that are both electrical and mechanical. Yet, reed relays are normally treated separately from most electromechanical relays. Reed relays are discussed later.

For many years, the electromagnetic relay (EMR) was the major element used to perform the logic described by a ladder diagram. In fact, the term ladder logic is often used in reference to the use of relays to perform a sequence or logic. With time, PLCs have been used more and more to perform the functions that had been achieved for years with the electromechanical relays. This has become the case particularly where a larger number of functions are to be performed. Yet, the electromechanical relays are still the better choice if there are only a few interlocks to be performed. In addition, some organizations prefer the use of EMRs to a PLC because troubleshooting can be quicker and more readily performed.

Some of the symbols associated with electromechanical control relays are shown in Figure 12.3. The Joint Industrial Council (JIC) symbols shown in the figure are used mostly in North America and the IEC symbols are used primarily in Europe.

One of the oldest types of relays, and perhaps best known, is commonly called the "control relay." Control relays go by a variety of other names including "machine tool relay," "industrial relay," and "general purpose relay." Control relays are available in numerous configurations. Some of the more common types are mentioned next.

*Cube relays*: Cube relays are also called "ice cube relays" by some manufacturers and "general purpose relays" or "industrial relay" by others. The term cube comes from the shape of the relays that is similar to that of some shapes

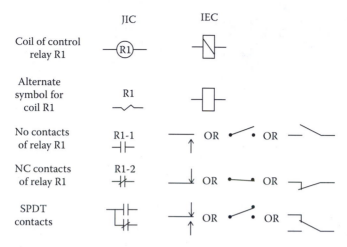

**FIGURE 12.3**
Control relay symbols.

of ice cubes. Although the devices are in the shape of an ice cube, they are nevertheless larger than a conventional ice cube and typically in the range of 1.7 in. × 1.7 in. × 2.4 in. high (4.3 cm × 4.3 cm × 6.0 cm high). Cube relays use the armature-type actuator and are available in several mounting configurations. One popular type of mounting uses a mounting base that is fixed in-place and that accepts conductor terminations. The relay is then plugged into the base. Two common types of plug-in mounts are the octal (also called socket or tube) and the spade. A plug-in type relay can be readily removed from its base and replaced with another plug-in relay in a matter of seconds as no conductors of a circuit need to be moved. Another type of relay mount is the "surface mount," which accommodates mounting of the relay on a flat sheet metal surface. With a surface mounted relay, conductors must be connected to terminals on the relay.

Compared to other types of relays, the cube relays are relatively inexpensive and are readily available worldwide. Popular models of cube relays are available in one pole to four pole configurations and normally open (NO), normally closed (NC), and double throw (DT). Coil power is in the range of 1.5–2.5 VA. Cube relays are available in a number of coil voltages. Contacts typically have a rating that is between 3000 and 7000 VA resistive. The relays are generally well suited to most control circuits as well as the control of motors up to 2 HP. The operating components of cube relays are protected to some extent by a dust cover, and in some models, the terminals are receded to minimize the potential for fingers to come in contact with live electrical parts. A cube-type relay of the plug-in type, with spade terminals, is shown in Figure 12.4 along with the matching plug-in base. Smaller versions of cube relays known as "miniature cube relays" are actually near the size of ice cubes. One popular model of a miniature cube relay measures 0.82 in W × 1.09 L × 1.46 high (2.0 cm × 2.7 cm × 3.7 cm high). The VA

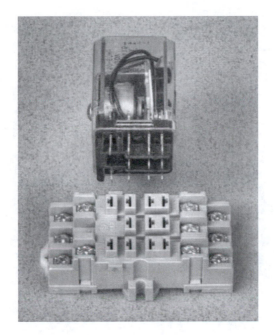

**FIGURE 12.4**
Cube relay with plug-in base. (Photograph by the author.)

capabilities of the miniature cube contacts are often comparable to that of the larger ice cube relays. The coil power consumption of the miniature cube relays is normally in the range of 1.5 VA.

*Machine tool relays*: The term machine tool relay is applied to a type of relay that has traditionally been used in the manufacturing industries. Machine tool relays are also called "industrial relays" or "heavy duty relays." The machine tool relays are generally considered as better suited to industrial and manufacturing requirements than the less-expensive cube relays. The automotive manufacturing industry in particular has been a regular user of this type of relay. Machine tool relays differ in appearance and construction from cube relays. Unlike cube relays, machine tool relays very often have field-replaceable contacts as well as replaceable coils. In most designs of machine tool relays, for example, a NO contact can generally be replaced in the field with a NC contact and vice versa. Some designs of machine tool relays are available with up to 12 poles. Many machine tool relays can be expanded in the field to add poles. In many designs, for example, a two pole relay can be readily changed to a three pole relay by merely adding contact "cartridges." Contacts in machine tool relays are enclosed to exclude to some extent the entrance of dust, debris, and liquids. Machine tool relays usually have a plunger-type actuator. A popular type of machine tool relay is shown in Figure 12.5.

*Open relays*: In most applications, relays of any design are enclosed within a cabinet. Sometimes, a cabinet that houses relays would be locked primarily

**FIGURE 12.5**
Machine tool relay. (Photograph by the author.)

for the purpose of preventing inexperienced persons from coming in contact with the voltage present on metallic components within the cabinet. A cabinet might also be locked to avoid malicious tinkering. Because of the protection afforded by a cabinet, some organizations use relays that are considered an "open" design. Open relays do not have covers or enclosures of the types used with cube relays or machine tool relays. Some models have terminals that are located atop the device and that are much more exposed than the terminals on either the cube relays or the machine tool relays. While the exposed terminals on open relays present a greater potential for shock to personnel who open the cabinet, the design more readily accommodates fabrication in a high-volume manufacturing process. The HVAC manufacturing industry in particular uses great numbers of the open relays. Open relays are available in a variety of contact configurations and ratings. Much to the credit of the open relay manufacturers, open relays are available with a variety of pilot duty contacts. Some designs of open relays use the plunger type actuators, whereas others use the armature actuators. (Pilot duty contacts are treated in detail later.)

A design of an open-type relays is shown in Figure 12.6. The relay is intended for surface mounting. It has screw-type terminals and an armature actuator. This particular model is called a "power relay" by the manufacturer. It has two poles that are NO and is rated 1.5 HP each pole.

*Latching relays:* The types of electromechanical relays mentioned earlier are described as "nonlatching relays." Nonlatching relays are said to have a normal, or shelf, position. For example, if a set of contacts is said to be NO, that set of contacts will be in the open position when the relay is either on a shelf

**FIGURE 12.6**
Open-type power relay. (Photograph by the author.)

or when the coil is in a de-energized electrical circuit. Whenever the relay's coil is energized, the contacts will be shifted to the opposite position—NO contacts will be closed and NC contacts will be opened. Latching relays have a different mode of operation. A latching relay remains in its last position, and it will remain in that position indefinitely until it is moved to the opposite position. A momentary pulse of power is required only to shift the relay's contacts from one position to an opposite position. Power is not necessary to hold the contacts in any of the two possible positions. Various means are used to hold contacts in-place. Some designs use magnets whereas other designs use an over-center spring. Obviously, latching relays require a logic circuit that is different from that which would be used for a nonlatching relay. Worldwide, there are far fewer installations of latching relays than nonlatching relays. Nevertheless, there are some control configurations to which latching relays are ideally suited. Latching relays are well suited to circuits that require contacts to be closed for long periods of time.

*Time delay relays*: Time delay relays are used in control circuits to momentarily postpose a control action. The most common version of time delay relays are of the electromechanical design. A set of contacts is moved by a timing device of some type. The timing mechanisms can be either mechanical or electronic. One type of mechanical timing mechanism that has been used for many years uses a pneumatic diaphragm. The two most commonly used types of time delay relays are the "on-delay" time delay relay and the "off-delay" time delay relay.

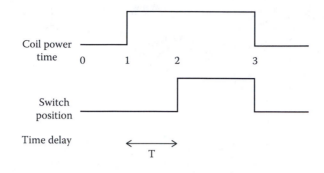

**FIGURE 12.7**
TDE relay.

*On-delay time delay relay:* The function of an on-delay time delay relay is to provide a delay after a control signal is received. When, say, voltage is applied to the coil of the relay, the relay's contacts do not shift immediately. Rather the contacts shift only after a predetermined time delay. The contacts could be NO, NC, single pole double throw (SPDT), or multipole. If voltage is removed from the coil, the contacts immediately shift to the normal position. If the contacts of a time delay relay are NO, the contact action would be normally open timed closed (NOTC). If the contacts are NC, the contact action would be normally closed timed open (NCTO). On-delay relays are often designated as time delay on energization (TDE) relays. The action of a TDE relay is best be described by a graph of the type shown in Figure 12.7. At time "0," there is no power applied to the relay's coil and the contacts are in the normal state. At time 1, power is applied to the coil and trimming begins but the contacts remain in-place. After time delay T elapses, the contacts shift from the normal position— at time 2. NO contacts close and NC contacts open. If power is removed from the coil, say at time 3, the contacts immediately return to the normal position.

*Off-delay time delay relay:* The function of an off-delay time delay relay is to provide a delay on de-energization. When voltage is applied to the coil of the relay, the contacts immediately shift from the normal position. The NO contacts close and the NC contacts open. The contacts could be NO, NC, SPDT, or multipole. The contacts will remain held in those positions until time T after power is removed from the relay's coil. Off-delay relays are often designated as time delay on de-energization (TDD) relays. The action of a TDD relay is described by the graph of Figure 12.8. At time "0," there is no power applied to the relay's coil and the contacts are in the normal state. At time 1, power is applied to the coil and the contacts immediately shift from the normal positions. At time 2, power is removed from the coil but the contacts remain in position until after time T when the contacts revert to

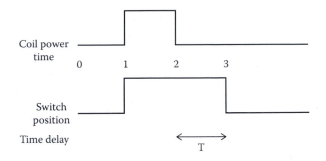

**FIGURE 12.8**
TDD relay.

the normal position. Except for some time delay relays with very short timing periods, TDD relays require a continuous power source in order to hold contacts in-place after the signal voltage has been removed.

The symbols used in ladder diagrams for the coils and contacts of TDE relays and TDD relays are shown in Figure 12.9. The TDE and TDD types of relays

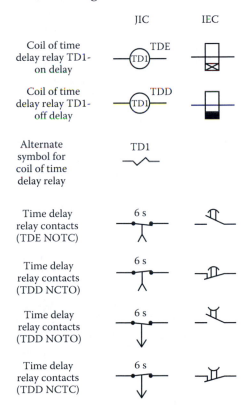

**FIGURE 12.9**
Time delay relay symbols.

are the most common types used in control circuits, but there are other less common types of relays available for a variety of special timing functions.

*Options and accessories*: Electromechanical relays are available with a number of options. Many designs of relays have manual operators that allow a person to manually shift the contacts from the normal position. Other relays have flags or small lights that indicate the position of a relay. Both of these features can be an aid in troubleshooting.

*Selecting an EMR*: The selection of an electromechanical relay should be carried out in recognition of a number of considerations. First, there is the circuit voltage that is a starting point. In North America, the circuit voltage for the control of most motors is 120 VAC. However, for many types of installations, the control voltage is 125 VDC. This is the case in nuclear plants and in the control of transmission and distribution gear. Many engineers in the electrical generating industry will argue that the 125 VDC power is better because a dc coil is a more dependable actuator. dc coils are less likely to chatter, burn out, or fail. Yet, it can be argued that a set of contacts can better interrupt an ac current than a dc current. The arc in dc contacts tends to linger, whereas an ac current automatically goes to zero value numerous times every second. Another significant factor is the speed of response. Protection relays are usually designed to accept and produce a dc signal. The reason is that dc devices, in solid-state designs, can respond in microseconds compared to the milliseconds characteristic of ac devices. When a fault of some sort appears, the fastest possible response is preferred.

Once the coil voltage has been determined for an electromechanical relay, the next consideration is the suitability of a prospective relay's contacts to the load. Literature pertinent to most relays will state a current capability, as "6 amps," at a particular voltage. Unless specifically stated, the current rating given for a relay may be assumed to be the relay's capability if used with the least demanding of all loads, namely, a resistive load. It is perhaps important to understand that the relay manufacturing business is a very competitive business and, so, relays must be fabricated to have very little extra capacity. A relay with an advertised capability of 6 A would be capable of reliably closing contacts with an inrush of 6 A, of maintaining a constant 6 A and of breaking a current of 6 A. The relay would be capable of somewhere in the region of 50,000–200,000 cycles in an average environment but only when cycling a pure resistive load that is no greater than 6 A. For reactive loads, a "6 amp" relay would have a much lower current capability.

For applications other than resistive loads, a relay's current interrupting capability is much less than the its resistive load rating. A typical set of recommended reductions in capability is shown in Table 12.1. Say, for

**TABLE 12.1**

Recommended Contact Current Capability

| Service | Percentage of Advertised Current Rating |
|---|---|
| For long contact life | |
| Resistive (e.g., heater) | 75 |
| Inductive | 40 |
| Capacitive | 75 |
| Motor | 20 |
| Incandescent | 10 |

example, that a relay with an advertised rating of 6 A is considered for the control of incandescent lighting. Its maximum current capability, according to Table 12.1, would be 10% of 6 A or 0.6 A.

While the contacts of a relay have a maximum capability, there is also a minimum capability that is often not taken into account. Consider, for example, the possible use of relay with a 6 A resistive, 120 VAC rating in a panel to operate another relay that has a 3 VA coil rating. The coil current would be 3/120 A or 0.025 A. The adjusted maximum capability of the contacts for an inductive load would be 0.40 × 6 or 2.4 A. The current draw of the coil would be (0.025/2.4) (100)% or 1.04% of the contacts rating. This would be considered a poor application. True, when initially installed, the relay and the circuit would function as expected. In time, however, there could be a problem. Basically, the contacts would have what is called inadequate wiping action. While excessive arcing between contacts can cause premature failure of contacts, inadequate arcing can likewise result in failure. A gentle arc between contacts is highly desirable for the reason that it creates an action that cleans the surfaces of the contacts. Without the wiping action, in time contacts very well may accumulate a film or dust that will prevent continuity.

Where low contact VAs are encountered, pilot duty contacts are recommended to enhance long contact life. Pilot duty contacts are specifically designed to prolong contact life in those applications where low currents are encountered. The contacts of pilot duty relays are much smaller than the regular relay contacts and have a high contact pressure. Many are made of silver or, at least, have silver plating. One popular relay on the market with pilot duty contacts has an advertised rating of 25 VA maximum and a 3 VA minimum. A relay with contacts of this description would be a good match with a relay that has a coil power between the minimum and maximum coil power draw. Current ratings should be certified as uncertified ratings might be lacking in capability. Certifications are discussed in Section 12.9.

## 12.3  Solid-State Relays

SSRs are increasingly being used in applications that not many years ago used only EMRs. Today, roughly one in every five relays manufactured is an SSR. SSRs are available in current ratings to at least 150 A and potentials to 600 V. As the name suggests, SSRs perform their function by means of electronic, or solid-state, components. More specifically, SSRs turn current on or off with thyristors. SSRs have a very long cycle life. Whereas EMRs might have a cycle life that is somewhere around 50,000–200,000 cycles, SSRs can easily be expected to withstand many millions of cycles. A typical SCR relay is shown in Figure 12.10.

Because of the absence of moving parts, SSRs are not affected by vibration or shock. SSRs are especially suited to the control of heaters. With a matching controller SSRs can send pulses of current to a heater, thereby maintaining the heater at a nearly constant temperature. The controlled medium would also be controlled at a more constant temperature.

Much as EMRs, SSRs also have limitations. In particular, SSRs are susceptible to high-voltage spikes. An EMR might very well be capable of withstanding voltage spikes that could be as high as 1000 V or more. High-voltage spikes of a certain high value can instantly destroy a SSR. So, in most applications precautions must be taken to protect an SSR against excessive voltage spikes. Surge protectors are one solution. When turned off, SSRs have a small leakage current that can pose a problem in some circuits. When current passes through a thyristor, heat of some magnitude is generated. For SSRs with a relatively low current rating, a firm mount to a vertical sheet metal

**FIGURE 12.10**
Solid-state relay. (Photograph by the author.)

surface is often sufficient to remove the heat that is generated. For higher current levels, SSRs must be attached to metallic heat sinks to maintain the relay within an acceptable temperature range.

## 12.4 Reed Relays

Reed relays function much as an electromechanical relay. A reed relay has a coil and a set of contacts that are shifted from one position to another when the coil is energized. In its most simple form, a reed relay consists of two ferromagnetic strips, or reeds, that are partly sealed within a glass capsule. A design of a single pole, NO reed relay is represented in Figure 12.11. A portion of each reed extends outside the capsule so that electrical connections may be made to that part of the reeds external to the capsule. Portions of the reeds within the capsule that act as contacts are usually plated, sometimes with gold. An electrical coil surrounds the glass tube. When the coil of a NO design is de-energized, there is a gap between the two reeds and the contacts are open. When voltage is applied to the coil a magnetic field is generated, drawing the reeds together and establishing contact closure. When the coil is de-energized, the reeds separate, opening the circuit. Usually, the capsule is either evacuated or charged with an inert gas to prolong contact life. The reeds of a reed relay are hermetically sealed, and for this reason, the contacts are impervious to contaminants, as dust or deleterious gases. Because of the hermetic design, some models of reed relays are approved for use in environments containing explosive gases or explosive dusts. The coils of reed relays accept only dc voltage. Designs of reed relays are available in NO, NC, and double pole configurations. Some models are also available as

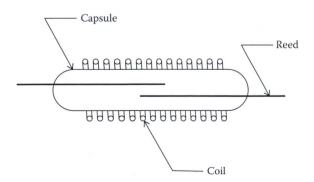

**FIGURE 12.11**
Reed relay.

latching relays. Reed relays respond more quickly than electromechanical relays and typically respond in approximately 1 ms. Most reed relays are encapsulated and intended for mounting on circuit boards.

## 12.5 Environmental Requirements

Relays used in conjunction with three-phase electrical power installations are very often exposed to elevated levels of gases or dusts that tend to deteriorate relay parts and contacts. A number of measures may be taken to help ensure that degradation of relays will be avoided or minimized. SSRs or reed relays are possible candidates as these designs have only the exterior terminals exposed to the ambient. Hermetic relays are another approach to protect the moving parts and contacts of relays.

Relays located within hazardous areas preset some unique requirements. Hazardous areas are defined as those areas that may contain flammable gases, liquids, fibers, or dusts. If a conventional EMR is located in a hazardous area and the contacts of that relay either close or open, an arc will be generated. That arc, if it is of an adequately high-energy level, can ignite gases and cause an explosion. In short, conventional EMRs are not suited for use in a hazardous area. Special precautions are needed when using arc-generating devices in a hazardous area. One viable approach is to use an enclosure that is suitable for hazardous areas. Both explosion proof enclosures and purged enclosures are suitable for hazardous areas. Some of the other possible alternatives for a hazardous area are

1. Hermetically sealed devices—A number of electrical devices, including switches, controls, and relays, are available in hermetic designs. A hermetic device has its contacts entirely contained within a sealed can. The seal excludes gases or dusts surrounding the relay from entering into the can. Hermetic relays are also a prudent choice for environments that contain gases that are not necessarily explosive but that over time may cause deterioration of relay parts or contacts.

2. Intrinsically safe (I.S.) devices—I.S. devices operate on the principle that an electric arc within a hazardous area will not necessarily cause an explosion. For ignition to occur, an electric arc must be of a minimal energy level. I.S. devices are made to operate below those energy levels. I.S. systems use barriers that are located outside a hazardous area. Wiring to devices within the hazardous area must pass through the I.S. barriers that limit voltages and currents within a hazardous area. As mentioned earlier, a minimum current level is recommended for relays to cause wiping of the contact surfaces. So, for I.S. applications, attention should be given to contact designs and currents with the dual objectives of both avoiding an explosion and prolonging contact life.

## 12.6 Enclosures

To enhance contact life, relays should be mounted in an enclosure of a type suitable for its environment. A variety of enclosures are available for housing relays. Some enclosures are relatively simple in design and suited only for the relatively benign surroundings of an indoor installation. Other enclosures are intended for more challenging environments as outdoors or hazardous locations. Most enclosures are fabricated of sheet steel but, to enhance the life of an enclosure, some enclosures are fabricated of stainless steel or fiberglass. To accommodate the mounting of relays, an enclosure should have a backplane that extends slightly from the rear of the enclosure. In North America, the requirements for enclosures are defined by the guidelines of NEMA. Some of the more commonly used types of NEMA enclosures are the types 1, 4, 12, 7, 9, and 10. Briefly these are defined as follows:

NEMA 1—General purpose indoor. Possibly vented. Prevents human contact with electrical parts when the door is closed. Excludes falling dirt. The least expensive of enclosures.

NEMA 4—Watertight. Excludes rain, snow, or sprayed water. Suitable for indoors or outdoors.

NEMA 12—Dust tight. Openings are sealed. Excludes dust. For indoor installations only.

NEMA 7, NEMA 10—Explosion proof. Suitable for explosive gases.

NEMA 9—Explosion proof. Suitable for explosive dusts.

Hazardous environments present some unique requirements for relays. The NEMA 7 and NEMA 9 enclosures are available to house conventional switches or relays. Housings are available in various sizes to contain either individual devices or groups of devices. These enclosures are mostly of a design that does not exclude explosive gases or dusts. Rather, the enclosures prevent the hot gases of an explosion within the enclosure from escaping to the surrounding area and thereby igniting an explosion outside the enclosure. Explosion proof enclosures are made of cast steel or aluminum and are designed with small flame paths. The principle of the design is that exploded gases within the enclosure will be cooled upon passing through the flame paths and therefore not be able to ignite gases outside the enclosure. To meet code, the enclosures must be of an approved design. Explosion proof enclosures tend to be heavy and expensive. Explosion proof enclosures also require maintenance by qualified personnel so as to keep flame paths clear and in the original configuration.

An acceptable alternative to an explosion proof enclosure is a purged enclosure. Codes accept a purged enclosure as a safe alternative in hazardous areas. A purged enclosure has relatively tight seals around doors and

**FIGURE 12.12**
NEMA 12 enclosure with motor controls. (Photo courtesy of State Motor and Control Solutions, St Louis, MO.)

openings. Air from an area outside the hazardous area is forced, through ducts, into the enclosure and the interior of the enclosure is maintained at a slight positive pressure with respect to its environment. A purged enclosure allows use of conventional electrical devices within the enclosure since explosive gases or dusts are not allowed to enter into the enclosure. Whenever a large number of electrical devices are involved, a purged enclosure is drastically less expensive than an explosion proof enclosure. Precautions are appropriate to guard against the possible situation in which purging of the enclosure might be interrupted at a time when explosive gases are in the proximity of the enclosure.

An IEC standard defines a set of criteria for enclosures that is different from the NEMA descriptions. While there is no identical match between the IEC and NEMA descriptions, a few of the enclosures are somewhat similar. Specifically, the IEC IP10 approximates the NEMA 1. The IEC IP56 approximates the NEMA 4 and the IEC IP52 approximates the NEMA 12. A NEMA 12 type enclosure (dust tight) is shown in Figure 12.12.

## 12.7 Protection Relays

Protection relays are a class of relay designed mostly to protect larger electrical gear as generators, motors, and transformers. Older protection relays were of the electromechanical design, but gradually these designs were replaced with faster acting microprocessor-based units. Protection relays are commonly identified by a device number as described in ANSI/IEEE

Standard C37.2. The device numbers run from "1" to "94." Some numbers also have a suffix. Each of the device numbers describe a specific kind of protective relay that is intended to detect a particular type of malfunction and provide an output that can be used to protect electrical gear.

One well-known manufacturer of protection relays recommends that for the protection of important motors below 1500 HP, approximately 13 protection relays should be used to ensure adequate protection. The manufacturer's list of recommendations includes the following device numbers: 27, 37, 38, 46, 49, 49R, 49S, 50, 57, 66, 86, and 87M. Following are some representative descriptions of the mentioned device numbers.

Device #2—Time delay starting.

Device #27—Undervoltage relay that operates when its input voltage is less than a predetermined value.

Device #37—Undercurrent or underpower relay that functions when the current or power flow decreases below a predetermined value.

Device #38—Bearing protective device. Provides interlock in the event of a high bearing temperature.

Device #46—Reverse-phase or phase-balance current relay that functions when the polyphase currents are of reverse phase sequence or when the polyphase currents are unbalanced or contain negative phase-sequence components above a given amount.

Device #47—Phase sequence or phase balance voltage relay.

Device #49—Machine or transformer thermal relay that functions when the temperature of a machine armature winding or other load carrying winding or element of a machine or power transformer exceeds a predetermined value.

Device #49R—Winding over temperature relay.

Device #49S—Locked rotor relay.

Device #50—Instantaneous overcurrent relay that functions instantaneously on an excessive level of current.

Device #57—Short circuit or grounding relay.

Device #59—Overvoltage relay.

Device #66—Notching or jogging relay that functions to allow only a specific number of operations within a given time of each other. Or, device can be used to energize a circuit periodically or for fractions of specified time periods.

Device #86—Lockout relay that is an electrically operated hand or electrically reset auxiliary relay that is operated upon the occurrence of abnormal conditions to maintain associated equipment out of service until it is reset.

Device 87M—Differential protective relay that functions on a percent-age, or phase angle, or other quantitative difference between two currents or some other electrical quantity.

Obviously, with the large number of protective relays available for the pro-tection of electrical gear, there are any number of possible combinations that might be used in any single application. In short, protective relays have an important role in the protection of electrical equipment of a great variety.

NOTE: Protection relays specifically suitable for motor protection are discussed in Section 9.2.3.

## 12.8 Contactors

A contactor is also called an "electromagnetic contactor," although it is most often merely called a "contactor." Contactors have a coil and a set of con-tacts arranged in a configuration that approximates what is found in control relays. When voltage is applied to the coil of a contactor, the contactor's con-tacts are closed to deliver power to the controlled electrical gear. The main difference between a relay and a contactor is that the latter is used to connect and interrupt higher levels of power. Typically, contactors control power to electrical gear as motors, heaters, lighting, and capacitor banks. Contactors are required in those applications where electrical power must be repeatedly applied and subsequently interrupted. Control of power to the very largest types of electrical gear is controlled by circuit breakers and not contactors. One significant difference between contactors and circuit breakers is that cir-cuit breakers are designed to interrupt short circuit currents but contactors do not have that capability.

Contactors are available in two-pole, three-pole, and four-pole configu-rations. Both enclosed and open designs are common. The open design is intended for mounting in a customer's enclosure. Most contactors are NO, although NC contactors are also available for specialized applications. Auxiliary contacts, which would have a lower current rating than the main contacts, are available with most contactors. The auxiliary contacts are often used for the purpose of interlocking in control circuits or for the purpose of remote indication of contactor position. Contactors are also classified as either "general purpose" or "definite purpose." Definite purpose contactors are mostly intended for the original equipment manufacturer market and are to be used in a specific, defined application. Definite purpose models are usually less expensive than the nearest equivalent general purpose design. In addition to the electromechanical types of contactors, there are solid-state models that are said to offer long service lives.

**FIGURE 12.13**
IEC contactor. (Photograph by the author.)

The use of contactors as specifically used in the control of motors is treated in detail in Chapter 9. For motor applications, NEMA rates contactors according to sizes 00, 0, and 1 to 9. IEC has its system of rating contactors for motor applications and those sizes range from AC-1 to AC-4. A typical IEC contactor is shown in Figure 12.13.

## 12.9 Terminals and Terminations

By common usage, a "terminal" could be a point on a device where a conductor is landed to complete an electrical connection. A terminal could also be a small component that is affixed to the end of a conductor to facilitate the connection of that conductor to a terminal on electrical gear. Conductors are "terminated" on terminals and a conductor can have a "wire termination." As is typical of many English-language words, the significance of the word "terminal" is best understood by its use in a sentence.

Relays and contactors are available with a variety of terminals where conductors may be landed. Some of the more common terminals are the screw type, quick-connect (also called push-on and fast-on) and the pressure type. Years ago, the screw-type terminal was the most prevalent type, but with time, the screw types were mostly replaced with other types.

The quick-connect type, which is a blade-type terminal, is favored on relays used in high-volume production. The terminal on the relays with quick-connect terminals are usually the female type. Quick-connect terminals are available in two sizes: 0.187 in. (4.7 mm) and 0.250 in. (6.3 mm). Otherwise, the pressure-type terminal has become the most popular type of terminal. Pressure types of terminals are available in a variety of configurations. In most cases, a pressure-type terminal consists of two small metallic tabs that are intended to receive a conductor between the tabs. After a conductor is inserted between the tabs, the turn of a screw brings the tabs together and applies pressure to hold the conductor and tabs in close contact.

Examples of relay terminals may be seen in the aforementioned illustrations. The relays in Figures 12.4 and 12.5 have pressure-type terminals, and the relays in Figures 12.6 and 12.10 have screw-type terminals. A relay with male quick-connect terminals is shown in Figure 12.14.

Although there are exceptions, conductors used to interconnect relays within a control panel are mostly stranded copper in sizes 16 AWG or 14 AWG. So, these are the types of conductors that must be landed on the electrical devices within a panel. If a conductor is to be landed on a screw terminal, it must have either a ring tongue-type termination or a split-ring-type wire terminal. A conductor intended for landing on a relay with male quick-connect terminals must have a matching female-type push-on terminal.

**FIGURE 12.14**
Relay with quick-connect terminals.

A conductor to be landed under a pressure-type terminal could have one of two types of ends. One method is to have the end of the conductor stripped of its insulation for approximately 3/8 in. (9.5 mm). The stripped portion would be inserted into the pressure-type terminal and a screw tightened to hold the conductor firmly in place and establish continuity. Some conductors intended for terminations in pressure-type terminals are terminated with a ferrule. The purpose of the ferrule is to keep all strands of copper within the ferrule so that stray strands of copper do not inadvertently make contact with metallic parts other than the target terminal. The ferrule also helps to ensure that terminations within a pressure-type terminal will be made correctly. The use of ferrules on conductor ends is especially favored where terminals are located in close proximity to one another.

As is characteristic to most electrical parts, terminals have ratings. This is true of the terminals located on electrical devices as well as terminals of the type used on conductors. For example, ring tongue terminals are restricted to the acceptable size of screw under which the terminal will be located as well as the size of conductor to which the terminal is attached. Pressure-type terminals are limited to the number and size of conductors that may be terminated at a single terminal. Terminals also have a current rating. Several types of conductor terminals are shown in Figure 12.15. In the top row are two split ring terminals and two quick-connect terminals. Two ring tongue

**FIGURE 12.15**
Various conductor terminals.

terminals are shown in the bottom row. The insulator color designates the approved conductor size. Yellow-colored terminals are for conductor sizes #10 AWG and #12 AWG. The blue colors are for #14 AWG and #16 AWG conductors and the red are for #16 AWG to #22 AWG conductors.

## 12.10  Certifications

In the evaluation of the capabilities of a prospective relay or contactor for an application, the device's certifications should be considered as an important part of the process. A certification is documentation that essentially confirms that a product has been tested according to a predefined protocol, and it has been found to meet the requirements of that protocol. In the United States, certifications are typically provided by a third part agency. One well-known testing laboratory is Underwriters Laboratories. UL is an agency that has offices in over 46 countries and is said to provide safety-related certification, validation, testing, inspection, auditing, advising, and training. If an electrical device bears the UL label, it may be assumed that the device has been examined and tested to confirm that the device is capable of the claimed capability. In Canada, a similar agency is the Canadian Standards Association (CSA).

Within the European Union, the "CE" marks are the more recognized method of identification for suitable electrical devices. According to the CE protocol, there is no third-party validation of a manufacturer's claims. If an electrical device bears the CE mark, it is understood that the claims of capabilities have been certified but that the claims come from only the manufacturer.

Electrical codes often require the use of "approved" devices. The accepted interpretation of "approved" is that the device has a certified electrical capability. In the United States, both UL and CSA marked devices are commonly accepted as "approved." Worldwide, there are numerous agencies that test and certify devices.

## 12.11  Common Relay-Associated Mistakes

Mistakes often made in the selection and application of relays are:

1. Failure to recognize that a relay's ac current rating is only its resistive rating and not its current capability when used with other, more demanding loads
2. Use of ac rated relay that has no dc rating in a dc application

3. Using a relay with a contact current below the contact's minimum rating

4. Using conductor terminations that are not within either the rating of the terminal or the rating of the conductor termination

5. Using a relay in a configuration that has the relay unnecessarily energized for long periods of time

A responsible selection of a relay will be made with the intent of using it in a circuit with components that will provide a long service life. This approach will require close attention to the factors that are involved. And, it goes without saying that efforts should be taken to avoid the mistakes often made in the selection of relays.

## Problems

1. What is the most common type of relay?

2. What are the two types of actuators used in electromechanical relays?

3. Which type of relay is best suited to manufacturing, the cube relay or the machine tool relay?

4. What are two types of plug-in bases commonly used with cube-type relays?

5. Are shading rings required in both the dc and ac coils used in relays? Explain.

6. Would it be fair to say that today control relays are being used more and more to replace PLCs?

7. If a relay has a stated current capacity of "10 amps" but no horsepower rating, what would be its current capability if used to control a motor?

8. Why is a contact wiping action necessary in a relay?

9. Is it true that solid-state relays are unaffected by shock or vibration? Explain.

10. Are any precautions needed to protect electromagnetic relays or SSRs against voltage spikes?

11. Which is the fastest to react to an input signal: electromagnetic relay or reed relay?

12. What are some of the methods that may be taken to use relays within a hazardous area?

13. What NEMA type of enclosure would be required to house relays in a flour mill where flammable particles of flour are common in the air?

14. Which IEC enclosure approximates the NEMA 4 enclosure?

15. Name some of the types of materials used to fabricate NEMA and IEC type enclosures?

16. Which protective device would be used to react instantaneously to a detected high current to a transformer?

17. What is the device number to sense a ground fault?

18. Where is a reversing contactor used?

19. Identify some methods of electrical certifications that are accepted in North America.

20. If a relay has a 10 A, 120 VAC rating would it be suitable for use at 2 A, 24 VDC?

# 13

## *Electrical Drawings*

Drawings are the universal language of those who deal with electrical power. At the conceptual stage of a project, drawings become the basis for the proposal of a new electrical system. When a new system has been approved, drawings become a large part of the criteria for the design and installation of the system. After installation, construction drawings are needed for the maintenance of an existing system. If an expansion of a system is under consideration, drawings of the system will be needed to determine how an expansion will be made and what modifications will be required.

In North America, drawings are normally prepared in one of five standard sizes of drawings. The North American standard sizes are tabulated in Table 13.1.

Outside North America, the IEC standard drawing sizes are commonly used. These sizes are summarized in Table 13.2.

There are many variations in the formats of electrical drawings. Also the symbols used in these drawings vary from country to country, from one agency to another, and from one firm to another. The purpose here is not to review all possible types of drawings and symbols. Rather, introductions are made of samples of the types of drawings used in the electrical industry. Representative electrical drawings are shown on the following pages. To illustrate the principles involved, relatively simple examples are cited. Primarily, the treated drawings are common to three-phase electrical systems. The accompanying sample drawings are necessarily of the size ANSI A. In practice, most of the electrical drawings represented here would be found on larger size sheets. Single-phase systems are generally more simple systems and single-phase drawings tend to be smaller. Control systems are necessarily single phase and, so, three-phase systems have single-phase control systems. Other than control systems, samples of single-phase systems are not presented in this chapter.

Common types of electrical drawings and their respective functions are listed in Table 13.3.

**TABLE 13.1**

North American Standard Drawing Sizes

| Drawing Size | Dimensions (in.) | Dimensions (mm) |
|---|---|---|
| ANSI A | 8.5 × 11 | 215.9 × 279.4 |
| ANSI B | 11 × 17 | 279.4 × 431.8 |
| ANSI C | 17 × 22 | 431.8 × 558.8 |
| ANSI D | 22 × 34 | 558.8 × 863.6 |
| ANSI E | 34 × 44 | 863.6 × 1117.6 |

**TABLE 13.2**

IEC Drawing Sizes

| Drawing Size | Dimensions (mm) |
|---|---|
| IEC A6 | 105 × 148 |
| IEC A5 | 148 × 210 |
| IEC A4 | 210 × 297 |
| IEC A3 | 297 × 420 |
| IEC A2 | 420 × 594 |
| IEC A1 | 594 × 841 |
| IEC A0 | 841 × 1189 |

**TABLE 13.3**

Common Types of Electrical Drawings

| Type of Drawing | Drawing Function |
|---|---|
| One line | Basic electrical sources, principal gear, some sizes, buses |
| Three line | Similar to one-line diagram but with more details and more lines |
| Logic | Criteria of a control system with logic symbols |
| Ladder | Components of an electrical control system and the interlocks |
| Panel layout | Physical positioning of electrical components within a panel |
| Cable | Routing and sizing of electrical cables |
| Raceway | Raceway and physical routing of conduits and cable trays |
| Connection | Terminations of conductors that interconnect electrical components |

## 13.1 One-Line Diagrams

Primary symbols used in one-line diagrams are shown in Figure 13.1. Most of the symbols in the table are in accordance with the recommended practices of IEEE as presented in Standard 141-1976.

By means of established notations and symbols, the one-line diagram describes the fundamentals of an electrical system. In a sense, the diagram

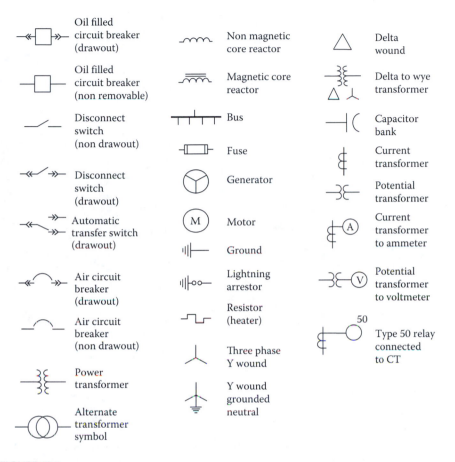

**FIGURE 13.1**
One-line symbols.

is the starting point for any new or proposed system. It sets the basis for the size and capability of an electrical system. A one-line diagram includes the principal electrical gear and the respective ratings of electrical gear. The diagram will show circuit breakers, transformers, bus bars, large motors, and in-house generators. Ratings as voltages, current, and VA ratings are included in the diagrams. In one-line diagrams, a single line represents a group of conductors. The drawing will represent, say, Conductors A, B, and C of a three-phase system with one line on the drawing. Thus, the title of the drawings. Traditionally, representations of the entry points to the system are entered either at the top of a drawing or at the left side. If the entry point is at the top of the drawing, energy flow proceeds toward the bottom of the diagrams. On drawings with the entry points at the left of the drawing, energy flow proceeds to the right side of

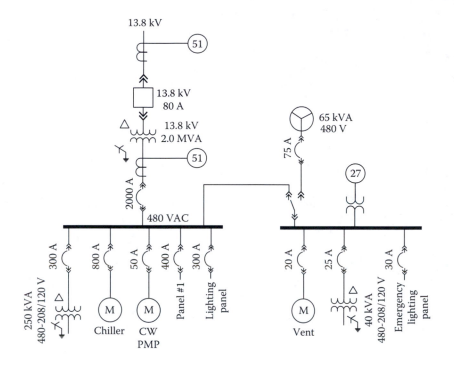

**FIGURE 13.2**
One-line diagram, "the hypothetical hospital."

the drawing. One-line diagrams intentionally avoid excessive lines and detail to better emphasize the fundamental concept of a system.

A typical one-line drawing is shown in Figure 13.2. The drawing is titled, "The hypothetical hospital." The drawing shows electrical power to a hospital originating from a 13.8 kV source, which is assumed to be supplied by a public utility. The electrical source connects first to an 80 A oil-filled circuit breaker and then to a 2.0 MVA, 13.8 kV/480 VAC transformer. The transformer and its breaker could be owned and supplied by the utility or it could be owned and provided by the hospital. Within the hospital, the main 480 bus is protected by a 2000 A air circuit breaker. The 480 bus provides power to a number of users within the facility. The largest single load is the building's water chiller that supplies chilled water to terminal air conditioning units. An in-house generator is available for periods when the 13.8 kV source may be unavailable. In the event of a utility outage, detector relay #27 senses the loss of power on the emergency 480 V bus. In the event of a loss of power from the utility, the automatic transfer switch will separate the emergency 480 bus from the primary bus and simultaneously start the in-house generator. The generator will be started to provide power to only the emergency bus.

## 13.2 Three-Line Diagrams

To a large extent, three-line diagrams continue where one-line diagrams stop. Three-line diagrams are prepared not so much to show the concept of a system. Rather, three-line drawings show more details than what is shown on the associated one-line drawing. For this reason, a three-line diagram will contain significantly more lines than the associated one-line diagram. Three-line diagrams will also be considerably larger than the associated one-line diagram. If a system is represented on, say, two sheets of a one-line drawing, there might be eight sheets for the associated three-line diagram. Three-line drawings will necessarily include the three conductors of a three-phase system. For example, if the system is a three-wire delta system, three conductors will be shown. For a four-wire system, the four conductors will be shown although the drawing will still be called a "three-wire drawing."

Three-line diagrams tend to use many of the symbols used in the development of one-line diagrams. In addition, three-line diagrams will use some of the symbols used in association with circuit protection. Many of the symbols contained in Figure 8.2 will be found in three-line diagrams.

A typical three-line diagram is presented in Figure 13.3. Shown is a three-line diagram for XYZ Chemical Co. The diagram might be characteristic of the electrical system for a small building on the premises of a chemical facility.

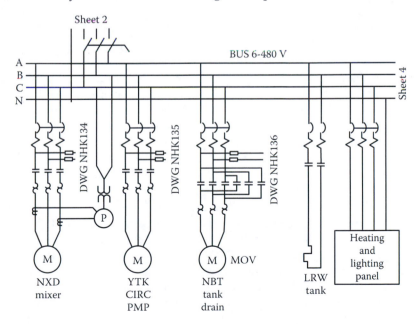

**FIGURE 13.3**
Three-line diagram XYZ Chemical Co.

The drawing shows a 480 VAC bus that is identified as "Bus 6." Power to the bus originates on another three-line diagram that is identified as "Sheet 2." Bus 6 could be contained, for example, within a motor control center located within the building. The diagram indicates that other gear served by Bus 6, but not shown on the current page, is shown on Sheet 4. The diagram shows Conductors A, B, C, and N of a four-wire system. A disconnect switch permits isolation of the MCC from its source. As shown, the disconnect switch interrupts only conductors A, B, and C but not N. Three-phase three-wire power is shown delivered to three motors within the building. The logic that determines operation of the motors is shown in, respectively, schematic (ladder) drawings NHK 134, –135, and –136. Power for the ladder diagrams, in each instant, is derived from Conductors A and B of the motor circuit but after the circuit breaker. Accordingly, power to the respective ladder logic is deactivated whenever the circuit breaker is open. One of the three motors powers a motor operated valve (MOV). The MOV is operated through a reversing contactor (shown in the diagram) that allows opening and closing of the valve. The other two motors in the diagram are activated through nonreversing contactors. Overhead lighting within the building is 277 VAC and is distributed through a heating and lighting panel. For this reason, the "N" conductor is extended to the heating and lighting panel. A single-phase immersion heater on Tank LRW is powered from Phases A and B. Here it may be seen that details not normally shown on the one-line diagram would be available in the associated three-line diagram.

## 13.3  Logic Diagrams

Logic diagrams present the basic criteria, expressed with a unique set of symbols, for the requirements of a control system. Very often logic diagrams are not developed by electrical personnel. Rather, the development might be conducted by, say, a mechanical engineer or a chemical engineer. The logic diagram then becomes a precise statement of what the originator wants the control system to do. While electrical personnel might not be involved in the development of a logic diagram, their interpretation and understanding of the diagram very well could be required.

Some organizations require development of a logic diagram prior to development of a ladder diagram. Following agreement on a logic diagram, development of the ladder diagram will often follow. Yet, many organizations rely entirely on ladder diagrams and do not use logic diagrams. In fact, many individuals who might have expertise with ladder diagrams might find logic diagrams to be a very foreign concept. It is significant to note that more and more PLCs (programmable logic controllers) are being used to perform control functions and many PLCs today can be programmed directly from a logic diagram. So, in many instances, a ladder diagram may likewise not be required to implement the controls of a system.

Logic diagrams use "gates" to produce a binary logic "1" or a logic "0." A logic "1" represents a specified condition and a "0" represents the opposite condition. However, there is no universally accepted system of symbols. Actually, a certain symbol might have a different significance to different organizations. So, care is needed to understand the significance of the symbols in a logic diagram. Binary logic is similar to the logic obtained with relay contacts since a relay contact can have only one of the two possible states, namely, "opened" or "closed." So, a ladder diagram using contacts that open and close follows readily after a logic diagram (Figure 13.4).

Some of the more commonly used logic gates are shown in Figure 13.5. The illustration also shows the equivalent ladder diagram symbols. It is sometimes easier for a person to understand the function of a gate through reference to the truth table associated with a specific logic symbol. To this end, Figure 13.4 presents the truth tables for each of the cited logic symbols. Whereas the gates of Figure 13.5 show only two inputs to the gates in all cases, except the "not" gate, more than two inputs can be accommodated by the gates. Logic diagrams are constructed with the input criteria to the left of the diagram and the outputs to the right. The logic gates, where the logic is performed, are positioned on the diagram between the inputs and the outputs.

A typical logic diagram using various logic gates is presented in Figure 13.6. Shown is a hypothetical version of what could be a control system for a centrifugal compressor. In order to start the motor of the unit's compressor, a set

| AND | | |
|---|---|---|
| A | B | C |
| 0 | 0 | 0 |
| 0 | 1 | 0 |
| 1 | 0 | 0 |
| 1 | 1 | 1 |

| OR | | |
|---|---|---|
| A | B | C |
| 0 | 0 | 0 |
| 0 | 1 | 1 |
| 1 | 0 | 1 |
| 1 | 1 | 1 |

| NOR | | |
|---|---|---|
| A | B | C |
| 0 | 0 | 1 |
| 0 | 1 | 0 |
| 1 | 0 | 0 |
| 1 | 1 | 0 |

| NAND | | |
|---|---|---|
| A | B | C |
| 0 | 0 | 1 |
| 0 | 1 | 1 |
| 1 | 0 | 1 |
| 1 | 1 | 0 |

| NOT | |
|---|---|
| A | C |
| 0 | 1 |
| 1 | 0 |

**FIGURE 13.4**
Truth table of logic gates.

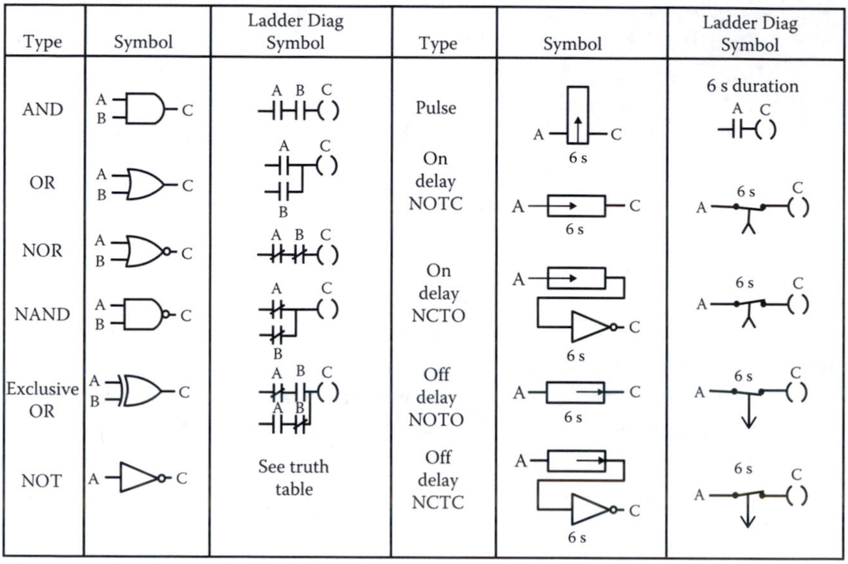

| Type | Symbol | Ladder Diag Symbol | Type | Symbol | Ladder Diag Symbol |
|---|---|---|---|---|---|
| AND | A, B → C | A B C ⊣⊢⊣⊢⊣⊢( ) | Pulse | A ⊓ C  6 s | 6 s duration  A C ⊣⊢( ) |
| OR | A, B → C | A C ⊣⊢( ) / B | On delay NOTC | A → C  6 s | 6 s  C |
| NOR | A, B → C | A B C ⊣⊢⊣⊢⊣⊢( ) | On delay NCTO | A → C  6 s | 6 s  C |
| NAND | A, B → C | A C / B | Off delay NOTO | A → C  6 s | 6 s  C |
| Exclusive OR | A, B → C | A B C ⊣⊢⊣⊢⊣⊢( ) / A B | Off delay NCTC | A → C  6 s | 6 s  C |
| NOT | A → C | See truth table | | | |

**FIGURE 13.5**
Logic diagram symbols.

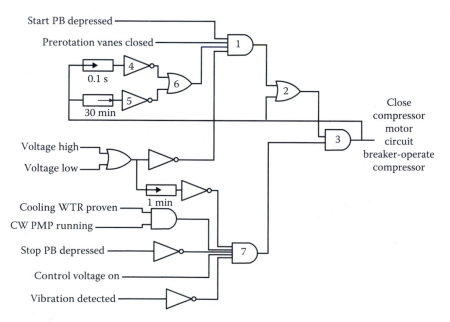

**FIGURE 13.6**
Typical logic diagram for centrifugal chiller.

of start permissives and a set of run permissives must be satisfied. The output of and gate 1 represents the start permissives. For the output of gate 1 to produce a logic 1, four conditions must be satisfied: (1) the start pushbutton must be depressed (thereby calling for a start), (2) the inlet prerotation vanes must be in the closed position (This is normal procedure to minimize the starting load on the motor.), (3) the voltage to the circuit breaker must be neither too high nor too low (This is also a practice to minimize overheating of the motor during starting.), and (4) the output of gate 6 must be a logic "1." There will be a logic "1" at the output of gate 6 during the following times: [1] Prior to the closure of the CB and up to 0.1 s after CB closure and [2] exclusive of the period from the opening of the CB to 30 min after opening of the CB. (The purpose of the 30 min period of time delay is to prevent a closure of the CB until 30 min after it has been opened. The 30 min delay is a common method of minimizing overheating of the motor due to frequent starts.)

## 13.4 Ladder Diagrams

Ladder diagrams are a type of electrical schematic drawing that represent the logic of an industrial control system with physical components. The term "ladder" comes from the practice of using two vertical lines on the sides of a drawing to represent the control system's power source and horizontal lines that contain the various representations of the system's components. Together, the vertical lines with the horizontal lines, or rungs, simulate the appearance of a ladder. Thus, the term "ladder."

If the ladder diagram is intended for a system that will use electromagnetic relays, the diagram will usually be developed using representations of control relays, contacts, time delay relays, as well as possibly other electromechanical devices. Ladder diagrams intended for programming PLCs generally use a set of symbols that differs from those used for a system that uses electromechanical devices. For example, PLC ladder diagrams mostly use parentheses and brackets instead of representations of contacts and relay coils. There are other differences as well. (In this textbook, the system using representations of electromechanical relay contacts and electromechanical relay coils is treated, but the PLC programming system of ladder diagrams is not.)

Ladder diagrams use a unique set of symbols although many of the symbols may be found in other types of drawings. Common symbols used in ladder diagrams are shown in Figure 13.7. The most common type of electromechanical relay identified in ladder diagrams is the "control relay." Control relays vary considerably in design but the most common type consist of a coil that, when electrically energized, moves an armature that in turn moves a set of contacts from one position to another position. Common

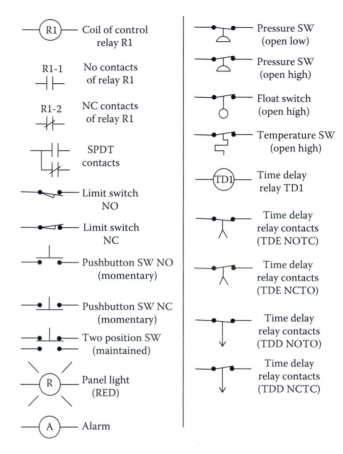

**FIGURE 13.7**
Ladder diagram symbols.

types of relays are NO, NC, SPST, and DPDT. Control relay could also be of the electronic type, which might have no moving parts whatsoever. Relays are treated in detail in Chapter 12.

A typical ladder diagram is shown in Figure 13.8. The ladder diagram shown was prepared with the assumption that the physical devices would be electromechanical devices. Nevertheless, a PLC could be programmed to design a control system that would perform the same results. The ladder diagram of Figure 13.8 performs the functions required by the logic diagram of Figure 13.7. By convention, components are shown in horizontal lines or rungs. The rungs of the ladder diagram are each numbered starting at the top of the diagram and proceeding downward. The locations of the contacts of relays, namely the rung number, are listed beside the coil of the relay. Usual practice is to underscore NC contacts but to avoid underscoring NO contacts.

The sequence of operation of the control system shown in Figure 13.8 is described as follows. The control system is for the control of a centrifugal

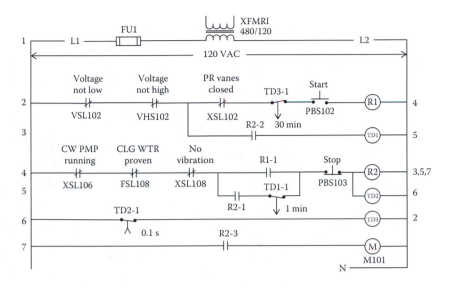

**FIGURE 13.8**
Ladder diagram for centrifugal chiller.

compressor. Note, however, that the control system was designed for the purposes of illustration and does not simulate any actual, existing control system. The symbol "M" represents the circuit breaker that would be closed to power the compressor motor. It is assumed that the circuit breaker had been in the open state for a period of time in excess of 30 min prior to someone depressing the "start" pushbutton. The 120 VAC that powers the control system is on and providing power to the control system. All of the control relays (R1 and R2) and time delay relays (TD1, TD2, and TD3) are in the deenergized state. Relay R1 is essentially the "start relay." All of the permissives to R1 must be satisfied before R1 can be energized when the start pushbutton is depressed thereby calling for a start. One of the start permissives is the condition of the voltage. The power voltage must be neither too high nor too low. The prerotation vanes must be in the closed position so that the motor would be started in the unloaded condition. Time delay relay TD2 is deenegized and contacts TD2-1 are open. Time delay relay TD3 is deenergized and contacts TD3-1 are closed.

If the start pushbutton is depressed with all of the start permissives normal, relay R1 will be energized and, provided the "run" permissives are satisfied, relay R2 will be energized. The permissives for energization of run relay R2 include (1) chilled water pump running, (2) cooling water to condenser running, and (3) absence of vibration. With both the start permissives and the run permissives satisfied, depressing the start pushbutton will energize R2. When R2 is energized, timer TD1 is energized through R2-2. Contacts TD1-1 are closed and R2 is sealed energized through R2-1. The compressor circuit breaker motor will be closed by R2-3 and the compressor started. When R2 is energized, TD2 is energized and 0.1 second later contacts TD3-1 are opened deenergizing start relay R1.

If the voltage to the compressor motor becomes either too high or too low while the compressor motor is operating, contacts TD1-1 will open after 1 min deenergizing R2 and stopping the compressor motor.

When R2 is deenergized, TD2 is likewise deenergized and contacts TD2-1 are opened deenergizing TD3. Contacts TD3-1 will remain open for a period of 30 min following stoppage of the compressor motor thereby prohibiting another start of the motor for 30 min. At 30 min, TD3-1 contacts close and permit a restart provided that all of the start and run permissives are satisfied. (Delaying a start after a stop of a large motor is often done to allow cooling of the motor for the purpose of enhancing motor life. Delaying motor restarts is a common practice in the control of large motors. The delays allow cooling of the insulation in immediate contact with the motor windings. In some control designs, a delay is imposed from the start of a motor rather than from the stoppage.)

## 13.5 Panel Layout Drawings

A panel layout drawing is used to locate electrical devices within a control panel. For new construction, the layout drawing shows where devices are to be mounted within a panel. Sometimes, identifying labels will be attached to identify individual devices within the panel. For an existing panel that, say, has been in use for many years, the layout panel drawing will assist maintenance personnel to correctly identify devices within the panel. After many years, labels that had been installed to identify components very often become illegible. Perhaps tags or labels were never used. In any event, the layout drawing will assist maintenance personnel to identify existing components within a panel. In this regard, the layout drawings can be very helpful as an aid in troubleshooting and locating specific devices.

A typical layout drawing is shown in Figure 13.9. As is common with electrical devices of all types, a sound practice is to allow for possible future additions. Accordingly, the layout of Figure 13.9 includes unused space for components that might be added at some future date.

## 13.6 Cable Drawings

Electrical cable drawings show the sizes as well as the origins and destinations of the cables of an installation. Each cable will be identified with a unique number. Often the cable number includes characters that describe

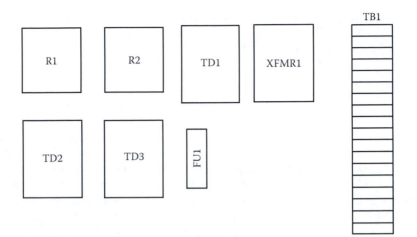

**FIGURE 13.9**
Panel layout drawing.

properties of the cable. Typical characters will describe the number of conductors in the cable and the AWG sizes of the conductors. Cable drawings vary in their composition and design. So, generalizations on the subject can be misleading. In fact, today most often cable sizing and routing is presented in spreadsheets that are more often called "schedules" rather than drawings. If an installation is to be made in compliance with a code, as the NEC, the sizing of cables will necessarily be in accordance with the requirements of the code. Whether by drawing or schedule, a cable's raceway routing will be shown in one of these documents.

## 13.7 Raceway

After electrical cables have been sized and the respective routes determined, the raceways (conduits and cable trays) will be sized to accommodate the cable selections. Raceways are sized much as cables so as to meet code requirements. Raceway drawings are often physical drawings that specify the exact physical routing of the raceways. In many instances, the exact physical locations must be resolved in advance because of the possible interference with other utilities as pipes, ducts, or lights. For more simple installations, the routing is often allowed to be at the discretion of the installation contractor. In this later case, there would be no drawing that shows the exact routing of conduits or trays.

## 13.8 Connection Drawings

Connection drawings serve to identify where, exactly, conductors are to be landed on electrical devices. As is the case with cables and conduits, many organizations use computer-generated schedules rather than drawings to designate where conductors are to be landed. Yet, the connection drawings still have some advantages. Specifically, connection drawings are prepared to represent the physical configuration of the devices where conductors are to be landed. Having a physical depiction of the device where a conductor is to be connected contributes to ensuring that a conductor will be landed on the proper terminal thereby avoiding subsequent errors and time-consuming troubleshooting.

A connection drawing, or an equivalent spreadsheet, is necessary for a new installation. Later, one of these documents will be helpful when replacements or modifications are needed. A connection drawing can show the connections that are to be made within a panel or it could show the connections between remotely located devices. Connection drawings are also called "interconnection drawings." A few rules are generally followed in the generation of connection drawings. One rule is that a line on the drawing represents a physical conductor. By this practice, less thought is required of the person installing the wiring. The rule is partly intended to avoid an excessive number of conductors landed under a single terminal, which could exceed a terminal's rating.

A typical connection drawing is shown in Figure 13.10. The drawing depicts a panel that includes the devices of Figure 13.9 and interconnections made in accordance with the latter diagram of Figure 13.8. As is very common, the panel includes a terminal strip to which field wiring may be connected. By this practice, field cables are landed on the single terminal strip rather than the individual components. Again, this practice is for the purpose of simplifying tasks that are to be done in the field. If cables are to be landed on the terminal strip rather than on the components within the panel, there would be less potential for error. One exception in the referenced drawing is the 480 VAC connections to the 480/120 transformer; it is best to keep the 480 VAC conductors at a distance from the 120 VAC conductors. Incoming cables are identified with a unique number. Note that the L2 side of the derived 120 VAC power to the panel must be connected to a system neutral that is designated by "N." The connection to neutral prevents the voltage within the panel from rising too high above ground potential. A connection to ground, or "G," would be disallowed by code as the L2 conductors are current carrying under normal operation. Of course, the enclosure in which the electrical components are located must be connected to an equipment ground.

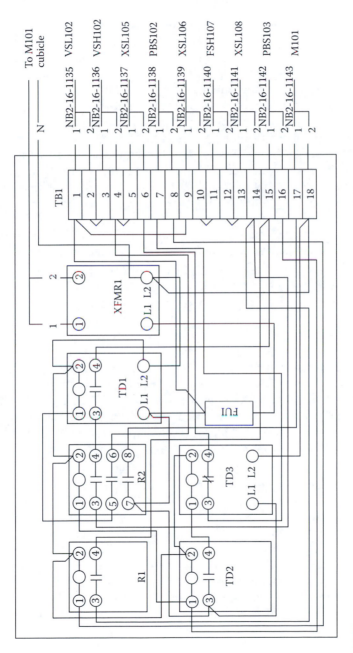

**FIGURE 13.10**
Connection drawing.

## Problems

1. Draw a one-line diagram that represents the electrical system at a building that is a 10-story, 80-unit condominium. The electric utility is supplying a 4160 V service to the building. The incoming service first comes to a disconnect switch and then a 200 A draw out oil circuit breaker. In the basement of the building, the incoming conductors connect to a power distribution center that contains within it a bus and cubicles for circuit breakers. Conductors lead from the power distribution center to five, 4160/208 transformers that are distributed throughout the building. A 40 A air circuit breaker is located in each of the lines that extend to the transformers. Conductors after the transformers are protected with 800 A air breakers that lead to distribution panels from which conductors would extend (not to be shown) to (5) panels within the respective condominiums that are numbered #1 through #5. The transformers are wired delta-to-grounded four-wire wye. (This configuration provides three-phase 208 voltage for air conditioners, ovens, and clothes driers, as well as 120 VAC single-phase voltage for convenience outlets.)

2. Draw a one-line diagram that represents an electrical system that provides power to three irrigation pumps. All three pumps are served from a common bus located in an outdoor motor control center. Service from the electric utility is by a 4160 VAC three-wire line. A 4160/480 transformer, near the motor control center, is provided at the site by the utility. The transformer is wired delta–delta and the secondary of the transformer is corner grounded. The motors are 10 HP, 15 HP, and 15 HP and are identified as M1, M2, and M3, respectively. The circuit breakers are sized 20, 30, and 30 A, respectively. Extensions to the motors are by two-wire cables. The incoming 480 V service connects first to a 100 A breaker located in the motor control center.

3. Draw a one-line diagram that represents an electrical system that powers a nitrogen gas–generating plant. The electric utility is supplying a 4160 V service to the plant. The incoming service is directed to a power distribution center that contains Bus #1. Bus #1 provides power to (1) motor M1, which is 2000 HP and has an associated 350 A circuit breaker, (2) motor M2, which is 3000 HP and has an associated 500 A circuit breaker, (3) a 4160/480 V –20 kVA transformer (delta to wye grounded) that provides 480 V power to Bus #2; the transformer has an associated disconnect switch and a 3 A fuse before the transformer, and (4) a 4160/208-20 kVA transformer (delta to wye grounded) for 208 power to a heating and lighting panel that provides utilities to a building on the property. Bus #2 provides power to three motors that are 5 HP, 10 HP, and 1 HP. The motors are identified as M3, M4, and M5, respectively. Each of the motors, M3, M4, and M5, served by Bus 2 are

protected with 20 A breakers. Both motor M1 and M2 have protective relays #27—under voltage, #46—reverse phase, 51—ac time overcurrent, and 59—overvoltage.

4. Draw a three-line diagram for the system described in Question #1.

5. Draw a three-line diagram for the system described in Question #2.

6. Draw a diagram that shows the logic for an outdoor fan that delivers outside air to an auditorium. Operation of the fan is to be from a manual start–stop control station. If there is a power outage, the controls shall call for cessation of operation. To avoid a burden on the electrical supply system, operation of the fan is not to be resumed upon the return of electrical power following an outage. The fan motor is started and operated whenever the (momentary) start pushbutton is depressed provided (1) motor voltage is present, (2) control voltage is present, and (3) the (momentary) stop pushbutton is not depressed. If the stop BP is depressed, the motor discontinues operation. If either the motor voltage or the control voltage is not present, operation of the motor is ceased. If the motor voltage and the control voltage return after the motor has been stopped, the motor will not resume operation until the start pushbutton is depressed. The inputs are (1) momentary start PB depressed, (2) motor voltage present, (3) control voltage present, and (4) momentary stop PB depressed. The output is operation of supply air fan motor.

7. Draw a diagram that shows the logic for operation of a large saw. To avoid an inadvertent start of the saw motor, operation will not commence until two pushbuttons are simultaneously depressed. The motor operation logic is to also include the following permissives: (1) motor voltage is present, (2) control voltage is present, and (3) the stop pushbutton is not depressed. If the stop BP is depressed, the motor discontinues operation. If either the motor voltage or the control voltage is not present, operation of the motor is ceased. If the motor voltage and the control voltage return after the motor has been stopped, the motor will not resume operation until the start pushbuttons are depressed. The inputs are (1) momentary start PB#1 depressed, (2) momentary pushbutton #2 depressed, (3) motor voltage present, (4) control voltage present, and (5) momentary stop PB depressed. The output is operation of saw motor.

8. Draw a diagram that shows the logic for a pump that delivers oil from a delivery truck to a holding tank. To help avoid accidental starts, the pump motor is started and operated whenever the (momentary) start pushbutton is depressed for a period of at least two seconds provided. Also operation is permitted only when (1) motor voltage is present, (2) control voltage is present, and (3) the (momentary) stop pushbutton is not depressed. If the stop BP is depressed, the motor discontinues

operation. If either the motor voltage or the control voltage is not present, operation of the motor is ceased. If both the motor voltage and the control voltage return after the motor has been stopped, the motor will not resume operation until the start pushbutton is depressed. The inputs are (1) momentary start PB depressed, (2) motor voltage present, (3) control voltage present, and (4) momentary stop PB depressed. The output is operation of the oil pump motor.

9. Draw a diagram that shows the logic for an outdoor air fan that delivers outside air to an auditorium. Operation of the fan is to be from a manual start–stop control station. If there is a power outage the controls shall call for cessation of operation. To avoid a burden on the electrical supply system, operation of the fan is not to be automatically resumed upon return of electrical power following an outage. The fan motor is started and operated whenever the (momentary) start pushbutton is depressed, provided motor start permissives are satisfied. The start permissive shall include (1) motor voltage is present and neither too high nor too low, (2) control voltage is present, and (3) the (momentary) stop pushbutton is not depressed. Once started, the motor shall continue to operate provided the motor voltage is present, the control voltage is present, and the motor voltage has been neither too high nor too low for a period of time in excess 1 s. If either the control voltage is not present, operation of the motor is ceased. If both the motor voltage and the control voltage return after the motor has been stopped, the motor will not resume operation until the start pushbutton is depressed. The inputs are (1) momentary start PB depressed, (2) motor voltage present, (3) control voltage present, (4) motor voltage not high, (5) motor voltage not low, and (6) momentary stop PB depressed. The output is operation of motor.

10. Draw a diagram that shows the logic for the exhaust fan for a battery charging room. The fan is intended to exhaust the hydrogen gas that is generated by the charging process. Continuous operation of the fan is considered critically important as its failure could result in a hazardous accumulation of explosive hydrogen gases within the battery room. The fan is fitted with a switch that opens whenever the fan is stopped. If the fan is not operating, the fan switch actuates a panel light on a remote panel. The inputs are (1) motor voltage present and (2) fan motor operating. The outputs are (1) exhaust fan operating and (2) red panel light activated if exhaust fan is not operating.

11. Draw a diagram to show the logic for control of a condensate day tank. The tank is located in a woman's dormitory at a university. The rooms in the dormitory are heated by steam on cold days and condensate from the heaters flows by gravity to the day tank that is located in the basement of the dormitory. When the level in the day tank rises to 5 ft to 1 in., the contacts of a high-level switch (LSH101) are opened and the condensate pump is started. When the level in the tank falls to 4 ft to

11 in., the contacts of LSH101 again close. When the level falls to 1 ft to 3 in., the contacts of low-level switch LSL101 are opened and the operation of the pump is stopped. The contacts of LSL101 are closed when the level rises to 1 ft to 5 in. A local control station permits operation of the condensate pump anytime the level in the tank is above the setting of LSL101. If the condensate level remains high for more than 2 min, a remote alarm is activated. The inputs are (1) motor voltage is present, (2) control voltage is present, (3) condensate level is high, (4) condensate level is low, (5) start pushbutton, and (6) stop pushbutton. The outputs are operation of the condensate pump and the activation of an alarm.

12. Draw a diagram to show the logic for a forced draft fan that delivers combustion air to a boiler. To minimize wear on the fan motor, an interlock prohibits starting the motor unless the air inlet vanes are proven in the closed position. After starting, the inlet vanes may be moved to an open position without halting fan operation. The owner asks that two panel lights be provided at the fan motor control station: (1) a green light to show that the permissives are satisfied and the fan motor is ready for starting and (2) a red light to indicate that the fan motor contactor has been closed calling for fan motor operation. When the fan motor is started, the green panel light is to be off. The inputs are (1) motor voltage is present, (2) control voltage is present, (3) inlet vanes proven closed, (4) start pushbutton, (5) stop pushbutton, and (6) fan motor contactor closed. The outputs are (1) demand for the operation of the forced draft fan, (2) green panel light indication that fan motor permissives are satisfied and motor is ready to start, and (3) red panel light indication that fan motor contactor is closed calling for fan motor operation.

13. Draw a ladder diagram for the logic required by Question #6. Assume that electromechanical devices are to be used and (1) the start PB is NO, identified as PB1; (2) the stop PB is NC, identified as PB2; (3) the control circuit is powered by the secondary of a 480/120 V transformer, identified as XFMR1; and (4) a fuse, identified as FU1, protects the circuit.

14. Draw a ladder diagram for the logic required by Question #9. Assume that electromechanical devices are to be used and (1) one start PB is NO, identified as PB1; (2) the stop PB is NC, identified as PB2; (3) the control circuit is powered by the secondary of a 480/120 V transformer, identified as XFMR1; and (4) a fuse, identified as FU1, protects the circuit. No relay is to be energized when the motor is not running.

15. Draw a layout diagram for the components assumed for the ladder diagram of above Question #13.

16. Draw a layout diagram for the components assumed for the ladder diagram of Question #14.

17. Draw a connection diagram for the ladder diagram of Question #13. (Assume terminal numbers for the devices.)

# *Appendix*

## A.1 Demonstration That Equation 5.1 Is the Equivalent of Equation 5.2

Equation 5.1 states

$$P^i = (v_{PK})(i_{PK})(\cos \theta_{SP})(\sin \omega t)^2 + (v_{PK})(i_{PK})(\sin \theta_{SP})(\sin \omega t)(\cos \omega t) \quad (5.1)$$

A common form of instantaneous power found in textbooks on the subject of three-phase power is

$$P^i = [(v_{PK})(i_{PK})/2]\cos \theta - [(v_{PK})(i_{PK})/2]\cos (2\omega t + \theta) \quad (5.2)$$

These two equations are equivalent as demonstrated next:

$$\cos (2\omega t + \theta) = \cos 2\omega t \cos \theta - \sin 2\omega t \sin \theta \quad (\text{Reference A.1})$$

Thus, Equation 5.2 becomes

$$P^i = [(v_{PK})(i_{PK})/2]\cos \theta - [(v_{PK})(i_{PK})/2][\cos 2\omega t \cos \theta - \sin 2\omega t \sin \theta]$$
$$P^i = [(v_{PK})(i_{PK})/2]\cos\theta - [(v_{PK})(i_{PK})/2][\cos 2\omega t \cos \theta] + [(v_{PK})(i_{PK})/2][\sin 2\omega t \sin \theta]$$

Since, $\cos 2\omega t = 1 - 2\sin^2 \omega t$

$$P^i = [(v_{PK})(i_{PK})/2]\cos \theta - [(v_{PK})(i_{PK})/2][1 - 2(\sin \omega t)^2][\cos \theta]$$
$$+ [(v_{PK})(i_{PK})/2][\sin 2\omega t \sin \theta]$$
$$P^i = [(v_{PK})(i_{PK})/2]\cos \theta - [(v_{PK})(i_{PK})/2]\cos \theta + [(v_{PK})(i_{PK})/2][2(\sin \omega t)^2 \cos \theta]$$
$$+ [(v_{PK})(i_{PK})/2][\sin 2\omega t \sin \theta]$$
$$P^i = [(v_{PK})(i_{PK})](\cos \theta)(\sin \omega t)^2 + [(v_{PK})(i_{PK})/2][\sin 2\omega t \sin \theta]$$
$$\sin 2\omega t = 2\sin \omega t \cos \omega t$$
$$P^i = [(v_{PK})(i_{PK})](\cos \theta)(\sin \omega t)^2 + [(v_{PK})(i_{PK})/2][(2\sin \omega t \cos \omega t)\sin \theta]$$
$$P^i = [(v_{PK})(i_{PK})](\cos \theta)(\sin \omega t)^2 + [(v_{PK})(i_{PK})][(\sin \theta)(\sin \omega t)(\cos \omega t)], \text{ Q.E.D.}$$

**NOTE:** Instantaneous current is described by the equation, $i^i = i_{PK} \sin (\omega t + \theta_{SP})$. If current leads voltage, then $\theta_L > 0$; and if current lags voltage, $\theta_L < 0$. However, many textbooks on the subject describe instantaneous current as $i_{PK} \sin (\omega t - \theta_{SP})$ in which case $\theta_L < 0$ for a leading current and $\theta_L > 0$ for a lagging current. If instantaneous current is assumed to be described by the expression $i_{PK} \sin(\omega t - \theta_{SP})$, then the corresponding equivalent equations to Equations 5.1 and 5.2 become, respectively,

$$P^i = (v_{PK})(i_{PK})(\cos \theta_{SP})(\sin \omega t)^2 - (v_{PK})(i_{PK})(\sin \theta_{SP})(\sin \omega t)(\cos \omega t) \quad \text{and}$$
$$P^i = [(v_{PK})(i_{PK})/2] \cos \theta - [(v_{PK})(i_{PK})/2] \cos(2\omega t - \theta)$$

## A.2 Table of Computed Values of Instantaneous Power for Example 5.1 (Table A.1)

$$P^i = A + B$$
$$A = (6720) \sin^2 \omega t; \quad B = -(6855.36) \sin \omega t \cos \omega t$$

## A.3 Confirmation that Total Power Is Equivalent to that of an Assumed Wye Circuit

In the textbook, it is stated that the power of either a delta circuit or an unknown circuit can be calculated by assuming the circuit to be a wye circuit. This assumption presents a very handy tool because, in many instances, only line parameters are available. No reference is provided at this time since the author independently realized this relationship.

While not a rigid mathematical proof, two corroborating calculations are presented in this appendix that supports the claim. The first step of the demonstration is to calculate the power of a delta circuit when the phase values of the delta circuit are known. Next, the line parameters of the delta circuit are calculated. It is then shown that if only the line currents were known, and not the phase currents, the total power consumption of the circuit may be determined from only the line parameters. This end is accomplished by assuming, according to the postulated method, that the user is solely a wye circuit. A second calculation is performed as support. These two calculations are shown: first, in Part 1 of this section and, second, in Part 2 of this section.

### A.3.1 Part 1 of Section A.3: First Demonstration That Power of an Unbalanced Delta Circuit Is Equivalent to That of an Assumed Wye Circuit

**Given**: An unbalanced delta circuit with the following properties:

Line potential: 480-3-60 V

$$I_{ab} = 5 \text{ A at PF} = 1.0$$
$$I_{bc} = 10 \text{ A at PF} = 0.9 \text{ lagging}$$
$$I_{ca} = 15 \text{ A at PF} = 0.8 \text{ leading}$$

Find line currents in conductors A, B, and C.

**TABLE A.1**

Example 5.1 Computed Values

| ($\omega t$) | Function A | Function B | $P^i = A + B$ |
|---|---|---|---|
| 0° | 0 | 0 | 0 |
| 10° | 202.63 | −1172.40 | −969.77 |
| 20° | 786.09 | −2203.40 | −1417.31 |
| 30° | 1680.00 | −2968.63 | −1288.63 |
| 40° | 2776.54 | −3375.80 | −599.26 |
| 50° | 3943.45 | −3375.80 | 567.64 |
| 60° | 5040.00 | −2968.63 | 2071.36 |
| 70° | 5933.90 | −2203.40 | 3730.50 |
| 80° | 6517.36 | −1172.40 | 5344.96 |
| 90° | 6720.00 | 0 | 6720.00 |
| 100° | 6517.36 | 1172.40 | 7689.77 |
| 110° | 5933.90 | 2203.40 | 8137.31 |
| 120° | 5040.00 | 2968.63 | 8008.63 |
| 130° | 3943.45 | 3375.80 | 7319.26 |
| 140° | 2776.54 | 3375.80 | 6152.35 |
| 150° | 1680.00 | 2968.63 | 4648.63 |
| 160° | 786.09 | 2203.40 | 2989.49 |
| 170° | 202.63 | 1172.40 | 1375.03 |
| 180° | 0 | 0 | 0 |
| 190° | 202.63 | −1172.40 | −969.77 |
| 200° | 786.09 | −2203.40 | −1417.31 |
| 210° | 1680.00 | −2968.63 | −1288.63 |
| 220° | 2776.54 | −3375.80 | −599.26 |
| 230° | 3943.45 | −3375.80 | 567.64 |
| 240° | 5040.00 | −2968.63 | 2071.36 |
| 250° | 5933.90 | −2203.40 | 3730.50 |
| 260° | 6517.36 | −1172.40 | 5344.96 |
| 270° | 6720.00 | 0 | 6720.00 |
| 280° | 6517.36 | 1172.40 | 7689.77 |
| 290° | 5933.90 | 2203.40 | 8137.31 |
| 300° | 5040.00 | 2968.63 | 8008.63 |
| 310° | 3943.45 | 3375.80 | 7319.26 |
| 320° | 2776.54 | 3375.80 | 6152.35 |
| 330° | 1680.00 | 2968.63 | 4648.63 |
| 340° | 786.09 | 2203.40 | 2989.49 |
| 350° | 202.63 | 1172.40 | 1375.03 |
| 360° | 0 | 0 | 0 |

**Solution**

$$\theta_{P-AB} = \cos^{-1}1.0 = 0$$
$$\theta_{P-BC} = \cos^{-1}0.9 = -25.84193°$$
$$\theta_{P-CA} = -\cos^{-1}0.8 = +36.86989°$$

These are the same line parameters that were assumed in Example 5.8. Calculations repeated in that section determined the line parameters corresponding to the assumed phase parameters as shown in the following summary.

| Line Current | Lead/Lag |
|---|---|
| $I_A = 19.6962506$ A | $\theta_{L-A/CA} = +42.5929°$ |
| $I_B = 14.413495$ A | $\theta_{L-B/AB} = +22.9263°$ |
| $I_C = 12.762278$ A | $\theta_{L-C/BC} = +55.6244°$ |

The power of the delta circuit is

$$P_{C-A} = V_P I_P \cos\theta_{P-A/CA} = (480)(15)(.8) = 5{,}760$$

$$P_{A-B} = V_P I_P \cos\theta_{P-B/AB} = (480)(5)(1) = 2{,}400$$

$$P_{B-C} = V_P I_P \cos\theta_{P-C/BC} = (480)(10)(.9) = 4{,}320$$

$$P_T = 5{,}760 + 2{,}400 + 4{,}320 = 12{,}480 \text{ W}$$

As postulated, assume that the user is a wye circuit instead of a delta circuit. Equation 5.21 is applicable:

$$P_T = V_L\left(1/\sqrt{3}\right)[I_{L-A}\cos(\theta_{L-A/CA} - 30°) + I_{L-B}\cos(\theta_{L-B/AB} - 30°)$$
$$+ I_{L-C}\cos(\theta_{L-C/BC} - 30°)] \qquad (5.21)$$

Assume $V_L = 480$ VAC

$$P_T = (480)\left(1/\sqrt{3}\right)\big[(19.6962506)\cos(42.5929° - 30°)$$
$$+ (14.41349)\cos(22.9263° - 30°)$$
$$+ (12.76227)\cos(55.6244° - 30°)\big]$$

$$P_T = (480)(1/\sqrt{3})\big[(19.696250)\cos 12.5929°$$

$$+(14.41349)\cos -7.0737°$$

$$+(12.76227)\cos 25.6244°\big]$$

$$P_T = (277.1281)\big[(19.696250)(.97594)$$

$$+(14.41349)(.992388)$$

$$+(12.76227)(.90165)\big]$$

$$P_T = (277.1281)\big[(19.22235)+14.30377+11.50710\big]$$

$$P_T = (277.1281)[45.03322] = 12,480.00 \text{ W}$$

Thus, it is seen that the power calculation yields the same value as that which was determined by adding the wattages of the three phases of a delta circuit. This calculation tends to confirm that the power of a delta circuit can be determined by assuming the user to be a wye circuit. A second, corroborating calculation is performed next as a second confirmation.

### A.3.2 Part 2 of Section A.3: Second Demonstration That Power of an Unbalanced Delta Circuit Is Equivalent to That of an Assumed Wye Circuit

Assume the following conditions for an unbalanced delta circuit:
  Potential: 480-3-60 V

$$I_{ab} = 10 \text{ A}, PF = .5, \quad \text{lagging}, \theta_{P-AB} = -60.0000°$$

$$I_{bc} = 20 \text{ A}, PF = .6, \quad \text{lagging}, \theta_{P-BC} = -53.1301°$$

$$I_{ca} = 30 \text{ A}, \ PF = .7, \quad \text{lagging}, \theta_{P-CA} = -45.5729°$$

$$P_{ab} = VI(PF) = (480)(10)(.5) = 2,400 \text{ W}$$

$$P_{bc} = VI(PF) = (480)(20)(.6) = 5,760 \text{ W}$$

$$P_{ca} = VI(PF) = (480)(30)(.7) = 10,080 \text{ W}$$

Total delta power: 18,240 W
  Use the stated currents for the delta circuit and calculate the power if the currents and voltages were descriptive of a wye circuit. The first step is to calculate line currents for the delta circuit.

Reference: Equations 4.4 through 4.9

Determine $I_A$:

$$I_{ba-x} = -I_{ab}\cos\theta_{P-AB} = -(10)(.5) = -5$$

$$I_{ca-x} = -I_{ca}(1/2)\left[(\sqrt{3})\sin\theta_{P-CA} + \cos\theta_{P-CA}\right]$$

$$= -(30)(1/2)\left[(\sqrt{3})\sin(-45.5729°) + \cos(-45.5729°)\right]$$

$$= -(30)(1/2)\left[(\sqrt{3})(-.714142843) + .700000\right]$$

$$= -(15)[-1.2369316 + .70000] = -(15)[-.5369316] = 8.05397531$$

$$I_A = I_{ba-x} + I_{ca-x} = -5 + 8.05397531 = 3.05397531$$

$$I_{ba-y} = -I_{ab}\sin\theta_{P-AB} = -(10)\sin(-60.00°) = 8.66025404$$

$$I_{ca-y} = I_{ca}(1/2)\left[(\sqrt{3})\theta_{P-CA} - \sin\theta_{P-CA}\right]$$

$$= (30)(1/2)\left[(\sqrt{3}) - 45.5729° - \sin(-45.5729°)\right]$$

$$= (15)\left[(\sqrt{3})(.7) - (-.714142843)\right] = (15)\left[1.21243 + (.71414)\right]$$

$$= (15)[1.9265784] = 28.89867612$$

$$I_{A-y} = I_{ba-y} + I_{ca-y} = 8.66025404 + 28.89867612 = 37.55893016$$

$$I_A = \left\{(I_{A-x})^2 + (I_{A-y})^2\right\}^{1/2} = \left\{(3.05397531)^2 + (37.55893016)^2\right\}^{1/2}$$

$$= 37.68288736\ A$$

$$\lambda = \sin^{-1}(I_{A-y} \div I_A) = \sin^{-1}(37.55893016 \div 37.68288736) = \sin^{-1}(.9967105)$$

$I_A$ is in Quadrant I.

$$\lambda = 85.3514173°$$

$$\theta_{L-A} = (\lambda - 120°) = (85.3514173° - 120°) = -34.6485827°$$

Determine $I_B$:

$$I_{ab-x} = I_{ab}\cos\theta_{P-AB} = (10)\cos - 60° = (10)(.5) = 5$$

$$I_{cb-x} = -I_{bc}(1/2)\left[(\sqrt{3})\sin\theta_{P-BC} - \cos\theta_{P-BC}\right]$$

$$= -(20)(1/2)\left[(\sqrt{3})\sin(-53.1301°) - \cos(-53.1301°)\right]$$

$$= -(10)\left[(\sqrt{3})(-.8000) - (.600)\right]$$

$$= -(10)\ [-1.38564 - (.60000)] = -(10)[-1.985640] = 19.856406$$

$$I_{B-x} = I_{ab-x} + I_{cb-x} = 5 + 19.85640 = 24.8564064$$

$$I_{ab-y} = I_{ab}\sin\theta_{P-AB} = (10)\sin - 60.000° = (10)(-.86602) = -8.6602540$$

$$I_{cb-y} = I_{bc}(1/2)\left[(\sqrt{3})\cos\theta_{P-CB} + \sin\theta_{P-CB}\right]$$

$$= (20)(1/2)\left[(\sqrt{3})\cos(-53.1301°) + \sin(-53.1301°)\right]$$

$$= (10)\left[(\sqrt{3})(.6) + (-.8)\right] = (10)\ [1.0392304 - .80000] = 2.392304$$

$$I_{B-y} = I_{ab-y} + I_{cb-y} = -8.660254 + 2.39230 = -6.267954$$

$$I_B = \left\{(I_{B-x})^2 + (I_{B-y})^2\right\}^{1/2} = \left\{(24.856406)^2 + (-6.267954)^2\right\}^{1/2} = 25.634511\ A$$

$$\theta_{L-B} = -\sin^{-1}(I_{B-y} \div I_B) = -\sin^{-1}(-6.267954 \div 25.634511)$$

$$= -\sin^{-1}(-.2445123)$$

$I_B$ is in Quadrant IV.

$$\sin^{-1}(-.2445) = -14.15301°$$

$$\theta_{L-B} = -14.15301°$$

Determine $I_C$:

$$I_{bc-x} = I_{bc}(1/2)\left[(\sqrt{3})\sin\theta_{P-BC} - \cos\theta_{P-BC}\right]$$

$$= (20)(1/2)\left[(\sqrt{3})\sin(-53.1301°) - \cos(-53.1301°)\right]$$

$$= (10)\left[(\sqrt{3})(-.80000) - (.60000)\right]$$

$$= (10)\left[-1.385640 - (-.60000) = (10)\right] - 1.98564 = -19.856406$$

$$I_{ac-x} = I_{ca}(1/2)\left[(\sqrt{3})\sin\theta_{P-CA} + \cos\theta_{P-CA}\right]$$

$$= (30)(1/2)\left[(\sqrt{3})\sin(-45.5729°) + \cos(-45.5729°)\right]$$

$$= (15)\left[(\sqrt{3})(-.7141428) + .70000\right] = (15)\left[-1.23693 + .70000\right]$$

$$= (15)[-.53693] = -8.053975$$

$$I_{C-x} = I_{bc-x} + I_{ac-x} = -19.856406 - 8.053975 = -27.91038$$

$$I_{ac-y} = -I_{ca}(1/2)\left[(\sqrt{3})\cos\theta_{P-CA} - \sin\theta_{P-CA}\right]$$

$$= -(30)(1/2)\left[(\sqrt{3})\cos(-45.5729°) - \sin(-45.5729°)\right]$$

$$= -(30)(1/2)\left[(\sqrt{3})(.70000) - (-.7141428)\right]$$

$$= -(15)\ [1.2124355 + .7141428] = -(15)[1.926578] = -28.898676$$

$$I_{bc-y} = -I_{bc}(1/2)\left[(\sqrt{3})\cos\theta_{P-Bc} + \sin\theta_{P-Bc}\right]$$

$$= -(20)(1/2)\left[(\sqrt{3})\cos(-53.130°) + \sin(-53.1301°)\right]$$

$$= -(10)\left[(\sqrt{3})(.60000) + (-.80000)\right]$$

$$= -(10)[1.039230 - (.80000)] = -(10)[.2392304] = -2.392304$$

$$I_{C-y} = I_{bc-y} + I_{ac-y} = -28.898676 - 2.392304 = -31.29098$$

$$I_C = \left\{(I_{C-x})^2 + (I_{C-y})^2\right\}^{1/2} = \left\{(-27.91038)^2 + (-31.29098)^2\right\}^{1/2} = 41.929878\ \text{A}$$

$$\varphi = \sin^{-1}(I_{C-y} \div I_C) = \sin^{-1}(-31.29098 \div 41.929878) = \sin^{-1}(-.746269)$$

$I_C$ is in Quadrant III.

$$\varphi = \sin^{-1}(-.746269) = 228.2682°$$

$$\theta_{L-C} = (\varphi - 240°) = (228.2682° - 240°) = -11.73176°$$

Following is a summary of the calculated values for the assumed delta circuit:

| Line Current | Lead/Lag |
|---|---|
| $I_A = 37.682887$ A | $\theta_{L-A} = -34.64858°$ |
| $I_B = 25.634511$ A | $\theta_{L-B} = -14.15301°$ |
| $I_C = 41.929878$ A | $\theta_{L-C} = -11.73176°$ |

Assume that the calculated line currents and the respective lead/lag angles are for a wye circuit and calculate the power.

$P_T = V_L (1/\sqrt{3}) [I_{L-A} \cos (\theta_{L-A/CA} - 30°) + I_{L-B} \cos (\theta_{L-B/AB} - 30°)$

$\quad + I_{L-C} \cos (\theta_{L-C/BC} - 30°)]$ Equation 5.21 is applicable. $\hspace{2em}$ (5.21)

$P_T = (480)(1/\sqrt{3})\big[(37.682887) \cos (-34.64858° - 30°)$

$\quad + (25.634511) \cos (-14.15301° - 30°) + (41.929878) \cos (-11.73176° - 30°)\big]$

$P_T = (277.1281)\big[(37.682887) \cos (-64.64858°)$

$\quad + (25.634511) \cos (-44.15301°) + (41.929878) \cos -41.73176°)\big]$

$P_T = (277.1281)[(37.682887)(.42817) + (25.634511)(.71748)$

$\quad + (41.929878)(.746269)]$

$P_T = (277.1281)\big[(16.13468) + (18.392248) + (31.290968)\big]$

$P_T = (277.1281)[65.81789] = 18{,}240.00$ W

The power calculation yields the same value as that which was calculated by adding the wattages of the three parts of a delta circuit. As was the case with the first calculation in Part 1, this second calculation likewise corroborates the statement that the power of a delta circuit can be determined by assuming the user to be a wye circuit.

The phasors representative of the voltages and currents in Part 2 calculation are shown and drawn to scale in Figure A.1.

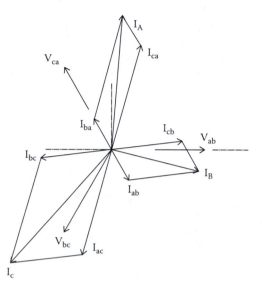

**FIGURE A.1**
Phasor diagram of Part 2.

## A.4 Proof of Two-Wattmeter and Three-Wattmeter Methods

In Section 7.2.2, it was stated that the two-wattmeter method is a valid method for measuring the total power of a three-phase three-wire circuit. It was also stated in Section 7.2.3 that the three-wattmeter method is also a valid method of measuring the total power of a three-phase three-wire circuit. It is demonstrated here that both are valid methods. In fact, proof of the two-wattmeter method follows from the proof of the three-wattmeter method. For this reason, the three-wattmeter method is treated first.

Reference is made to Figure 7.9 where the negative leads of the three wattmeters are shown joined together. The positive leads of the respective meters connected to the phase in which the current coils are located. It is demonstrated that the voltage of point "x" of Figure 7.9 is immaterial and can be at any level.

As mentioned in Chapter 5, the expression for instantaneous power in a single-phase circuit is

$$P^i = v^i \cdot i^i$$

where
  $v^i$ is the instantaneous value of voltage
  $i^i$ is the instantaneous value of current

The average power throughout a specific period of time for each phase is $P = (1/T)\int v^i \cdot i^i dt$ evaluated throughout the period from $t_1$ to $t_2$.

In Figure 7.9, the three single-phase wattmeters are configured to measure the currents of all three lines, that is, phase A, phase B, and phase C. The total of the readings of the three meters is

$$P = (1/T)\int v^i_{A-X} \cdot i^i_A dt + (1/T)\int v^i_{B-X} \cdot i^i_B dt + (1/T)\int v^i_{C-X} \cdot i^i_C dt$$

$$P = (1/T)\int \left[ v^i_{A-X} \cdot i^i_A + v^i_{B-X} \cdot i^i_B + v^i_{C-X} \cdot i^i_C \right] dt$$

Consider a delta circuit of the configuration represented in Figure 1.14.

$$i^i_A = i^i_{C-A} - i^i_{A-B}$$

$$i^i_B = i^i_{A-B} - i^i_{B-C}$$

$$i^i_C = i^i_{B-C} - i^i_{C-A}$$

Then,

$$\left[ v^i_{A-X} \cdot i^i_A + v^i_{B-X} \cdot i^i_B + v^i_{C-X} \cdot i^i_C \right]$$

$$= \left[ v^i_{A-X} \cdot (i^i_{C-A} - i^i_{A-B}) + v^i_{B-X} \cdot (i^i_{A-B} - i^i_{B-C}) + v^i_{C-X} \cdot (i^i_{B-C} - i^i_{C-A}) \right]$$

$$= \left[ i^i_{A-B}(v^i_{B-X} - v^i_{A-X}) + i^i_{B-C}(v^i_{C-X} - v^i_{B-X}) + i^i_{C-A}(v^i_{A-X} - v^i_{C-X}) \right]$$

According to Kirchhoff's voltage law, the sum of the voltages around a circuit must equal to zero. Accordingly, it is recognized that

$$v^i_{A-B} = v^i_{B-X} - v^i_{A-X}$$

$$v^i_{B-C} = v^i_{C-X} - v^i_{B-X} \quad \text{and}$$

$$v^i_{C-A} = v^i_{A-X} - v^i_{C-X}$$

It follows that

$$P = (1/T)\int [v^i_{A-B} \cdot i^i_{A-B} + v^i_{B-C} \cdot i^i_{B-C} + v^i_{C-A} \cdot i^i_{C-A}] dt$$

evaluated from $t_1$ to $t_2$.

This expression is the equivalent to the power that would be determined by measuring the currents and voltages of each of the three phases, that is, phases A–B, B–C, and C–A, and adding the three readings to obtain a total. Thus, the three-wattmeter method is proven valid for a delta circuit. A similar proof can be established for a wye circuit.

In a wye connected circuit, as with the delta circuit, the expression for instantaneous power is

$$P^i = v^i \cdot i^i$$

Let $v^i_x$ be the elevation of the neutral point voltage, which is the voltage of the common negative leads of the three wattmeters:

$$v^i_{A-X} = v^i_{A-N} + v^i_X$$

$$P = (1/T)\int v^i_{A-N} \cdot i^i_A \, dt + (1/T)\int v^i_{B-N} \cdot i^i_B \, dt + (1/T)\int v^i_{C-N} \cdot i^i_C \, dt$$

$$P = (1/T)\int [v^i_{A-N} \cdot i^i_A + v^i_{B-N} \cdot i^i_B + v^i_{C-N} \cdot i^i_C] dt$$

With the three wattmeters connected so that the negative leads are common and at voltage "x" above the neutral voltage,

$$P = (1/T)\int \left[ v^i_{A-X} \cdot i^i_A + v^i_{B-X} \cdot i^i_B + v^i_{C-X} \cdot i^i_C \right] dt$$

$$v^i_{A-X} = v^i_{A-N} + v^i_X$$

$$v^i_{A-X} \cdot i^i_A = i^i_A \left( v^i_{A-N} + v^i_X \right)$$

$$v^i_{B-X} \cdot i^i_B = i^i_B \left( v^i_{B-N} + v^i_X \right)$$

$$v^i_{C-X} \cdot i^i_C = i^i_C \left( v^i_{C-N} + v^i_X \right)$$

$$\left[ v^i_{A-X} \cdot i^i_A + v^i_{B-X} \cdot i^i_B + v^i_{C-X} \cdot i^i_C \right]$$

$$= \left[ i^i_A (v^i_{A-N} + v^i_X) + i^i_B (v^i_{B-N} + v^i_X) + i^i_C (v^i_{C-N} + v^i_X) \right]$$

$$= \left[ i^i_A v^i_{A-N} + i^i_{B-N} v^i_B + i^i_{C-N} v^i_C \right] + v^i_X \left[ i^i_A + i^i_B + i^i_C \right]$$

According to Kirchhoff's current law,

$$i^i_A + i^i_B + i^i_C = 0$$

Then,

$$P = (1/T)\int [v^i_{A-X} \cdot i^i_A + v^i_{B-X} \cdot i^i_B + v^i_{C-X} \cdot i^i_C] dt =$$

$$P = (1/T)\int v^i_{A-N} \cdot i^i_A \, dt + (1/T)\int v^i_{B-N} \cdot i^i_B \, dt + (1/T)\int v^i_{C-N} \cdot i^i_C \, dt$$

Thus, it is apparent that the three-wattmeter method as applied to a wye circuit provides the same results that would be obtained by measuring the power of each phase and then adding the three measurements.

Here, it was demonstrated that the three-wattmeter method is applicable to either a delta or a wye circuit. Therefore, it may be concluded that the three-wattmeter method as depicted in Figure 7.9 is applicable to any form of a three-phase circuit, delta or wye.

Given that the three-wattmeter method is valid for a three-phase three-wire circuit, it is demonstrated next that the two-wattmeter method is likewise a valid method of measuring power in a three-phase three-wire circuit. As mentioned earlier with respect to the three-wattmeter method, the three negative leads may be placed at any voltage. Obviously, if the negative leads were placed common to one of the phase lines, that is, phase A, phase B, or phase C, the wattmeter that measures the current in that phase would read "0." In other words, the meter with the "0" readings can be removed and the total power of the circuit would be the total of the readings on the remaining two wattmeters. Thus, the two-wattmeter method is also a valid means to measure the total power of a three-phase three-wire circuit.

NOTE: Some authors on the subject of three-phase power state that the two-wattmeter method is valid only for balanced circuits. In fact, the method is applicable to both balanced and unbalanced circuits. The following calculations corroborate this position.

### A.4.1 Computations to Corroborate the Two-Wattmeter Method for Three-Phase Three-Wire Circuits

The computations of this appendix corroborate that the two-wattmeter method is applicable to a three-wire three-phase circuit. According to the two-wattmeter method, only two currents of a three-wire three-phase circuit need to be measured in order to determine the power consumption of the circuit. The currents to be measured may be currents A–B, B–C, or C–A. Potentials are to be measured according to a specific order that is determined by the currents that are selected for the measurement.

#### A.4.1.1 Part 1 of Section A.4

This part considers the scenario in which two wattmeters are configured as depicted in Figure 7.5.

The configuration of Figure 7.5 requires measurement of currents A and B as well as voltage C–A and voltage C–B. The currents and voltages of Part 2 of Section A.3 are assumed. The phasors for the voltages and currents are represented in Figure A.1.

The voltage leads of the two meters must be arranged as described in Figure 7.5 with care taken to position the positive and negative leads as shown.

For this computation, assume that meter #1 has its voltage connections with the negative lead on phase C and the positive lead on phase A. Meter #2 has the negative voltage lead on phase C and the positive lead on phase B.

As demonstrated in the example of Part 2 of Section A.3, the following line values were determined:

| Line Current | Lead/Lag |
|---|---|
| $I_A = 37.682887$ A | $\theta_{L-A} = -34.64858°$ |
| $I_B = 25.634511$ A | $\theta_{L-B} = -14.15301°$ |

Since only two meters are used and only the currents in conductors A and B are measured, current $I_C$ and its lead/lag angle $\theta_{L-C}$ are not required and are not repeated here.

Reference is made to Figure A.2, which shows the respective currents and the pertinent leads/lags of currents $I_A$ and $I_B$. Meter #1 measures $I_A$, $V_{CA}$, and $\theta_{L-A}$. Meter #2 measures $I_B$, $V_{CB}$, and $\theta_{L-B}$. As determined in Section A.3,

$$I_A = 37.682887 \text{ A}$$

$$\theta_{L-A} = -34.64858° \quad \text{and}$$

$$V_{CA} = 480 \text{ VAC}$$

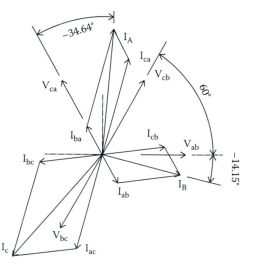

**FIGURE A.2**
Phasor diagram of Part 1.

Thus, the reading of meter #1 would be

$$W_1 = V_{CA}I_A\cos\theta_{L-A}$$

$$W_1 = (480)(37.682887)\cos-34.64858°$$

$$W_1 = 14879.9936 \text{ W} \approx 14,880 \text{ W}$$

Meter #2 measures $I_B$, $V_{cb}$, and the angle between these two parameters. As illustrated in Figure 7.10, the angle of lag between $I_B$ and $V_{cb}$ is $\theta_{L-B} + 60°$. Since $\theta_{L-B} = -14.15301°$, the angle of lag between $I_B$ and $V_{cb}$ is $60° + 14.15301°$ or $+74.15302°$.

Therefore, the reading of meter #2 would be

$$W_2 = (480)(25.634511)\cos-74.15302°$$

$$W_2 = 3359.999 \text{ W} \approx 3360 \text{ W}$$

The net measured wattage then is

$$W_T = W_1 + W_2$$

$$W_T = 14,880 \text{ W} + 3,360 \text{ W} = 18,240 \text{ W}$$

Thus, the sum of the measured wattage is in agreement with the calculated net wattage as determined in Part 2 of Section A.3. This computation corroborates the two-wattmeter theory as applicable to a three-wire three-phase circuit.

### A.4.1.2 Part 2 of Section A.4

This part considers the scenario in which the wattmeters are arranged as depicted in Figure 7.6.

The configuration of Figure 7.6 requires measurement of currents A and C as well as voltages B–C and B–A. The currents and voltages of Part 2 of Section A.3 are assumed. The phasors for the voltages and currents are represented in Figure A.1.

The voltage leads of the two meters would be arranged as described in Figure 7.6 with care taken to position the positive and negative leads as shown.

For this computation, assume that meter #1 has its voltage connections with the negative lead on phase B and the positive lead on phase C. Meter #2 has the negative voltage lead on phase B and the positive lead on phase A.

As demonstrated in the example of Part 2 of Section A.3, the following line values were determined:

| Line Current | Lead/Lag |
|---|---|
| $I_A = 37.682887$ A | $\theta_{L-A} = -34.64858°$ |
| $I_C = 41.929878$ A | $\theta_{L-C} = -11.73176°$ |

Since only two meters are used and only the currents in conductors A and C are measured, current $I_B$ and its lead/lag angle $\theta_{L-B}$ are not required and are not repeated here.

Reference is made to Figure A.3, which shows the currents and the pertinent leads/lags of currents $I_A$ and $I_C$. Meter #1 measures $I_C$, $V_{BC}$, and $\theta_{L-C}$; meter #2 measures $I_A$, $V_{BA}$, and $\theta_{L-A}$. As determined in Section A.3,

$$I_C = 41.929878 \text{ A}$$

$$\theta_{L-C} = -11.73176°, \quad \text{and}$$

$$V_{BC} = 480 \text{ VAC}$$

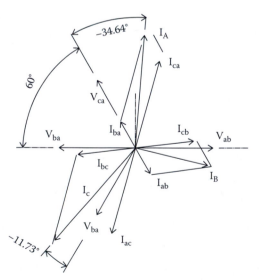

**FIGURE A.3**
Phasor diagram of Part 2.

Thus, the reading of meter #1 would be

$$W_1 = V_{CA}I_A\cos\theta_{L-C}$$

$$W_1 = (480)(41.929878)\cos-11.73176°$$

$$W_1 = 19{,}705.907\ W \approx 19{,}706\ W$$

Meter #2 measures $I_A$, $V_{ba}$, and the angle between these two parameters. As illustrated in Figure 7.11, the angle of lag between $I_A$ and $V_{ca}$ is $(\theta_{L-A} + 60°)$. Since $\theta_{L-A} = -34.64°$, the angle of lag between $I_B$ and $V_{ba}$ is $-94.64°$.

Therefore, the reading of meter #2 would be

$$W_2 = (480)(37.682887)\cos-94.64858°$$

$$W_2 = -1465.907\ W \approx -1466\ W$$

The net measured wattage then is

$$W_T = W_1 + W_2$$

$$W_T = 19{,}706\ W - 1{,}466\ W = 18{,}240\ W$$

Thus, the sum of the measured wattage is in exact agreement with the calculated net wattage as determined in the example of Part 2 of Section A.3. This computation corroborates the two-wattmeter theory as applicable to a three-wire three-phase circuit.

### A.4.1.3 Part 3 of Section A.4

This part considers the scenario in which two wattmeters are arranged in a configuration similar to that of Figure 7.5 except that the current measurements are in phase B and phase C and the potential measurements are of A–B and A–C. The phasors for the voltages and currents are represented in Figure A.1. (There is no figure to represent the arrangement.)

For this computation, assume that meter #1 measures the current in phase B and has its voltage connections with the negative lead on phase A and the positive lead on phase B. Meter #2 measures the current in phase C and has the negative voltage lead on phase A and the positive lead on phase C.

As demonstrated in the example of Part 2 of Section A.3, the following line values were determined:

| Line Current | Lead/Lag |
|---|---|
| $I_B = 25.634511$ A | $\theta_{L-B} = -14.15301°$ |
| $I_C = 41.929878$ A | $\theta_{L-C} = -11.73176°$ |

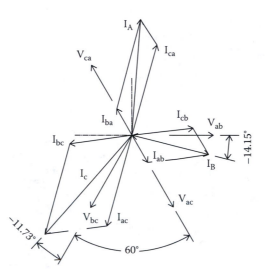

**FIGURE A.4**
Phasor diagram of Part 3.

Since only two meters are used and only the currents in conductors B and C are measured, current $I_B$ and its lead/lag angle $\theta_{L-B}$ are not required and are not repeated here.

Reference is made to Figure A.4, which shows the currents and the pertinent leads/lags of currents $I_B$ and $I_C$. Meter #1 measures $I_B$, $V_{AB}$, and $\theta_{L-B}$. Meter #2 measures $I_C$, $V_{AC}$, and $\theta_{L-C}$. As determined in Section A.3,

$$I_B = 25.634511 \text{ A}$$

$$\theta_{L-B} = -14.15301°$$

$$V_{AB} = 480 \text{ VAC}$$

Thus, the reading of meter #1 would be

$$W_1 = V_{ab}I_B\cos\theta_{L-B}$$

$$W_1 = (480)(25.634511)\cos-14.15301°$$

$$W_1 = 11931.07 \text{ W} \approx 11931 \text{ W}$$

Meter #2 measures $I_C$, $V_{ac}$, and the angle between these two parameters. As illustrated in Figure 7.12, the angle of lag between $I_C$ and $V_{ac}$ is $\theta_{L-C} + 60°$. Since $\theta_{L-AC} = -11.73176°$, the angle of lag between $I_C$ and $V_{ac}$ is $\theta_{L-C} = -71.73176°$.

Therefore, the reading of meter #2 would be

$$W_2 = (480)(41.929878)\cos(-71.73176°)$$

$$W_2 = 6308.92 \text{ W} \approx 6309 \text{ W}$$

The net measured wattage then is

$$W_T = W_1 + W_2$$

$$W_T = 11{,}931 \text{ W} + 6{,}309 = 18{,}240 \text{ W}$$

Thus, the sum of the measured wattage is in exact agreement with the calculated net wattage as determined in Section A.3. This computation corroborates the two-wattmeter theory as applicable to a three-wire three-phase circuit.

## A.5 Table of Typical Three-Phase Power Values versus ωt for One Cycle

The values tabulated later are for the assumed conditions of Example 5.2. Column I is phase A–B power. Column II is phase B–C power. Column III is phase C–A power. Column IV is the total three-phase power, that is, column I plus column II plus column III.

| | I | II | III | IV |
|---|---|---|---|---|
| (ωt) | Phase A–B Power | Phase B–C Power | Phase C–A Power | Total Power |
| 0° | 0 | 2071.36 | 8008.63 | 10,080.00 |
| 10° | −969.77 | 3730.50 | 7319.26 | 10,080.00 |
| 20° | −1417.31 | 5344.96 | 6152.35 | 10,080.00 |
| 30° | −1288.63 | 6720.00 | 4648.63 | 10,080.00 |
| 40° | −599.26 | 7689.77 | 2989.49 | 10,080.00 |
| 50° | 567.64 | 8137.31 | 1375.03 | 10,080.00 |
| 60° | 2071.36 | 8008.63 | 0 | 10,080.00 |
| 70° | 3730.50 | 7319.26 | −969.77 | 10,080.00 |
| 80° | 5344.96 | 6152.35 | −1417.31 | 10,080.00 |
| 90° | 6720.00 | 4648.63 | −1288.63 | 10,080.00 |
| 100° | 7689.77 | 2989.49 | −599.26 | 10,080.00 |
| 110° | 8137.31 | 1375.03 | 567.64 | 10,080.00 |
| 120° | 8008.63 | 0 | 2071.36 | 10,080.00 |
| 130° | 7319.26 | −969.77 | 3730.50 | 10,080.00 |

*(Continued)*

| ($\omega t$) | I Phase A–B Power | II Phase B–C Power | III Phase C–A Power | IV Total Power |
|---|---|---|---|---|
| 150° | 4648.63 | −1288.63 | 6720.00 | 10,080.00 |
| 160° | 2989.49 | −599.26 | 7689.77 | 10,080.00 |
| 170° | 1375.03 | 567.64 | 8137.31 | 10,080.00 |
| 180° | 0 | 2071.36 | 8008.63 | 10,080.00 |
| 190° | −969.77 | 3730.50 | 7319.26 | 10,080.00 |
| 200° | −1417.31 | 5344.96 | 6152.35 | 10,080.00 |
| 210° | −1288.63 | 6720.00 | 4648.63 | 10,080.00 |
| 220° | −599.26 | 7689.77 | 2989.49 | 10,080.00 |
| 230° | 567.64 | 8137.31 | 1375.03 | 10,080.00 |
| 240° | 2071.36 | 8008.63 | 0 | 10,080.00 |
| 250° | 3730.50 | 7319.26 | −969.77 | 10,080.00 |
| 260° | 5344.96 | 6152.35 | −1417.31 | 10,080.00 |
| 270° | 6720.00 | 4648.63 | −1288.63 | 10,080.00 |
| 280° | 7689.77 | 2989.49 | −599.26 | 10,080.00 |
| 290° | 8137.31 | 1375.03 | 567.64 | 10,080.00 |
| 300° | 8008.63 | 0 | 2071.36 | 10,080.00 |
| 310° | 7319.26 | −969.77 | 3730.50 | 10,080.00 |
| 320° | 6152.35 | −1417.31 | 5344.96 | 10,080.00 |
| 330° | 4648.63 | −1288.63 | 6720.00 | 10,080.00 |
| 340° | 2989.49 | −599.26 | 7689.77 | 10,080.00 |
| 350° | 1375.03 | 567.64 | 8137.31 | 10,080.00 |
| 360° | 0 | 2071.36 | 8008.63 | 10,080.00 |

NOTE: The values in columns I, II, and III are shown to two significant places but were actually calculated to four significant places.

The computations of instantaneous power (which determined a continuous power level of exactly 10,080.00 W) can be verified with Equation 5.13, which is applicable to three-phase delta circuits using phase parameters. The equation states

$$P = 3V_P I_P \cos \theta_P$$

Using the given parameters,

$$V_P = 480 \text{ VAC}$$

$$I_P = 10 \text{ A}$$

$$PF = \cos \theta_P = 0.70$$

$$P = 3V_P I_P \cos \theta_P$$

$$P = 3(480)(10)(.7) = 10,080.00 \text{ W}$$

Thus, the computation of power using Equation 5.13 confirms the computations of instantaneous power.

## A.6 Values of Sine Wave with Peak of 14.142 A: Example 5.9

| $(\omega t)$ (°) | Position | $(i) = \text{Sin } \omega t$ | $(i)^2$ |
|---|---|---|---|
| 0 | 1 | 0 | 0 |
| 5 | 2 | .0871 | .00759 |
| 10 | 3 | .1736 | .03015 |
| 15 | 4 | .25881 | .06698 |
| 20 | 5 | .34202 | .11697 |
| 25 | 6 | .42261 | .17860 |
| 30 | 7 | .50000 | .25000 |
| 35 | 8 | .57357 | .32898 |
| 40 | 9 | .64278 | .41317 |
| 45 | 10 | .70710 | .50000 |
| 50 | 11 | .76604 | .58682 |
| 55 | 12 | .81915 | .67101 |
| 60 | 13 | .866025 | .75000 |
| 65 | 14 | .90630 | .82139 |
| 70 | 15 | .93969 | .88302 |
| 75 | 16 | .96592 | .93301 |
| 80 | 17 | .98480 | .96984 |
| 85 | 18 | .99619 | .99240 |
| 90 | 19 | 1.0000 | 1.00000 |
| | | Total | 9.49993 |

In this table, values of $(i)^2$ are computed for the equation $f(\omega t) = \sin(\omega t)$ in order to calculate the value of root mean square of the function $\sin(\omega t)$ within the range $(\omega t) = 0° - (\omega t) = 90°$. The following method of calculation is used:

$$f(X)\text{rms} = \left\{ [1/n] \left[ X_1^2 + X_2^2 + \cdots X_n^2 \right] \right\}^{1/2}$$

## A.7 Computations of Instantaneous Power: Example 5.10

| ($\omega t$) (°) | Point Number | ($v$)(volts) | ($i$)(amps) | $v \cdot i$ |
|---|---|---|---|---|
| 2.5 | 1 | 14.80491 | 10.5212 | 155.7654 |
| 7.5 | 2 | 44.3020 | 11.3048 | 500.8250 |
| 12.5 | 3 | 73.4620 | 12.0023 | 881.7172 |
| 17.5 | 4 | 102.0629 | 12.6085 | 1286.8671 |
| 22.5 | 5 | 129.8870 | 13.1188 | 1703.9644 |
| 27.5 | 6 | 156.7226 | 13.5292 | 2120.3365 |
| 32.5 | 7 | 182.3655 | 13.8366 | 2523.3326 |
| 37.5 | 8 | 206.6204 | 14.0388 | 2900.7060 |
| 42.5 | 9 | 229.3029 | 14.1341 | 3240.9931 |
| 47.5 | 10 | 250.2402 | 14.1218 | 3533.8519 |
| 52.5 | 11 | 269.2730 | 14.0020 | 3730.3848 |
| 57.5 | 12 | 286.2565 | 13.7757 | 3943.4056 |
| 62.5 | 13 | 301.0614 | 13.4446 | 4047.6554 |
| 67.5 | 14 | 313.5751 | 13.0111 | 4079.9569 |
| 72.5 | 15 | 323.7022 | 12.4786 | 4039.3644 |
| 77.5 | 16 | 331.3658 | 11.8511 | 3927.0737 |
| 82.5 | 17 | 336.5075 | 11.1335 | 3746.5093 |
| 87.5 | 18 | 339.0882 | 10.3311 | 3503.1581 |
| 92.5 | 19 | 339.0882 | 9.4500 | 3204.4135 |
| 97.5 | 20 | 336.5075 | 8.4971 | 2859.3528 |
| 102.5 | 21 | 331.3658 | 7.4795 | 2478.4611 |
| 107.5 | 22 | 323.7022 | 6.4049 | 2073.3112 |
| 112.5 | 23 | 313.5751 | 5.2817 | 1656.2138 |
| 117.5 | 24 | 301.0614 | 4.1182 | 1239.8413 |
| 122.5 | 25 | 286.2565 | 2.9234 | 836.8459 |
| 127.5 | 26 | 269.2730 | 1.7063 | 459.4719 |
| 132.5 | 27 | 250.2402 | 0.4762 | 119.1858 |
| 137.5 | 28 | 229.3029 | −0.7573 | −173.6730 |
| 142.5 | 29 | 206.6204 | −1.9853 | −410.2061 |
| 147.5 | 30 | 182.3655 | −3.1981 | −583.2269 |
| 152.5 | 31 | 156.7226 | −4.3865 | −687.4776 |
| 157.5 | 32 | 129.8870 | −5.5416 | −719.7912 |
| 162.5 | 33 | 102.0629 | −6.6545 | −679.1858 |
| 167.5 | 34 | 73.4620 | −7.7168 | −566.8947 |
| 172.5 | 35 | 44.3020 | −8.7203 | −386.3301 |
| 177.5 | 36 | 14.8049 | −9.6575 | −142.9789 |
| 180.0 | — | | | |
| | | | Total | 60443.20 |

In this Appendix, the value of root mean square power, P, is calculated by determining the area under the curve of instantaneous power from the equations for instantaneous current, ($i$), and instantaneous voltage, ($v$).

## A.8 Values of Nonlinear Current of Example 5.11

| $(\omega t)$ (°) | Position | $(i)$ | $(v)$ | $P$ |
|---|---|---|---|---|
| 0 | | 0 | 0 | 0 |
| 2.5 | 1 | .907 | 14.804 | 13.427 |
| 7.5 | 2 | 2.446 | 44.302 | 108.362 |
| 12.5 | 3 | 4.832 | 73.462 | 354.968 |
| 17.5 | 4 | 4.830 | 102.062 | 492.959 |
| 22.5 | 5 | 5.789 | 129.887 | 751.915 |
| 27.5 | 6 | 6.626 | 156.722 | 1,038.439 |
| 32.5 | 7 | 7.296 | 182.365 | 1,330.535 |
| 37.5 | 8 | 7.825 | 206.620 | 1,616.801 |
| 42.5 | 9 | 8.196 | 229.302 | 1,879.359 |
| 47.5 | 10 | 8.444 | 250.240 | 2,113.026 |
| 52.5 | 11 | 8.588 | 269.273 | 2,312.516 |
| 57.5 | 12 | 8.621 | 286.256 | 2,467.812 |
| 62.5 | 13 | 8.779 | 301.061 | 2,643.014 |
| 67.5 | 14 | 9.183 | 313.575 | 2,879.559 |
| 72.5 | 15 | 9.946 | 323.702 | 3,219.540 |
| 77.5 | 16 | 11.532 | 331.365 | 3,821.301 |
| 82.5 | 17 | 12.887 | 336.507 | 4,336.565 |
| 87.5 | 18 | 13.839 | 339.088 | 4,692.638 |
| 90.0 | | 14.142 | — | |
| | | | Total | 36,072.736 |

# Symbols, Acronyms, and Equations

## B.1 Symbols

Following is a summary of the symbols used in this textbook:

$\Delta$—symbol for (three-phase) delta loads

$\Phi$—symbol for "phase" (e.g., $3\Phi$ = three-phase)

$v^i$—instantaneous voltage (volts)

$v^i_X$—value of instantaneous voltage $v^i$ at time X (V)

$v^i_{XY}$—instantaneous value of voltage measured from "X" to "Y" (V)

$v_{PK}$—peak value of instantaneous voltage $v^i$ (V)

$V(t)$—voltage expressed as a function of time (rms volts)

$V(t)_{ab}$—voltage A–B expressed as a function of time (rms volts)

$V$—(absolute) numerical value of dc voltage or (rms) ac voltage (V)

$V_L$—line voltage (rms volts—used in reference to a three-phase source)

$V_{XY}$—voltage vector V with positive direction from X to Y

$\overline{V_{XY}}$—ac voltage with positive direction measured from "X" to "Y" (rms volts or voltage vector)

$V_{L-XY}$—line voltage (rms volts—measured from X to Y)

$V_P$—voltage (rms volts—used in reference to a phase of a three-phase load)

$V_{P-XY}$—phase voltage (rms volts—measured from X to Y)

$i^i$—instantaneous current (A)

$i^i_t$—value of instantaneous current $i^i$ at time $t$ (A)

$i_{PK}$—peak value of instantaneous current $i^i$ (A)

$I$—(absolute) numerical value of dc current or (rms) ac current (A)

$I(t)$—current expressed as a function of time (rms amps)

$I(t)_{ab}$—current A–B expressed as a function of time (rms amps)

$I_L$—line current (rms amps) (used in reference to a three-phase source)

$I_{XY}$—current vector I with positive direction from X to Y

$\overline{I_{XY}}$—ac current with positive direction measured from X to Y (rms amps or current vector)

$I_{L-X}$—line current in conductor X (rms amps)

$I_P$—phase current (rms amps) (used in reference to a three-phase load)

$I_{P-XY}$—current in phase X to Y (rms amps)

$L$—general representation of an electrical load (which could be resistive, capacitive, inductive, or any combination thereof)

$R$—electrical resistance (ohms)

P—electrical power (W)

$P_{XY}$—electrical power in circuit X toY (W)

$P^i$—instantaneous power (W)

$t$—time (s)

$t_X$—time at X (s)

f—frequency (Hz)

$\theta_{SP}$—single-phase lead/lag angle between current and voltage (degrees)

$\theta_L$—general representation of line lead/lag angle between line current and line voltage (degrees or radians)

$\theta_P$—phase lead/lag angle between phase current and phase voltage (degrees or radians) (for leading current, $\theta_P > 0$; for lagging current, $\theta_P < 0$)

$\theta_{L-A}$—lead/lag angle of line current A with respect to line voltage $V_{ca}$

$\theta_{L-B}$—lead/lag angle of line current B with respect to line voltage $V_{ab}$

$\theta_{L-C}$—lead/lag angle of line current C with respect to line voltage $V_{bc}$

$\theta_{L-CA}$—lead/lag of line current A with respect to line voltage C–A

$\theta_{L-AB}$—lead/lag of line current B with respect to line voltage A–B

$\theta_{L-BC}$—lead/lag of line current C with respect to line voltage B–C

$\theta_{P-A/AD}$—lead/lag of phase current A with respect to line voltage A–D

$\theta_{P-B/BD}$—lead/lag of phase current B with respect to line voltage B–D

$\theta_{P-C/CD}$—lead/lag of phase current C with respect to line voltage C–D

PF—power factor = $\cos \theta_P$ (for balanced delta or balanced wye loads)

$\omega$—$2\pi f$ (radians)

## B.2 Acronyms

Acronyms are used extensively in electrical drawings, manuals as well as in the spoken word. An acronym is defined as a word formed from the first (or first few) letters of a series of words. Following is a summary of common electrical-related acronyms used in this textbook.

| Acronym | Defined |
| --- | --- |
| A, a | Amperes |
| ac | Alternating current |
| AHJ | Authority having jurisdiction |
| AIR | Ampere interrupting rating |
| AMI | Advanced metering infrastructure |
| AWG | American wire gauge |
| BTU | British thermal unit |
| CB | Circuit breaker |
| CT | Current transformer |
| dc | Direct current |

*(Continued)*

| Acronym | Defined |
|---|---|
| DIN | Deutsches Institut für Normung |
| DPDT | Double-pole double-throw (switch) |
| DPST | Double-pole single-throw (switch) |
| DR | Demand response |
| DT | Double throw |
| FLA | Full load amperage |
| FLP | Florida light and power |
| GFD | Ground fault detector |
| HGR | High ground resistor |
| HP | Horsepower |
| HV | High voltage |
| ICCB | Insulated case circuit breaker |
| ICEA | Insulated Cable Engineers Association |
| IEC | International Electromechanical Commission |
| IEEE | Institute of Electrical and Electronic Engineers |
| IPP | Independent power producer |
| JIC | Joint Industrial Council |
| kV | Kilovolts (1000 V) |
| kVA | Kilovolts-amps (1000 volts-amps) |
| kW | Kilowatts (1 kW = 1000 watts) |
| LRA | Locked rotor amps |
| LV | Low voltage |
| LVPCB | Low-voltage power circuit breaker |
| ma | Milliamps (0.001 A) |
| ms | Milliseconds (0.001 s) |
| MCCB | Molded case circuit breaker |
| MKS | Meter-kilogram-second |
| MPT | Main power transformer |
| MV | Medium voltage |
| MW | Megawatt |
| N | Neutral (conductor) |
| NC | Normally closed (contacts) |
| NCTO | Normally closed timed to open (contacts) |
| NEC | National electrical code |
| NFPA | National Fire Protection Association |
| NGR | Neutral ground resistor |
| NO | Normally open (contacts) |
| NOTC | Normally open timed to close (contacts) |
| OEM | Original equipment manufacturer |
| OCPD | Overcurrent protective device |
| P | Power |
| PB | Push button (switch) |
| PC | Personal computer |

*(Continued)*

| Acronym | Defined |
|---------|---------|
| PLC | Programmable logic controller |
| PF | Power factor |
| R | (Electrical) Resistance |
| SI | Système international |
| SOX | Sulfur oxide (emissions compounds—$SO_2$ or $SO_3$) |
| SPST | Single-pole single-throw (switch) |
| SPDT | Single-pole double-throw (switch) |
| SU | Step up (transformer) |
| TCC | Time–current characteristics |
| TDD | Time delay on deenergization (time delay relay) |
| TDE | Time delay on energization (time delay relay) |
| TOU | Time of (day) use |
| USA | United States of America |
| V, v | Volts |
| VFD | Variable frequency drive |
| VAC | Volts alternating current |
| VDC | Volts direct current |

## B.3 Equations

Following is a summary of the equations used in this textbook:

Instantaneous voltage expressed as a function of time:

$$v^i = (v_{PK})\sin \omega t \tag{1.1}$$

Instantaneous current expressed as a function of time:

$$i^i = i_{PK} \sin(\omega t + \theta_{SP}) \tag{1.2}$$

Single-phase RMS voltage expressed as a function of time:

$$V(t) = V \sin(\omega t) \tag{1.3}$$

$$I(t) = I \sin(\omega t + \theta_{SP}) \tag{1.4}$$

where
   I(*t*) is the current expressed as a function of time (rms amps)
   I is the numerical value of current (rms)
   $\theta_{SP}$ is the angle of lead or angle of lag (radians) (current with respect to voltage in a single-phase circuit)
   for a lagging power factor, $\theta_{SP} < 0$
   for a leading power factor, $\theta_{SP} > 0$

For single-phase circuits (Figure 1.12): The magnitude and lead/lag of line current $I_a$ resulting from the addition of line current $I_1$ at lag/lead angle $\theta_1$ (to line voltage) and line current $I_2$ at lead/lag angle $\theta_2$ (to line voltage) and ... line current $I_n$ at lead/lag angle $\theta_n$ (to line voltage):

$$|I_a| = \{(I_{a-x})^2 + (I_{a-y})^2\}^{\frac{1}{2}} \tag{1.5}$$

where

$$I_{a-x} = I_1 \cos\theta_1 + I_2 \cos\theta_2 + \cdots I_n \cos\theta_n, \text{ and}$$

$$I_{a-y} = I_1 \sin\theta_1 + I_2 \sin\theta_2 + \cdots I_n \sin\theta_n$$

$$\theta_a = \sin^{-1}(I_{a-y} \div I_a)$$

Power consumption of a balanced, linear three-phase wye or balanced, linear three-phase delta load:

$$P = \left(\sqrt{3}\right) V_L I_L \cos\theta_P \tag{1.6}$$

Equation for the relationship of phase current and line current lead/lag angle in a balanced delta and a balanced or unbalanced wye circuit:

$$\theta_P = \theta_L - 30° \tag{1.7}$$

where
   $\theta_L$ is the line lead/lag angle between line current and line voltage (degrees or radians) (for lagging current, $\theta_L < 0$; for leading current, $\theta_L > 0$)
   $\theta_P$ is the phase lead/lag angle between phase current and phase voltage (degrees or radians) (for lagging current, $\theta_P < 0$; for leading current, $\theta_P > 0$)

Specifically,
  For balanced delta circuits:

$$\theta_{P-CA} = \theta_{L-A/CA} - 30°,$$

where
  $\theta_{P-CA}$ is the lead/lag of current in phase C–A with respect to voltage C–A
  $\theta_{L-A/CA}$ is the lead/lag of current in conductor A with respect to voltage C–A

$$\theta_{P-AB} = \theta_{L-B/AB} - 30°,$$

where
  $\theta_{P-AB}$ is the lead/lag of current in phase A–B with respect to voltage A–B
  $\theta_{L-B/AB}$ is the lead/lag of current in conductor B with respect to voltage A–B

$$\theta_{P-BC} = \theta_{L-C/BC} - 30°,$$

where
  $\theta_{P-BC}$ is the lead/lag of current in phase B–C with respect to voltage B–C
  $\theta_{L-C/BC}$ is the lead/lag of current in conductor C with respect to voltage B–C

For balanced or unbalanced wye circuits:

$$\theta_{P-A/AD} = \theta_{L-A/CA} - 30°,$$

where
  $\theta_{P-A/AD}$ is the lead/lag of current in phase A–D with respect to voltage A–D
  $\theta_{L-A/CA}$ is the lead/lag of current in conductor A with respect to voltage C–A

$$\theta_{P-B/BD} = \theta_{L-B/AB} - 30°,$$

where
  $\theta_{P-B/BD}$ is the lead/lag of current in phase B–D with respect to voltage B–D
  $\theta_{L-B/AB}$ is the lead/lag of current in conductor B with respect to voltage A–B

$$\theta_{P-C/CD} = \theta_{L-C/BC} - 30°,$$

where
  $\theta_{P-C/CD}$ is the lead/lag of current in phase C–D with respect to voltage C–D
  $\theta_{L-C/BC}$ is the lead/lag of current in conductor C with respect to voltage B–C

Impedance:

$$\underline{Z} = \underline{V}/\underline{I} \tag{1.8}$$

Admittance:

$$\underline{Y} = \underline{I}/\underline{Z} \tag{1.9}$$

Line current of a balanced three-phase delta load:

$$I_L = \left(\sqrt{3}\right)I_P \tag{4.1}$$

For balanced three-phase circuits, delta, or wye. The magnitude and lead/lag of line current $I_B$ resulting from the addition of line current $I_1$ at lag/lead angle $\theta_1$ (to line voltage) and line current $I_2$ at lead/lag angle $\theta_2$ (to line voltage) and up to …line current $I_n$ at lead/lag angle $\theta_n$ (to line voltage). (Figure 4.9). Although Figure 4.9 shows a delta load the equation is applicable to any combination of balanced loads.

$$|I_B| = \{(I_{B-x})^2 + (I_{B-y})^2\}^{\frac{1}{2}} \tag{4.2}$$

where

$$I_{B-x} = I_1 \cos\theta_1 + I_2 \cos\theta_2 + \cdots I_n \cos\theta_n, \text{ and}$$

$$I_{B-y} = I_1 \sin\theta_1 + I_2 \sin\theta_2 + \cdots I_n \sin\theta_n$$

$$\theta_B = \sin^{-1}(I_{B-y} \div I_B)$$

In a balanced three-phase circuit, the three-line currents, $I_A$, $I_B$, and $I_C$, are equal in absolute value. Therefore,

$$\underline{I}_B = \underline{I}_A = \underline{I}_C$$

Phase voltage in a balanced three-phase wye load or an unbalanced three-phase load with a grounded neutral:

$$V_L = \left(\sqrt{3}\right)V_P \tag{4.3}$$

## Equations 4.4 through 4.9

Equations for calculating line currents when the phase currents and respective leads/lags in an unbalanced delta circuit are known.

With reference to Figure 1.14—where $I_{ab}$ is the current in phase a–b; $I_{bc}$ is the current in phase b–c; $I_{ac}$ is the current in phase a–c; $\theta_{P-AB}$ is the lead/lag of current in phase A–B; $\theta_{P-BC}$ is the lead/lag of current in phase B–C; $\theta_{P-CA}$ is the lead/lag of current in phase C–A; $I_A$, $I_B$, and $I_C$ are the line currents in, respectively, conductors A, B, and C; $\theta_{L-A}$, is the lead/lag of the line current in phase A; $\theta_{L-B}$ is the lead/lag of the line current in phase B; and $\theta_{L-C}$ is the lead/lag of the line current in phase C.

$$|I_A| = \left\{ (I_{A-x})^2 + (I_{A-y})^2 \right\}^{\frac{1}{2}} \qquad (4.4)$$

$$\theta_{L-A} = (\lambda - 120°) \qquad (4.5)$$

where

$I_A$ is the current in line A

$\theta_{L-A}$ is the lead (lag) of current $I_A$ with respect to line voltage $V_{ca}$

$I_{ba-x} = -I_{ab} \cos \theta_{P-AB}$

$I_{ca-x} = -I_{ca}(1/2)\left[ \left( \sqrt{3} \right) \sin \theta_{P-CA} + \cos \theta_{P-CA} \right]$

$I_{A-x} = I_{ba-x} + I_{ca-x}$

$I_{ba-y} = -I_{ab} \sin \theta_{P-AB}$

$I_{ca-y} = I_{ca}(1/2)\left[ \left( \sqrt{3} \right) \cos \theta_{P-CA} - \sin \theta_{P-CA} \right]$

$I_{A-y} = I_{ba-y} + I_{ca-y}$

$\lambda = \sin^{-1}(I_{A-y} \div I_A)$

Valid range of $\theta_{P-AB}$ and $\theta_{P-CA}$: $\pm 90°$; valid range of $\theta_{L-A}$: $+120°$ to $-60°$

$$|I_B| = \left\{ (I_{B-x})^2 + (I_{B-y})^2 \right\}^{\frac{1}{2}} \qquad (4.6)$$

$$\theta_{L-B} = \sin^{-1}(I_{B-y} \div I_B) \qquad (4.7)$$

where

$I_B$ is the current in line B

$\theta_{L-B}$ is the lead (lag) of current $I_B$ with respect to line voltage $V_{ab}$

$I_{ab-x} = I_{ab} \cos \theta_{P-AB}$

$I_{cb-x} = -I_{bc}(1/2)\left[(\sqrt{3})\sin \theta_{P-BC} - \cos \theta_{P-BC}\right]$

$I_{B-x} = I_{ab-x} + I_{cb-x}$

$I_{ab-y} = I_{ab} \sin \theta_{P-AB}$

$I_{cb-y} = I_{bc}(1/2)\left[(\sqrt{3})\cos \theta_{P-BC} + \sin \theta_{P-BC}\right]$

$I_{B-y} = I_{ab-y} + I_{cb-y}$

Valid range of $\theta_{P-AB}$ and $\theta_{P-CB}$: $\pm 90°$; valid range of $\theta_{L-B}$: $+120°$ to $-60°$

$$|I_C| = \left\{(I_{C-x})^2 + (I_{C-y})^2\right\}^{\frac{1}{2}} \tag{4.8}$$

$$\theta_{L-C} = (\varphi - 240°) \tag{4.9}$$

where

$|I_C|$ is the current in line C

$\theta_{L-A}$ is the lead (lag) of current $I_C$ with respect to line voltage $V_{bc}$

$I_{bc-x} = I_{bc}(1/2)\left[(\sqrt{3})\sin \theta_{P-BC} - \cos \theta_{P-BC}\right]$

$I_{ac-x} = I_{ca}(1/2)\left[(\sqrt{3})\sin \theta_{P-CA} - \cos \theta_{P-CA}\right]$

$I_C = I_{bc-x} + I_{ac-x}$

$I_{bc-y} = -I_{bc}(1/2)\left[(\sqrt{3})\cos \theta_{P-BC} + \sin \theta_{P-BC}\right]$

$I_{ac-y} = -I_{ca}(1/2)\left[(\sqrt{3})\cos \theta_{P-CA} - \sin \theta_{P-CA}\right]$

$I_{C-y} = I_{bc-y} + I_{ac-y}$

$\varphi = \sin^{-1}(I_{C-y} \div I_C)$

Valid range of $\theta_{P-BC}$ and $\theta_{P-AC}$: $\pm 90°$; valid range of $\theta_{L-C}$: $+120°$ to $-60°$

## Equations 4.10 through 4.15

Equations for determining line currents in an unbalanced delta circuit when the loads on all three phases are resistive.

With reference to Figure 1.14—where $I_{ab}$ is the current in phase a–b; $I_{bc}$ is the current in phase b–c; $I_{ac}$ is the current in phase a–c; $\theta_{P-AB}$ is the lead/lag of current in phase A–B; $\theta_{P-BC}$ is the lead/lag of current in phase B–C; $\theta_{P-CA}$ is the lead/lag of current in phase C–A; $I_A$, $I_B$, and $I_C$ are the line currents in, respectively, conductors A, B, and C; $\theta_{L-A}$, is the lead/lag of the line current in phase A; $\theta_{L-B}$ is the lead/lag of the line current in phase B; and $\theta_{L-C}$ is the lead/lag of the line current in phase C.

$$|I_A| = \left\{ (I_{A-x})^2 + (I_{A-y})^2 \right\}^{\frac{1}{2}}$$  (4.10)

$$\theta_{L-A} = (\lambda - 120°)$$  (4.11)

where

$I_A$ is the current in line A

$\theta_{L-A}$ is the lead (lag) of current $I_A$ with respect to line voltage $V_{ca}$

$I_{A-x} = -I_{ab} - (1/2)I_{ca}$

$I_{A-y} = I_{ca}\left(\sqrt{3}/2\right)$

$\lambda = \sin^{-1}\left(I_{A-y} \div I_A\right)$

$$|I_B| = \left\{ (I_{B-x})^2 + (I_{B-y})^2 \right\}^{\frac{1}{2}}$$  (4.12)

$$\theta_{L-B} = \sin^{-1}(I_{B-y} \div I_B)$$  (4.13)

where

$I_B$ is the current in line B

$\theta_{L-B}$ is the lead (lag) of current $I_B$ with respect to line voltage $V_{ab}$

$I_B = -I_{ab} - (1/2)I_{bc}$

$I_{B-y} = I_{bc}\left(\sqrt{3}/2\right)$

$$|I_C| = \left\{ (I_{C-x})^2 + (I_{C-y})^2 \right\}^{\frac{1}{2}}$$  (4.14)

$$\theta_{L-C} = (\varphi - 240°) \tag{4.15}$$

where

$I_C$ is the current in line C

$\theta_{L-C}$ is the lead (lag) of current $I_C$ with respect to line voltage $V_{bc}$

$I_{C-x} = (1/2)I_{ca} - (1/2)I_{bc}$

$I_{C-y} = I_{bc}\left(\sqrt{3}/2\right) - I_{ca}\left(\sqrt{3}/2\right)$

$(\varphi) = \sin^{-1}(I_{C-y} \div I_C)$

## Equations 4.16 through 4.21

Equations for determining the currents in a feeder that delivers power to two or more three-phase loads any of which could be balanced or unbalanced. Reference is made to Figure 4.27. Conductor A-1 is the conductor connecting common conductor A to phase A of load #1; A-2 is the conductor connecting conductor A to phase A of load #2. $I_A$, $I_B$, and $I_C$ are the line currents in, respectively, feeder conductors A, B, and C, $\theta_{L-A}$ is the lead/lag of the line current in phase A, $\theta_{L-B}$ is the lead/lag of the line current in phase B, and $\theta_{L-C}$ is the lead/lag of the line current in phase C.

$$|I_A| = \left\{(I_{A-x})^2 + (I_{A-y})^2\right\}^{\frac{1}{2}} \tag{4.16), and}$$

$$\theta_{L-A} = (\kappa - 120°) \tag{4.17}$$

where

$\kappa = \sin^{-1}(I_{A-y} \div I_A)$

$I_{A-x} = \Sigma(I_{A1-x} + I_{A2-x} + I_{A3-x} \cdots + I_{AN-x})$

$I_{A-y} = \Sigma(I_{A1-y} + I_{A2-y} + I_{A3-y} \cdots + I_{AN-y})$

$I_{A1-x} = -I_{A1}(1/2)\left[\left(\sqrt{3}\right)\sin\theta_{A1} + \cos\theta_{A1}\right]$

$I_{A2-x} = -I_{A2}(1/2)\left[\left(\sqrt{3}\right)\sin\theta_{A2} + \cos\theta_{A2}\right] \ldots$ to

$$I_{AN-x} = -I_{AN}(1/2)\left[\left(\sqrt{3}\right)\sin\theta_{AN} + \cos\theta_{AN}\right]$$

$$I_{A1-y} = I_{A1}(1/2)\left[\left(\sqrt{3}\right)\cos\theta_{A1} - \sin\theta_{A1}\right]\ldots\text{to}$$

$$I_{AN-y} = I_{AN}(1/2)\left[\left(\sqrt{3}\right)\cos\theta_{AN} - \sin\theta_{AN}\right]$$

$I_{A1}$ is the line current in branch of conductor A to load #1

$I_{A2}$ is the line current in branch of conductor A to load #2 ... to

$I_{AN}$ is the line current in branch of conductor A to load #$N$

$\theta_{A1}$ is the lead/lag of line current $I_{A-1}$ with respect to line voltage $V_{ca}$

$\theta_{A2}$ is the lead/lag of line current $I_{A-2}$ with respect to line voltage $V_{ca}$ ... to

$\theta_{AN}$ is the lead/lag of line current $I_{A-N}$ with respect to line voltage $V_{ca}$

$$|I_B| = \left\{(I_{B-x})^2 + (I_{B-y})^2\right\}^{\frac{1}{2}}, \quad \text{and} \tag{4.18}$$

$$\theta_{L-B} = \sin^{-1}(I_{B-y} \div I_B) \tag{4.19}$$

where

$I_{B-x} = \Sigma(I_{B1-x} + I_{B2-x} + I_{B3-x} \ldots + I_{BN-x})$

$I_{B-y} = \Sigma(I_{B1-y} + I_{B2-y} + I_{B3-y} \ldots + I_{BN-y})$

$I_{B1-x} = I_{B1}\cos\theta_{B1}$ ... to

$I_{BN-x} = I_{BN}\cos\theta_{BN}$

$I_{B1-y} = I_{B1}\sin\theta_{B1}$ ... to

$I_{BN-y} = I_{BN}\sin\theta_{LN}$

$I_{B1}$ is the line current in branch of conductor B to load #1

$I_{B2}$ is the line current in branch of conductor B to load #2... to

$I_{BN}$ is the line current in branch of conductor B to load #$N$

$\theta_{B1}$ is the lead/lag of line current $I_{B1}$ with respect to line voltage $V_{ca}$

$\theta_{B2}$ is the lead/lag of line current $I_{A2}$ with respect to line voltage $V_{ca}$ ... to

$\theta_{BN}$ is the lead / lag of line current $I_{AN}$ with respect to line voltage $V_{ca}$

$I_{B1}$ is the line current in branch of conductor B to load #1

$I_{B2}$ is the line current in branch of conductor B to load #2 ... to

$I_{BN}$ is the line current in branch of conductor B to load #$N$

$\theta_{B1}$ is the lead/lag of line current $I_{B-1}$ with respect to line voltage $V_{ab}$

$\theta_{B2}$ is the lead/lag of line current $I_{B-2}$ with respect to line voltage $V_{ab}$ ... to

$\theta_{BN}$ is the lead/lag of line current $I_{B-N}$ with respect to line voltage $V_{ab}$

$$|I_C| = \left\{(I_{C-x})^2 + (I_{C-y})^2\right\}^{\frac{1}{2}} \quad \text{and} \tag{4.20}$$

$$\theta_{L-C} = \zeta - 240° \tag{4.21}$$

$\zeta = \sin^{-1}(I_{C-y} \div I_C)$

$I_{C-x} = \Sigma(I_{C1-x} + I_{C2-x} + I_{C3-x} \cdots + I_{CN-x})$

$I_{C-y} = \Sigma(I_{C1-y} + I_{C2-y} + I_{C3-y} \cdots + I_{CN-y})$

$I_{C1}$ is the current in branch of conductor C to load #1

$I_{C2}$ is the current in branch of conductor C to load #2... to

$I_{CN}$ is the current in branch of conductor C to load #N

$I_{C1-x} = I_{C1}(1/2)\left[(\sqrt{3})\sin\theta_{C1} - \cos\theta_{C1}\right]$

$I_{C2-x} = I_{C2}(1/2)\left[(\sqrt{3})\sin\theta_{C2} - \cos\theta_{C2}\right]...$to

$I_{CN-x} = I_{CN}(1/2)\left[(\sqrt{3})\sin\theta_{CN} - \cos\theta_{CN}\right]$

$I_{C1-y} = -I_{C1}(1/2)\left[(\sqrt{3})\cos\theta_{C1} + \sin\theta_{C1}\right]$

$I_{C2-y} = -I_{C2}(1/2)\left[(\sqrt{3})\cos\theta_{C2} + \sin\theta_{C2}\right]$ ...to

$I_{CN-y} = -I_{CN}(1/2)\left[(\sqrt{3})\cos\theta_{CN} + \sin\theta_{CN}\right]$

$\theta_{C1}$ is the lead / lag of line current $I_{C1}$ with respect to line voltage $V_{bc}$

$\theta_{C2}$ is the lead/lag of line current $I_{C2}$ with respect to line voltage $V_{bc}$...to

$\theta_{CN}$ is the lead/lag of line current $I_{CN}$ with respect to line voltage $V_{bc}$

## Equation 5.22

Equation that defines root mean square of a known function:

$$f(t)\ \text{rms} = \left\{1/(T_2 - T_1)\int_{T_1}^{T_2}[f(t)]^2\,dt\right\}^{\frac{1}{2}}. \tag{5.22}$$

## Equation 5.23

Equation for determining root mean square from known values of a function sampled at uniformly spaced distances:

$$f(X)\text{rms} = \left\{[1/n]\left[X_1^2 + X_2^2 + \cdots X_n^2\right]\right\}^{1/2} \tag{5.23}$$

**Equation B.1**: Guidelines for readily determining the arc sin (sin⁻¹) of angles
$\lambda$, $\varphi$, $\kappa$, or $\zeta$.

Some of the computations of this textbook require the calculation of the arc
sin (sin⁻¹) of angles $\lambda$, $\varphi$, $\kappa$, and $\zeta$ in order to determine the displacement angle
of current vectors in the CCW direction from the positive abscissa. Unless
a person is well experienced and well practiced in calculating the sin⁻¹ of
angles, it is easy to make errors in the performance of this exercise. Presented
here are simple guidelines that may be used as an assist to correctly and simply determine the value of the sin⁻¹ of an angle in any quadrant.

Reference is made to Figure B.1 that depicts a vector "I" in quadrant I.
Vector I is also the hypotenuse of a triangle with vertical component "y" and
horizontal component "x," and vector I is at an angle "a" from the positive
abscissa. By definition, the sin of angle a is y/I, and the arc sin of y/I (sin⁻¹ y/I)
is the angle a. While dealing with angles and values pertinent to quadrant I,
little confusion usually ensues. Confusion usually comes about when computing values in quadrant II, quadrant III, and quadrant IV. In quadrant II,
the sine of I is the sin of angle "b." Similarly in quadrant III, the sine of vector I
is the sine of angle "c," and in quadrant IV, the sin of I is the sine of angle "d."

If, say, angle b in quadrant II is 120° and a handheld calculator is used to
compute the sin of b, the calculator would correctly indicate the true sin of
b to be 0.866. However, if the same calculator is used to compute the sin⁻¹ of
0.866, the calculator indicates the angle to be 60° since the calculator does
not "know" that the angle under consideration is in quadrant II. Therein is
the potential for mistaken calculations. Following are some simple rules of
thumb that will prove helpful to accurately determine the values of the sin⁻¹
of any angle in any quadrant.

1. Given: The value of sin⁻¹ of vector I, which is in quadrant I, II, III, or
   IV.

2. Determine: The angle of I measured CCW from the positive abscissa
   (i.e., $\lambda$, $\varphi$, $\kappa$, or $\zeta$)

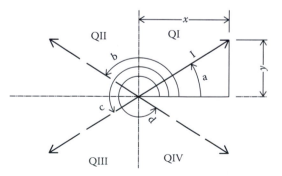

**FIGURE B.1**
Aid in determining arc sin of an angle.

3. Let $M = \left|\sin^{-1}\lambda\right|, \left|\sin^{-1}\varphi\right|, \left|\sin^{-1}\kappa\right|,$ or $\left|\sin^{-1}\zeta\right|$.

4. Compute M by calculator or table. (The calculated value of M will be between 0° and 90°.)

5. From Table B.1, determine the quadrant in which I is located.

6. Calculate the value of $\lambda$, $\varphi$, $\kappa$, or $\zeta$ per fourth row of Table B.1.

Summary of three-phase power equations: Three-phase delta, wye, or mixed circuit power using line parameters is shown in Tables B.2 through B.4.

**TABLE B.1**

Determination of Quadrant

| Value of x | Value of y | Quadrant | Value of $\lambda$, $\varphi$, $\kappa$, or $\zeta$ |
|---|---|---|---|
| $\geq 0$ | $\geq 0$ | I | $= M$ |
| $\leq 0$ | $\geq 0$ | II | $= 180° - M$ |
| $\leq 0$ | $\leq 0$ | III | $= 180° + M$ |
| $\geq 0$ | $\geq 0$ | IV | $= 360° - M$ |

**TABLE B.2**

Summary of Three-Phase Power Equations: Three-Phase Wye Circuit

| Type Circuit | Equation Number[a] | Phase Parameters[b] Line Parameters |
|---|---|---|
| Balanced resistive | 5.6 | $P = 3V_PI_P$ |
| | 5.7 | $P = \sqrt{3}\,V_L\,I_L$ |
| Balanced inductive or capacitive | 5.8 | $P = 3V_PI_P\cos\theta_P$ |
| | 5.9 | $P = \sqrt{3}\,V_L\,I_L\,\cos(\theta_L - 30°)$ |
| Unbalanced resistive | 5.14 | $P = V_P[I_A + I_B + I_C]$ |
| | 5.15 | $P = (1/\sqrt{3})V_L[I_A\cos(\theta_{L-A/CA} - 30°)$ |
| | | $\qquad + I_B\cos(\theta_{L-B/AB} - 30°) + I_C\cos(\theta_{L-C/BC} - 30°)]$ |
| Unbalanced inductive or capacitive | 5.16 | $P = V_P[I_A\cos\theta_{P-AD} + I_B\cos\theta_{P-BD} + I_C\cos\theta_{P-CD}]$ |
| | 5.17 | $P = (1/\sqrt{3})V_L[I_A\cos(\theta_{L-A/CA} - 30°)$ |
| | | $\qquad + I_B\cos(\theta_{L-B/AB} - 30°) + I_C\cos(\theta_{L-C/BC} - 30°)]$ |

[a] The equation number is the same as the associated paragraph number.
[b] In a wye circuit, $V_P = \left(1/\sqrt{3}\right)V_L$.

**TABLE B.3**

Summary of Three-Phase Power Equations: Three-Phase Delta Circuit

| Type Circuit | Equation Number[a] | Phase Parameters[b] Line Parameters |
|---|---|---|
| Balanced resistive | 5.10 | $P = 3V_P I_P$ |
| | 5.11 | $P = \sqrt{3}\, V_L I_L$ |
| Balanced inductive or capacitive | 5.12 | $P = 3V_P I_P \cos\theta_P$ |
| | 5.13 | $P = \sqrt{3}\, V_L I_L \cos(\theta_L - 30°)$ |
| Unbalanced resistive | 5.18 | $P = V_P[I_{P-AB} + I_{P-BC} + I_{P-CA}]$ |
| | 5.19 | $P = \left(1/\sqrt{3}\right)V_L\left[I_A \cos(\theta_{L-A/CA} - 30°)\right.$ |
| | | $\left. + I_B \cos(\theta_{L-B/AB} - 30°) + I_C \cos(\theta_{L-C/BC} - 30°)\right]$ |
| Unbalanced inductive or capacitive | 5.20 | $P = V_P\left[I_{A-B} \cos\theta_{P-AB} + I_{B-C} \cos\theta_{P-BC}\right.$ |
| | | $\left. + I_{C-A} \cos\theta_{P-CA}\right]$ |
| | 5.21 | $P = \left(1/\sqrt{3}\right)V_L\left[I_A \cos(\theta_{L-A/CA} - 30°)\right.$ |
| | | $\left. + I_B \cos(\theta_{L-B/AB} - 30°) + I_C \cos(\theta_{L-C/BC} - 30°)\right]$ |

[a] The equation number is the same as the associated paragraph number.
[b] In a delta circuit $V_P = V_L$.

**TABLE B.4**

Summary of Three-Phase Power Equations: Three-Phase Mixed Circuit

| Type of Circuit | Equation |
|---|---|
| Balanced resistive | $P = \sqrt{3}\, V_L I_L$ |
| Balanced inductive or capacitive | $P = \sqrt{3}\, V_L I_L \cos(\theta_L - 30°)$ |
| Unbalanced resistive | $P = \left(1/\sqrt{3}\right)V_L\left[I_A \cos(\theta_{L-A/CA} - 30°)\right.$ $\left. + I_B \cos(\theta_{L-B/AB} - 30°) + I_C \cos(\theta_{L-C/BC} - 30°)\right]$ |
| Unbalanced inductive or capacitive | $P = \left(1/\sqrt{3}\right)V_L\left[I_A \cos(\theta_{L-A/CA} - 30°)\right.$ $\left. + I_B \cos(\theta_{L-B/AB} - 30°) + I_C \cos(\theta_{L-C/BC} - 30°)\right]$ |

# References

1.1 J.W. Hammond, *Charles Proteus Steinmetz—A Biography*, The Century Co. New York/London, 1924, pp. 195–227.

1.2 G.C. Blalock, *Principles of Electrical Engineering*, 3rd edn., McGraw-Hill Book Company, New York, 1950, p. 270.

1.3 T.L. Floyd, *Principles of Electric Circuits*, 6th edn., Prentice Hall, Upper Saddle River, NJ, 2000, p. 627.

1.4 T.L. Floyd, *Principles of Electric Circuits*, 6th edn., Prentice Hall, Upper Saddle River, NJ, 2000, pp. 430–455.

1.5 T.L. Floyd, *Principles of Electric Circuits*, 6th edn., Prentice Hall, Upper Saddle River, NJ, 2000, p. 892.

1.6 V. Del Toro, *Electrical Engineering Fundamentals*, 2nd edn., Prentice Hall, Upper Saddle River, NJ, 1986, pp. 305–314.

2.1 U.S. Energy Information Administration, *Coal Overview*, January 2013.

2.2 L. Ward, Going with the flow, *Wall Street Journal,* November 11, 2013.

2.3 Global Energy Network Institute, *National Energy Grid Canada*, January 2014.

3.1 Merriam-Webster, Inc., *Webster's Ninth New Collegiate Dictionary*, Merriam-Webster, Inc., Springfield, MA, 1990.

3.2 National Institute of Safety and Health, Publication 98-131, *Worker Deaths by Electrocution*, May 1998.

3.3 IEEE Std 141-1993, *Recommended Practice for Electrical Power Distribution for Industrial Plants (Red Book)*, Paragraph 4.4.1.2.

4.1 J.W. Nilsson, S.A. Riedel, *Electric Circuits*, 6th edn., Prentice Hall, Upper Saddle River, NJ, 2000, p. 553.

4.2 J.W. Nilsson, S.A. Riedel, *Electric Circuits*, 6th edn., Prentice-Hall, Upper Saddle River, NJ, 2000, p. 548.

7.1 New York Public Services Commission Bulletin, *A Primer on Smart Metering*, Fall 2013.

7.2 Maryland Public Service Commission, *Approved Electric Submeters*, COMAR 20.25.01.04A.(3), 2010.

8.1 IEEE Std 242-2001, *Recommended Practices for Protection and Coordination of Industrial and Commercial Power Systems (Buff Book)*.

8.2 Eaton Corporation, *Short Circuit Current Rating and Available Fault Currents*, August 27, 2013.

8.3 GE Publication GET3550F, *Short Circuit Current Calculations for Industrial and Commercial Power Systems*, 1989.

8.4 UL Standard 489, *Molded Case Circuit Breakers and Circuit Breaker Enclosures*.

8.5 IEEE Std C37.13, *Standard for Low Voltage AC Power Circuit Breakers Used in Enclosures*, 2008.

9.1 C. Wester, *Motor Protection Principles*, GE Multilin Publication, 2014.

9.2 USA National Electric Code, 2002, National Fire Protection Association, Article 430.

9.3 NEMA Standard MG 1-2003, *Motors and Generators*.

A.1 R.S. Burlington, *Handbook of Mathematical Tables and Formulas*, 3rd edn., Handbook Publishers, Inc., Sandusky, OH, 1954, p. 18.

# Index